NATO ASI Series

Advanced Science Institutes Series

A series presenting the results of activities sponsored by the NATO Science Committee, which aims at the dissemination of advanced scientific and technological knowledge, with a view to strengthening links between scientific communities.

The Series is published by an international board of publishers in conjunction with the NATO Scientific Affairs Division

A Life Sciences B Physics	Plenum Publishing Corporation London and New York
C Mathematical and Physical Sciences D Behavioural and Social Sciences E Applied Sciences	Kluwer Academic Publishers Dordrecht, Boston and London
F Computer and Systems Sciences G Ecological Sciences H Cell Biology	Springer-Verlag Berlin Heidelberg New York London Paris Tokyo Hong Kong

Series H: Cell Biology Vol. 46

The ASI Series Books Published as a Result of Activities of the Special Programme on CELL TO CELL SIGNALS IN PLANTS AND ANIMALS

This book contains the proceedings of a NATO Advanched Research Workshop held within the activities of the NATO Special Programme on Cell to Cell Signals in Plants and Animals, running from 1984 to 1989 under the auspices of the NATO Science Committee.

The books published as a result of the activities of the Special Programme are:

Vol. 1: **Biology and Molecular Biology of Plant-Pathogen Interactions.** Edited by J.A. Bailey. 1986.
Vol. 2: **Glial-Neuronal Communication in Development and Regeneration.** Edited by H.H. Althaus and W. Seifert. 1987.
Vol. 3: **Nicotinic Acetylcholine Receptor: Structure and Function.** Edited by A. Maelicke. 1986.
Vol. 4: **Recognition in Microbe-Plant Symbiotic and Pathogenic Interactions.** Edited by B. Lugtenberg. 1986.
Vol. 5: **Mesençhymal-Epithelial Interactions in Neural Development.** Edited by J.R. Wolff, J. Sievers, and M. Berry. 1987.
Vol. 6: **Molecular Mechanisms of Desensitization to Signal Molecules.** Edited by T.M. Konjin, P.J.M. Van Haastert, H. Van der Starre, H. Van der Wel, and M.D. Houslay. 1987.
Vol. 7: **Gangliosides and Modulation of Neuronal Functions.** Edited by H. Rahmann. 1987.
Vol. 9: **Modification of Cell to Cell Signals During Normal and Pathological Aging.** Edited By S. Govoni and F. Battaini. 1987.
Vol. 10: **Plant Hormone Receptors.** Edited by D. Klämbt. 1987.
Vol. 11: **Host-Parasite Cellular and Molecular Interactions in Protozoal Infections.** Edited by K.-P. Chang and D. Snary. 1987.
Vol. 12: **The Cell Surface in Signal Transduction.** Edited by E. Wagner, H. Greppin, and B. Millet. 1987.
Vol. 19: **Modulation of Synaptic Transmission and Plasticity in Nervous Systems.** Edited by G. Hertting and H.-C. Spatz. 1988.
Vol. 20: **Amino Acid Availability and Brain Function in Health and Disease.** Edited by G. Huether. 1988.
Vol. 21: **Cellular and Molecular Basis of Synaptic Transmission.** Edited by H. Zimmermann. 1988.
Vol. 23: **The Semiotics of Cellular Communication in the Immune System.** Edited by E.E. Sercarz, F. Celada, N.A. Mitchison, and T. Tada. 1988.
Vol. 24: **Bacteria, Complement and the Phagocytic Cell.** Edited by F.C. Cabello and C. Pruzzo. 1988.
Vol. 25: **Nicotinic Acetylcholine Receptors in the Nervous System.** Edited by F. Celementi, C. Gotti, and E. Sher. 1988.
Vol. 26: **Cell to Cell Signals in Mammalian Development.** Edited by S.W. de Laat, J.G. Bluemink, and C.L. Mummery. 1989.
Vol. 27: **Phytotoxins and Plant Pathogenesis.** Edited by A. Graniti, R.D. Durbin, and A. Ballio. 1989.
Vol. 31: **Neurobiology of the Inner Retina.** Edited by R. Weiler and N.N. Osborne. 1989.
Vol. 32: **Molecular Biology of Neuroreceptors and Ion Channels.** Edited by A. Maelicke. 1989.
Vol. 33: **Regulatory Mechanisms of Neuron to Vessel Communication in the Brain.** Edited by F. Battaini, S. Govoni, M.S. Magnoni, and M. Trabucchi. 1989.
Vol. 35: **Cell Separation in Plants: Physiology, Biochemistry and Molecular Biology.** Edited by D.J. Osborne and M.B. Jackson. 1989.
Vol. 36: **Signal Molecules in Plants and Plant-Microbe Interactions.** Edited by B.J.J. Lugtenberg. 1989.
Vol. 39: **Chemosensory Information Processing.** Edited by D. Schild. 1990.
Vol. 41: **Recognition and Response in Plant-Virus Interactions.** Edited by R.S.S. Fraser. 1990.
Vol. 43: **Cellular and Molecular Biology of Myelination.** Edited by G. Jeserich, H. H. Althaus, and T. V. Waehneldt. 1990.
Vol. 44: **Activation and Desensitization of Transducing Pathways.** Edited by T. M. Konijn, M. D. Houslay, and P. J. M. Van Haastert. 1990.
Vol. 45: **Mechanism of Fertilization: Plants to Humans.** Edited by B. Dale. 1990.
Vol. 46: **Parallels in Cell to Cell Junctions in Plants and Animals.** Edited by A. W. Robards, W. J. Lucas, J. D. Pitts, H. J. Jongsma, and D. C. Spray. 1990.

Parallels in Cell to Cell Junctions in Plants and Animals

Edited by

A. W. Robards

Institute for Applied Biology
The University of York, York Y01 5DD, U.K.

W. J. Lucas

Department of Botany, University of California, Davis CA 95616, USA

J. D. Pitts

Beatson Institute for Cancer Research, Wolfson Laboratory
for Molecular Pathology, Garscube Estate, Switchback Road,
Bearsden, Glasgow G61 1BD, U.K.

H. J. Jongsma

Department of Physiology, University of Amsterdam, Meibergdreef 15,
1105 AZ Amsterdam, The Netherlands

D. C. Spray

Department of Neuroscience, Albert Einstein College of Medicine
of Yeshiva University, 1300 Morris Park Ave, Bronx, NY 10461, USA

Springer-Verlag Berlin Heidelberg New York
London Paris Tokyo Hong Kong
Published in cooperation with NATO Scientific Affairs Division

Proceedings of the NATO Advanced Research Workshop on Parallels in Cell to Cell Communication in Plants and Animals held in York, England, July 2–7, 1989

ISBN-13:978-3-642-83973-3 e-ISBN-13:978-3-642-83971-9
DOI: 10.1007/978-3-642-83971-9

Library of Congress Cataloging-in-Publication Data.
NATO Advanced Research Workshop on Parallels in Cell to Cell communication in Plants and Animals (1989 : York, England) Parallels in cell to cell junctions in plants and animals / edited by A. W. Robards . . . [et al.]. p. cm. —(NATO ASI series. Series H, Cell biology ; vol. 46)
"Proceedings of the NATO Advanced Research Workshop on Parallels in Cell to Cell Communication in Plants and Animals, held in York, England, July 2–7, 1989"—T.p. verso. Includes index. ISBN-13:978-3-642-83973-3
1. Cell interaction—Congresses. 2. Cell junctions—Congresses. 3. Plasmodesmata—Congresses. I. Robards, A. W. (Anthony William) II. Title. III. Series. QH604.2.N36 1989
574.87'5—dc20 90-9917

Softcover reprint of the hardcover 1st edition 1990

2131/3140-543210 – Printed on acid-free-paper

PREFACE

Functionally similar but structurally different permeable intercellular junctions provide routes for direct cell-to-cell signalling in both plants and animals. These are the plasmodesmata and gap junctions respectively. The physiology of junctional communication is better understood in plants while the technologies for investigation have been better developed in animal systems. The Proceedings of the Advanced Research Workshop recorded in this book explore the elements of similarity and difference produced by the parallel evolution of these junctional signalling systems and discuss the most recent methods for examining the physiological functions and regulation of intercellular communication.

There is increasingly strong evidence of both molecular as well as functional similarities between plasmodesmata and gap junctions. For example, the close correspondence in the size of molecules allowed to pass through plasmodesmata to those conducted through gap junctions has been demonstrated in a number of systems. Even more interesting is the discovery that animal gap junction proteins cross-react immunologically with some proteins in plant cells. Thus the molecular construction and function of these crucially important ultrastructural cell components is now open to a concerted research effort to understand how cells, both plant and animal, facilitate and regulate intercellular transport.

This book not only provides a contemporary review on the subject of intercellular communication, it also acts as a signpost for future work and for where interdisciplinary collaboration will have an important part to play in new developments in this important field. The rôle of communicating junctions in development, disease or simply the normal life of a plant or animal is of fundamental importance whether in the application of cell biology to medicine or to agriculture. Modern research is often at its most effective at the frontiers between conventional subject disciplines. Nowhere is this more clearly demonstrated than in the present volume where the value, indeed necessity, of using a wide range of experimental tools is amply demonstrated.

Table of contents

Structural and molecular diversity of gap junctions
R Dermietzel and T K Hwang 1

A structural analysis of the gap junctional channel and the 16K protein
M E Finbow, P Thompson, J Keen, P Jackson, E Eliopolous, L Meagher and J B C Findlay 13

Molecular heterogeneity of gap junctions in different mammalian tissues
K Willecke, H Hennemann, K Herbers, R Heynkes, G Kozjek, J Look, R Stutenkemper, O Traub, E Winterhager and B Nicholson 21

Biochemical and structural properties of the protein constituent of junctional domains in eye lens fiber plasma membranes
E L Benedetti, I Dunia, S Manenti and H Bloemendal 35

Junctional communication: The role of communication compartments in complex multicellular organisms
J D Pitts 53

Electrophysiological properties of gap junction channels
D C Spray 63

Cardiac gap junctions: Gating properties of single channels
H J Jongsma and M B Rook 87

Patterns of gap junctional permeability in developing insect tissues
S Caveney 101

Gap junctional intercellular communication and carcinogenesis
H Yamasaki 115

Progressive restrictions in gap junctional communication during development
F Serras and J A M van den Biggelaar 129

The neck region of plasmodesmata: General architecture and some functional aspects
P Olesen and A W Robards 145

Immunological investigations of relatedness between plant and animal connexins
A Xu, S Meiners and M Schindler 171

Secondary formation of plasmodesmata in cultured cells - Structural and functional aspects
J Monzer 185

Distribution of plasmodesmata in leaves. A comparison of *Cananga odorata* with other species using different measures of plasmodesmatal frequency
D G Fisher 199

Plasmodesmatal structure and function in nectaries
E P Eleftheriou 223

Analytical studies of dye-coupling between plant cells
E B Tucker 239

Function and localisation of movement proteins of tobacco mosaic virus and red clover mottle virus
K Tomenius 249

Plasmodesmata-virus interaction
W J Lucas, S Wolf, C M Deom, G M Kishore and R N Beachy 261

Subject Index 275

STRUCTURAL AND MOLECULAR DIVERSITY OF GAP JUNCTIONS

R Dermietzel and T K Hwang
Institute of Anatomy
University of Regensburg
UniversitätsstraBe 31
Fed Rep Germany

Introduction

Among the various membrane contacts (desmosomes,tight junctions,zonulae occludentes) gap junctions or nexuses have generated widespread interest during the last decade. One of the reasons for this increasing effort in gap junction research is a list of obvious functions with which these types of membrane contacts are considered to be involved. They are endowed with multi-facetted functions, albeit mostly speculative, including such diverse cell biological aspects as: a) regulation of differentiation during embryo- and organogenesis (Wolpert, 1978) b) co-ordination of cell proliferation (Dermietzel et al. 1987), c) co-operation of metabolic activities (Meda et al. 1982) and d) cell-to-cell propagation of electrical impulses (Bennett, 1977). The different aspects of possible gap junction function, which imminently harbour the connotation of signal trafficking from one cell to the next, led to the term cell-to-cell communication (Loewenstein, 1981) being coined. Being aware that different forms of cell-to-cell communication exist, such as hormonal and nervous communication, this specialised form of gap junction-mediated signal transfer can be regarded as "direct intercellular signal exchange".

The specificity of this type of communication is that it bypasses the extracellular space by transporting messenger molecules from a cytoplasmic source to the adjacent cytoplasmic interior via trans-membranous channels. These channels, called connexons, are joined in mirrored symmetry with connexons in the membrane of the adjacent cell thereby bridging the extracellular space. Thus the unitary cellular compartments which had been considered to represent individual units are connected by a network of routes which create a functional cellular syncytium. The importance of such a general intercellular communicative system can be appreciated if one considers the widespread occurrence of gap junctions among the living phyla. From primitive organisms like mesozoa (Revel, 1988) to the complex nervous system of higher vertebrates (Bennett, 1966) gap junctions are ubiquitous. Only few cell populations like skeletal muscle cells (which represent functional syncytia per se) and circulating blood cells do not possess gap junction channels.

NATO ASI Series, Vol. H 46
Parallels in Cell to Cell Junctions in Plants and Animals
Edited by A. W. Robards et al.

Communication between the cytoplasmic compartment of contiguous cells is also common in higher plants (Marchant, 1976).Here the intercellular connections are provided by tubular structures, called plasmodesmata, which offer a route of intercellular traffic for large and small molecules (Gunning, 1976). The critical molecular weight (M.W.) for the passage of hydrophilic molecules through plasmodesmata has been established by fluorescent dye injection and is reported to be about 1 kDa. This approaches the exclusion M.W. as described for the gap junction channel (Gunning, 1983). In addition,there are a number of functional similarities such as the control of channel gating by pH and/or Ca2+ (Gunning, 1983) which indicates that there are at least some common physiological properties for both types of intercellular channels. Recently antigenic cross-reactivity of soybean and root cell homogenates against a mono-specific anti-liver gap junction antibody was reported, suggesting a more evolutionary similarity than hitherto anticipated (Meiners and Schindler, 1987).

To better understand the similarities and disparities of both types of communicating channels it is appropriate to offer some detailed information on the structure and composition of the gap junction channels. Detailed information on the structural organization and composition of plasmodesmata are given later in this monograph.

Structural features of gap junctions: The gap junction channels are aligned in a unique paracrystalline pattern. Until now it is not surprising that morphologists have played a pioneering role in the initial discovery of gap junctions. Besides the classical description of electrical synapses in Mauthner cells of gold fish by Robertson (1963), the work of Revel and Karnovsky (1967) describing the occurrence of regular membrane differentiation between heart and liver cells by lanthanum staining, are historical milestones in elucidating the structure of gap junctions. The feature which emerged from conventional electron microscopy is that gap junctions are characterized by a close regional apposition of adjacent plasma membranes separated by a gap 2 nm in width (Fig. 1a). This led explorers to coin the term "gap junction" despite the fact that the gap itself is obviously an artefact of the staining procedure, commonly applied, to enhance the contrast in electron microscopy. In fact, by special staining techniques, particulate subunits spanning the extracellular space can be rendered visible.

Structural evaluation of gap junctions was made feasible with the introduction of freeze-fracturing technique into membrane research (Fig. 1b). According to freeze-fracture data, the gap junction consists of intra-membranous particles, aligned in a highly ordered pattern. The degree of order, however, seems to be variable. It has been suggested that this might reflect the actual physiological state of the channels, i.e. high resistance state providing a more ordered crystalline arrangement and low resistance state manifesting in a less well-ordered arrangement

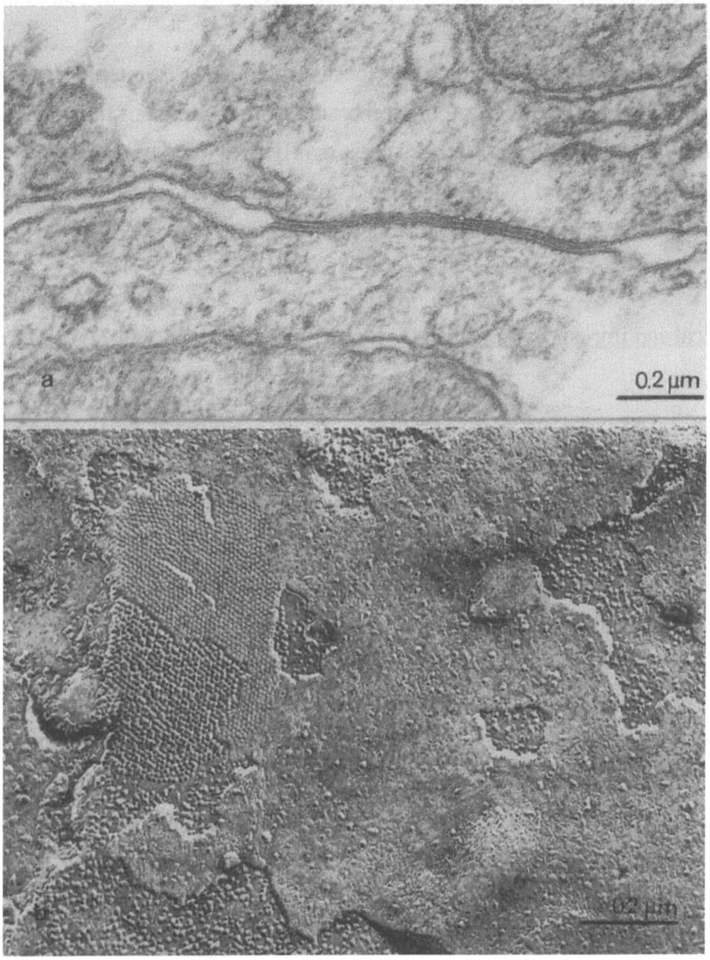

Fig. 1 a) Ultrathin section of a gap junction between two hepatocytes.
b) Corresponding freeze-fracture micrograph of a gap junction showing P- and E-face aspects of the connexons

of the particles (Peracchia and Peracchia, 1981). However, considerable differences existing in the crystalline packing between junctions in various organs even under uncoupling conditions which render such a general acceptance of this interpretation unlikely (Hanna et al. 1985).

A more detailed analysis of the structure of the gap junction channel was obtained by low angle X-ray diffraction (Makowski et al. 1977) and Fourier transformation of a series of tilted electron micrographs (Unwin and Zampighi, 1980). According to these data, the channel is composed of single subunits, called connexins, which are arranged hexamerically around a central pore with a maximum diameter of 1.2-1.5 nm. The hemi-channels spanning one plasma membrane are docked with the extracytoplasmic extremities of their corresponding counterparts in the adjacent plasma membrane.

Recent data obtained by cDNA cloning of the liver and heart connexins (see below) are in agreement with this model predicting four membrane-spanning units of the connexins with two highly conserved extracytoplasmic loops, presumably providing the docking sites. By site specific proteolysis the C- and N-termini of the connexins were found to be orientated towards the cytoplasm (Zimmer et al., 1987; Hertzberg et al., 1988).

Immunochemical and immunocytochemical evidence indicate the existence of a connexin family

A better refinement of the molecular composition of gap junctions was rendered possible when monospecific antibodies became available. These allowed the qualitative and quantitative detection of gap junction proteins in various tissues. The successful production of antibodies depended basically on the improvement of isolation techniques utilizing liver tissue as a source for gap junctions (Henderson et al., 1979; Hertzberg and Gilula, 1980). Thus, besides the lens fibre tissues which has also been frequently used for isolating gap junctions; but which will not be considered here, the initial data on the chemical composition of gap junctions refer to that of liver. The protein predominant in gap junction preparations from murine liver on sodium dodecyl sulphate - polyacrylamide gel electrophoresis (SDS-PAGE) gels had a relative molecular weight of 26-28 kDa. Apart from this major protein two further bands were consistently found. One of which migrated at a M.W. of about 45 kDa, thought to be the dimeric form of the 26-28 kDa protein and the other, found to be about 2-4 times less abundant than the 26-28 kDa protein in murine liver, at 21 kDa. Affinity purified polyclonal antibodies raised against the 26-28 kDa protein and the 21 kDa protein yielded a positive reaction with both the appropriate band and the 45 kDa band on western blots (Traub et al. 1982). By indirect immunofluorescence a discrete spotlike staining of membrane domains was discernible on liver cryostat sections (Fig. 2a). Immunoelectron microscopy evidently documented the 26-28 kDa protein as being localized within the gap junctional plasma membrane domains (Fig. 2b). Screening of diverse tissues with an anti-26 kDa antibody indicated that the same or cross-reacting homologous protein is present in various tissues, preferentially but not exclusively of epithelial origin (Dermietzel et al., 1984). However, some tissues known to be well-provided with gap junctions, e.g. heart cells and granulosa cells of the ovary did not cross-react with the anti-26 kDa antibody.

These findings strongly suggested that besides the 26 kDa protein of liver gap junctions other connexins would also occur. Through Edman degradation (Nicholson et al., 1985) and the successful cloning of a heart gap junction cDNA predicting a junctional protein with a molecular weight of 43 kDa (see below), initial immunocytochemical findings of a diversity in gap junction

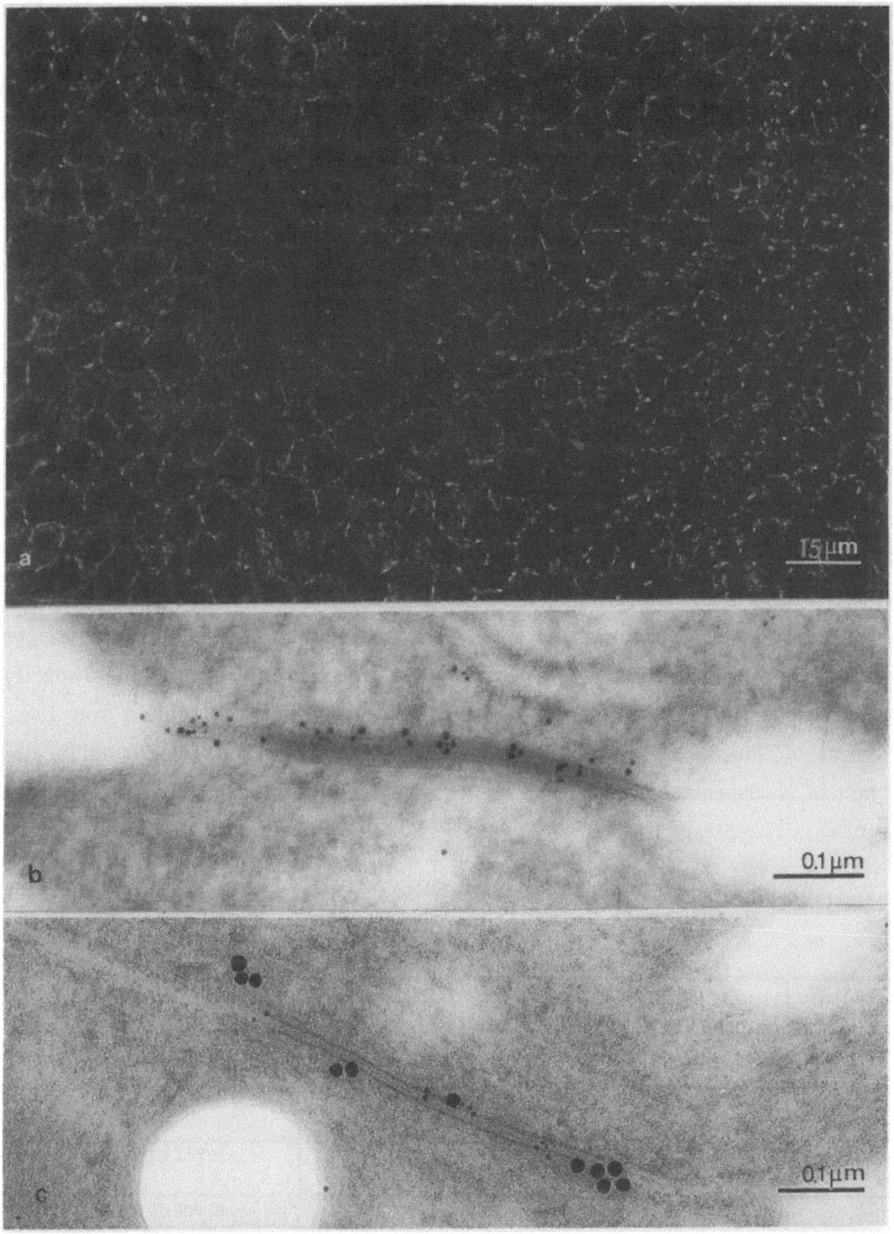

Fig. 2 a) Immunofluorescence of liver cryostat section stained with anti-connexin 32.
b) Immunogold labelling of a junctional domain from liver with anti-connexin 32.
c) Double immunolabelling indicating the co-existence of connexin 32 (large gold beads) and connexin 26 (small gold beads) within the same gap junctional plaque

proteins have recently been corroborated on a molecular level. Using the same immunochemical and immunocytochemical approaches in conjunction with N-terminal microsequencing, the 21 kDa protein (which had long been thought to be a contaminant of the preparation), has been shown to be second liver gap junction protein (Nicholson et al., 1987; Traub et al., 1989). Both proteins co-localize within the same gap junctional plaques of murine hepatocytes as recently evidenced by immunofluorescence and immunoelectron microscopy (Traub et al., 1989) (Fig. 2c). The message emerging from these discoveries is that gap junctions are composed of a family of antigenetically homologous proteins that are expressed in varying amounts in cell types of different origins.

Additional non-immunological evidence indicating the existence of related, but non-identical connexins in different tissues has been obtained by the successful cloning of the cDNA of the rat liver 26-28 kDa protein (Paul, 1986) and its human analogue (Kumar and Gilula, 1986).

Beyer et al. (1987), using the rat liver clone as a probe under low stringency hybidization conditions, were able to clone the cDNA for a heart and a lens connexin. The molecular weights predicted from the cDNA for the liver, the heart and the lens varied considerably from the molecular weight in their electrophoretic mobility. This neccessitated the introduction of a nomenclature system (Beyer et al. 1987) which uses the generic term connexin to be assigned to the protein family and the predicted molecular mass obtained from the cDNA to specify the different members of the gap junction family. The major 26-28 kDa protein from liver gap junctions would thus be referred to as connexin 32, the 21 kDa as connexin 26 and the 43 kDa heart protein as connexin 43. A list of different connexins as defined by their cDNAs is given in Table 1.

<u>Table 1</u>

<u>Connexins</u>	<u>Source</u>	<u>Species</u>	<u>Authors</u>
32	liver	rat/human	Paul (1986) Kumar et al. (1986)
26	liver	rat	Nicholson and Zhang (1988)
43	heart	rat	Beyer et al. (1987)
46	lens	rat	Beyer et al. (1988)
38	embryo	Xenopus	Ebihara et al. (1989)
30	embryo	Xenopus	Gimlich et al. (1988)

Diversification and time dependent expression of connexins

The question of tissue specific expression of connexins can be clarified either by immunocytochemical/immunochemical or molecular biological (Northern blot/in situ hybridization) analysis.

Both methods, the immunochemical as well as the molecular biological approaches have their advantages and limitations. While the former technique allows cell-specific detection and rapid screening of connexins, its major draw-back being concerned with immuno-crossreactivity, especially when polyclonal antibodies with "broad-clonality" or unselected monoclonal antibodies are utilized. Site-specific antibodies to non-homologous sequences are best suited to overcome this problem (Goodenough et al., 1988).

The detection of mRNAs, on the other hand, by the latter molecular biological approach is hampered by similar difficulties, i.e. cross-hybidization. The application of high stringency conditions is an effective way to minimize cross-hybridization. However, even under these conditions a positive reaction does not necessarily guarantee the probe to be identical to the mRNA it detects, especially in cases where high homology in the gene copies exist. In addition, in-situ hybridization gives only a limited amount of information with regard to the cell-specific presence of the particular RNA.

By compiling different data on tissue specific distribution of connexins, mRNAs from the literature and an immunocytochemical screening of a wide range of diverse organs using three different connexin antibodies we obtained a fairly accurate list of the abundances and cell-specific expression of the different connexins.

Two affinity purified polyclonal and one monoclonal antibody to the liver gap junction proteins connexin 32 and connexin 26, and a site-specific anti-connexin 43 antibody (Beyer et al., 1989) were used. A complete list of the expression sites is given in Table 2. Our results show that there is a remarkable diversification of connexin expression in different tissues. From the 14 tissues and cell cultures screened, 6 showed predominance of a single connexin. The rest expressed the presence of more than one connexin. As regards the monotypic expression of connexins, however, one has to take into account that our probes covered only a selected fraction of the connexins known to date. This warrants a re-examination of the expression pattern in the near future.

Through the phenomenon of co-expression, there seem to be predominant correlations indicating a preferred co-existence of connexin 32 and connexin 26 in tissue of epithelial origin. However, connexin 43 is also expressed in some epithelial derivatives including that of the epithelium of proximal tubules and epidermis. Interestingly, connexin 43 immuno-reactivity, occurring in conjunction with that of connexin 32 and 26, as evidenced in epithelial tissue of the kidney, showed an entirely different immuno-staining pattern. Most of the anti- 43

Table 2 Expression of connexins in diverse tissues (A) and cell cultures (B) and relative abundance of connexin mRNAs[a]

		anti-connexin 26	anti-connexin 32 (mRNAs[a])	anti-connexin 43 (mRNAs[a])
A)	liver rat	+	+++ (+++)	-/+ (-)
	liver mouse	++	+++ (ND)	-/+ (ND)
	heart	-	- (-)	+++ (+++)
	pancreas exocrine	+++	+++ (ND)	(ND)
	kidney	+++	++ (+++)	++ (++)
	endometrium	N.D.	+++	-
	myometrium	N.D.	++	++ (+++)b
	neuron	-	+/- (++)[c]	ND (-)[c]
	astrocytes	-	-	+++
	oligo-dendrocytes	-	+++	-
	epidermis	++	- (ND)	+++ (ND)
B)	fibroblasts rat	+/-	-	++
	WBL cells	-	-	+++
	Marshall cells	-	-	+++
	3T3 fibroblasts	-	-	+/-

immunoreactivity in epithelial tissues are obviously derived from interstitial fibroblasts. Until now we have no evidence of a co-expression between connexin 32 and connexin 43 although there is good evidence of connexin 26 and connexin 32.

The described observations strongly favours the neccessity of a thorough, cell-specific determination of connexin expression which should also include immuno-electron-microscopical detection. Our data on the expression of connexins in brain tissues (Table 2) gives a plausible example how this task can be resolved. By applying cell-specific markers for neuronal as well as glial cells, we obtained evidence for a highly diversified expression of connexins in the brain at a cellular level. Immunoreactivity to the heart connexin 43 occurs in cells, staining positively to the astrocyte specific marker - glial fibrillary acid protein (GFAP). The liver connexin 32 is also present in cells where myelin basic protein (MBP) occurs. It also occurs in a small fraction of nEn (neuron specific enolase) positive cells in the central nervous system. Connexin 26 on the other hand, is neither expressed in glial nor neuronal tissues of the adult brain, but abundantly present in leptomeningeal cells.

Furthermore, during embryonic and postpartal maturation of brain tissues, a differential expression of the three connexins (26, 32, 43) was encountered, showing the predominance of connexin 26 and connexin 43 in the neuroepithelium of embryonic brains, with a reduction in connexin 26 approaching zero level around the 3rd week after parturition. Connexin 32, on the other hand, shows a reversed order of appearance, i.e. an increase at the time when connexin 26 is at its minimum.

These data strongly suggest that besides the obvious diversification of the connexins there seems to be a time dependent expression, at least in developing tissues, a finding that has been corroborated by the recent detection of embryo specific mRNA in Xenopus (Gimlich et al., 1988; Ebihara et al., 1989).

The discovery of a time-dependent expression in connexins, however, is not a recent one. Variation in the concentration of connexins has also been reported to be under operational control under different physiological influences such as cellular proliferation (Dermietzel et al., 1984), gland cell secretion and hormonal regulation (Meda et al., 1987).

---

a) mRNA data obtained from Beyer et al. (1988)
b) No differentiation on uterus tissues done
c) No differentiation on brain tissues done
N.D. not determined
\+ relative abundance of immuno/mRNAs reactivity
\+ weak
++ medium
+++ strong
\- negative

Does the structural diversity of connexins also indicate functional differences in the gap junction channels?

Speaking in terms of the functional significance in the gap junction channels, we are still concerned with phenomenological clues. The permissive function of the gap junction channels which allow the bi-directional exchange of ions and small metabolites from cell-to-cell is routinely assayed by dye- and/or electrical coupling. However, besides a definitive "all or none" concept concerning the coupling event these techniques say little about the functional and molecular properties of the channels.

The crucial issue arising from the data given above is: Does the molecular diversity in gap junctional composition reflect some functional differences in channel properties? To date there is little evidence indicating an interdependency between the molecular composition of the channels and their functional characteristics. Double-patch clamp recordings of single channels show that there are differences in the conductance of heart gap junction channels and liver gap junction channels (Burt and Spray, 1988; Spray et al., 1986). Differences in the pharmacological properties of the different channels also seem to exist (see D. Spray, this Volume).

In conclusion, immunochemical, and molecular biological evidence strongly indicate that gap junctions are composed of a family of related but non-identical proteins. Diverse tissues seem to express their specific sets of connexins which vary in their qualitative and quantitative composition according to their cellular composition and actual functional and/or developmental stage.

A major goal in future gap junction research would be a convergent application of biochemical, immunochemical, molecular biological and physiological techniques to unravel the functional meaning of the aforementioned diversity of connexins.

References

Bennett, M.V.L. (1966) Physiology of electrotonic junctions. Ann. New York Acad. Sci. 509-539

Bennett, M.V.L. (1977) In: Handbook of Physiology, Section I: The Nervous System, ed. Kandel E.R. (Williams and Wilkins, Baltimore, MD) pp. 357-416

Beyer E.C., Paul D.L. and Goodenough D.A. (1987) Connexin 43: A protein from rat heart homologous to gap junction protein from liver. J. Cell Biol. 105: 2621-2629

Beyer E.C., Kistler J., Paul D.L. and Goodenough D.A. (1989) Antisera directed against connexin 43 peptides react with a 43 kDa protein localized to gap junctions in myocardium and other tissues. J. Cell Biol. 108: 595-605

Burt J.M. and Spray D.C. (1988) Iontropic agents modulate gap junctional conductance between cardiac myocytes. Am. J. Physiol. 254: H 1206-H 1210

Dermietzel R., Leibstein A., Frixen U. Janssen-Timmen U, Traub O. and Willecke K. (1984) Gap junctions in several tissues share antigenic determinants with liver gap junctions. EMBO J. 3: 2261-2270

Dermietzel R., Yancey S.B., Traub O., Willecke K. and Revel J-P. (1987) Major loss of the 28,000 dalton protein of gap junction in proliferating hepatocytes. J. Cell Biol. 105: 1925-1934
Ebihara L., Beyer E.C., Swenson K.I., Paul D.L. and Goodenough D.A. (1989) Cloning and expression of a Xenopus embryonic gap junction proteins. Science 243: 1194-1195
Gimlich R.L., Kumar N., and Gilula N.B. (1988) Sequence and developmental expression of mRNA coding for a gap junction protein in Xenopus. J. Cell Biol. 107: 1065-1073
Goodenough D.A., Paul D.L., Jesaitis L., (1988) Topological distribution of 2 connexin 32 antigenic sites in intact and split rodent hepatic gap junctions. J. Cell Biol. 107: 1817-1824
Gunning B.E.S. (1976) Intercellular communication in plants: Studies on plasmodesmata. Springer-Verlag, New York, pp. 1-13
Gunning B.E.S. and Overall R.L. (1983) Plasmodesmata and cell-to-cell transport in plants. Bioscience 33: 260-265
Hanna R.B., Ornberg R.L. and Reese T.S (1985) In: Gap Junction (eds. Bennett, M.V.L. and Spray D.C.) Cold Spring Harbor 1985, pp. 23-32
Henderson D., Eibl H. and Weber K. (1979) Structure and biochemistry of mouse hepatic gap junctions. J. Mol. Biol. 132: 193-218
Hertzberg E.L. and Gilula N.B. (1979) Isolation and characterization of gap junctions from rat liver. J. Cell Biol. 254: 2138-2147
Hertzberg E.L., Disher R.M., Tiller A.A., Zhou Y. and Cook R.G. (1988) Topology of the Mr 27,000 liver gap junction protein. J. Biol. Chem. 264: 19105-19111
Kumar N.M. and Gilula N.B. (1986) Cloning and characteri-zation of rat liver cDNAs coding for a gap junction protein. J. Cell Biol. 103: 767-776
Loewenstein W.R. (1981) Junctional intercellular communications. The cell-to-cell membrane channel. Physiol. Rev. 61: 829-913
Makowski L., Caspar D.L.D., Phillips W.C. and Goodenough D.A. (1977) Gap junction structure I. Correlated electron microscopy and x-ray diffraction. J. Cell Biol. 74: 605-628
Marchant H.J. (1976) In: Intercellular Communication in Plants: Studies in Plasmodesmata (eds. Gunning B.E.S. and Robards A.W.), Springer-Verlag, New York, pp. 59-80
Meda P., Kohen E., Kohen C., Rabinowitch A. and Orci L. (1982) Direct communication of homologous and heterologous endocrine islet cells in culture. J. Cell Biol. 92: 221-226
Meda P., Bruzzone R., Chanson M., Bosco D., Orci L. (1987) Gap junctional coupling modulates the secretion of exocrine pancreas. Proc. Natl. Acad. Sci. U.S.A. 84: 4901-4903
Meiners S. and Schindler M. (1987) Immunological evidence for gap junction polypeptide in plant cells. J. Biol. Chem. 262: 951-953
Nicholson B.J., Gros D.B., Kent S.B.H., Hood L.E. and Revel J-P. (1985) The Mr28,000 gap junction proteins from rat heart and liver are different but related. J. Biol. Chem. 260: 6514-6517
Nicholson B.J., Dermietzel R., Teplow D., Traub O., Willecke K. and Revel J-P. (1987) Two homologous protein components of hepatic gap junctions. Nature 329: 732-734
Paul D.L. (1986) Molecular cloning of cDNA for rat liver gap junction protein. J. Cell Biol. 103: 123-134
Peracchia C. and Peracchia L. (1981) Gap junction dynamics: Reversible effects of hydrogen ions. J. Cell Biol. 87: 719-723
Revel J-P. and Karnovsky M.J. (1967) Hexagonal arrays of subunits in intercellular junctions of the mouse heart and liver. J. Cell Biol. 33: C7-C18
Revel J-P. (1988) The oldest multicellular animal and its junctions. In: Gap Junctions (eds. Hertzberg E.L. and Johnson R.G.) Alan R. Liss. Inc. New York.
Robertson J.D. (1963) The occurrence of a subunit pattern in the unit membranes of club endings in Mauthner cell synapses in goldfish brains. J. Cell Biol. 19: 210-232
Spray D.C., Saez J.C. Brosius D., Bennett M.V.L., and Hertzberg E.L. (1986) Isolated liver gap junctions gating of transjunctional current is similar to that in intact pairs of rat hepatocytes. Proc. Natl. Acad. Sci. U.S.A. 83: 5494-5497
Traub O., Janssen-Timmen U., Druege P., Dermietzel R., Willecke K. (1982) Immunological properties of gap junction protein from mouse liver. J. Cell Biochem. 19: 27-44

Traub O., Look U., Dermietzel R., Bruemmer F., Huelser D. and Willecke K. (1989) Comparative characterization of the 21 kDa and 26 kDa gap junction proteins in murine liver and cultured hepatocytes. J. Cell Biol. 108: 1039-1051

Unwin P.N.T. and Zampighi G. (1980) Structure of the junction between communicating cells. Nature (Lond.) 283: 545-549

Wolpert L. (1978) Gap junctions: Channels for communications in development. In: Intercellular junctions and synapses (eds. Feldmann J. et al.), Chapman & Hall, London, pp. 83-94

Zimmer D.B., Green C.R., Evans W.H. and Gilula N.B. (1987) Topological analysis of the major protein in isolated intact rat liver gap junctions and gap junction-derived single membrane structures. J. Biol. Chem. 262: 7751-7763

A STRUCTURAL ANALYSIS OF THE GAP JUNCTIONAL CHANNEL AND THE 16K PROTEIN

Malcolm E. Finbow, Paul Thompson*, Jeff Keen, Phillip Jackson* Elias Eliopolous+, Liam Meagher and John B.C. Findlay*
Beatson Institute for Cancer Research
Garscube Estate
GLASGOW G61 1BD
Scotland

Introduction

Membrane proteins have been the focus of much attention in recent years. Advances in the handling of membrane proteins, in protein sequencing technology and in molecular biology have been coupled with high resolution imaging techniques and chemical labelling studies. This has resulted in detailed pictures of a number of membrane proteins including the photoreaction centre (Huber, 1989), acetyl choline receptor (Unwin *et al*, 1988) and rhodopsin (Findlay & Pappin, 1986). The knowledge which has been gained from these proteins provides a useful framework for the analysis of the organisation of other membrane protein complexes.

The gap junction was one of the first membrane complexes to be isolated (Benedetti and Emmelot, 1968). These junctions are composed of aggregates of channel particles in one membrane which align end-to-end with channel particles in the apposing membrane. Each channel contains a centrally located pore and a pair of channels provides a conduit for direct communication between the cytoplasms of neighbouring cells. Low resolution models of the individual channels have been available for some time (Caspar *et al*, 1977; Makowski *et al*, 1977) and functional studies have defined their permeability properties (see Pitts & Finbow, 1986). Yet despite this information and the large number of biochemical studies, the protein composition has been a point of controversy over many years.

Two types of protein have been proposed as the sole constituent of gap junctions, the connexins (MW 26k-45k) (see chapters by Willecke *et al* and Dermietzel *et al*, this volume) and a protein of Mr 16,000 (16k protein; Finbow *et al*, 1983, 1984; Buultjens *et al*, 1988).

Departments of Biochemistry* and Biophysics+, University of Leeds, Leeds LS2 9JT

NATO ASI Series, Vol. H 46
Parallels in Cell to Cell Junctions in Plants and Animals
Edited by A. W. Robards et al.

The complete sequences of several connexins are known and they have no homology with the 16k protein. Furthermore, recent evidence suggest they are components of different membrane complexes (Finbow & Hertzberg, unpublished results). It is therefore not known whether the models of gap junctional channels obtained from preparations of rodent liver gap junctions (Unwin & Zampighi, 1979; Unwin & Ennis, 1984) were isolated structures with a connexin component, a 16k protein component or a mixture of both. This makes structural correlations difficult on the presently available evidence. One approach to resolving this difficulty is to isolate the structures formed by only one of the two proteins. The gap junctional preparations from hepatopancreas of the marine lobster *Nephrops norvegicus* offer such an opportunity. Particularly pure preparations can be made in high yield from this tissue which are enriched in only the 16k protein (Finbow *et al*, 1984). As yet no connexins or connexin-related molecules have been identified in arthropods.

The Structure of Gap Junctions from the Hepatopancreas of *Nephrops norvegicus*

The hepatopancreas of crustaceans is a particularly rich source of gap junctions which may comprise upwards of several percent of the total surface area of the columnar epithelial cells, the major cell-type (Lane *et al*, 1986). The junctions appear very similar to those of other species consisting, of two apposed plasma membranes with an overall width of 15-18 nm. *En face* views after infiltration with electron dense extracellular tracers reveal the characteristic hexagonally packed 8-10 nm particles. These particles are also seen in freeze-fracture but unlike their vertebrate counterparts, the particles partition preferentially to the E-face rather than the P-face. However, there is considerable phyletic pleomorhophism in this property and it may reflect differences in lipid composition or other attachments rather than inherent differences of the channel proteins.

The isolated gap junctions from *Nephrops* hepatopancreas retain much of their *in situ* structure and appear morphologically similar to those isolated from rodent liver. The overall thickness is 14-15 nm suggesting a unit length of the channel in each membrane is 7-7.5 nm (Leitch & Finbow, unpublished results). Negative staining reveals the hexagonal array of 7-8 nm particles each of which contains a central pore that is better penetrated by cationic than anionic negative stains (Baker *et al*, 1985; Finbow *et al*, 1984; Buultjens *et al*, 1988; Finbow *et al*, unpublished results). The minor differences in thickness and particle dimensions before and after isolation are to be expected as different electron microscopic techniques are used and the methods of extraction (e.g. detergent and trypsin treatment) required to isolate pure fractions may cause loss of loosely associated material.

The 16k protein is the major protein in these preparations from the hepatopancreas and immunological approaches have recently confirmed the 16k protein as a component of the isolated gap junctional structures and of gap junctional areas in membrane preparations (Buultjens *et al*, 1988; Leitch & Finbow, unpublished results). The isolated structures are particularly resistant to proteases and *in situ* so is the 16k protein suggesting it is largely buried in the membrane. Thus, any projected model of the 16k protein must take into account this feature.

The Structure of the 16k Protein Channel Complex

Chemical sequencing of the 16k proteins from *Nephrops* hepatopancreas and mouse liver is nearing completion and this shows a high degree of conservation (>85% identity). Computer predictions and molecular modelling suggest the 16k protein forms four extended alpha helices, the smallest being no less than 24 residues in length, that pass through the membrane at a tilt in a manner analagous to the helices of the photoreaction centre (Huber, 1989). The helical regions are also tilted with respect to each other and have a diamond-shaped arrangement when viewed in cross-section. They are joined by short loops and are capable of being tightly packed in the membrane. In this disposition, most of the protein is buried in the membrane consistent with the protease studies.

Two of the helices (1 and 4) contain single, but opposite charged residues (Lys and Glu) which appear to point into the central region between the four helices and may be involved in salt bridging. All other charged residues are located at the ends of the helices and in loop regions on either side of the membrane in hydrophylic enviroments.

Diffraction analysis of electron micrographs of negatively stained gap junctions from the *Nephrops* hepatopancreas show hexagonal symmetry (Buultjens *et al* 1988). Indeed, in well stained areas individual channels can be seen to be composed of a ring of six subunits surrounding a central pore (Finbow *et al*, unpublished results). Such an arrangement can be achieved with the model of the 16k protein. Aromatic interactions can take place between helices 2 and 4 of adjacent 16k protein subunits to produce an oligomeric complex. This arrangement also results in the hydrophobic surfaces of helix 3 facing outwards into the membrane and the relatively polar (e.g. ser, ala and gly) face of helix 1 forming the wall of a central pore. The pore is likely to be water filled and its relatively large diameter (1-2nm) would not have any obvious selectivity other than size. That is, it has features expected of the pore of a gap junctional channel. Similar pores can be created using long ampiphatic peptides composed of a leu-leu-ser-ser repeat (Lear *et al*, 1988).

The orientation of the 16k protein in the membrane is of importance in identifying which regions of the molecule may be involved in protein-protein interactions. Pronase treatment of intact gap junctions does not alter the migration of the 16k protein on SDS-PAGE but does remove a small blocked N-terminal peptide (Finbow *et al*, unpublished results). The removal of this shows it is located on an exposed surface, most likely the equivalent of the cytoplasmic surface of the intact gap junction. This being so, the two loops between helices 1 and 2 and between helices 3 and 4 would lie on the equivalent of the extracellular face and perhaps offer sites of interaction between apposing channels. This orientation is supported by the finding that the single cysteine residue present in the second of these two loops, is inefficiently labelled by iodoacetamide in the intact structure (Thompson, unpublished results) indicating it is somehow masked, as might occur at a site of interaction .

The Gap Junctional Channel of Vertebrates

It is now possible to return to the original question of which protein is responsible for the accepted model of the rodent liver gap junctional channel (Unwin & Ennis, 1984; Unwin & Zampighi, 1979; Caspar *et al*, 1977; Makowski *et al*, 1977) by comparing this model with the structure of the *Nephrops* gap junctions and with the computer derived model of the channel formed from the *Nephrops* 16k protein. This comparison shows the main features of the two forms are very similar including the hexagonal symmetry, the channel dimensions, and the disposition of the protein in the membrane conferring the well documented insensitivity of gap junctions to proteases. The electron diffraction studies on the rodent form also show the cross-sectional area of each subunit in the membrane region is sufficient to accomodate four alpha helices (Unwin, 1987) and there is evidence that these pass through at a tilt (Unwin & Ennis, 1985). Furthermore, the calculation of mass of protein per channel from the models of the rodent gap junctional channel in the light of more recent data from the photoreaction centre suggests a subunit molecular weight of about 16kD. For example, an average of 25 residues are required for each transmembrane helix of the photoreaction centre (Huber, 1989). Given four helices, the subunit mass in the membrane is about 11,000 daltons (2,750 daltons per helix). All the published models show about two thirds of the total mass is in the membrane and the total predicted mass is therefore close to 16kD.

The marked similarity between the *Nephrops* and rodent forms of gap junctions suggests a common structural component. As described above, a closely related 16k protein is present in gap junctional isolates from rodent liver. It seems likely that this protein provides a substantial part of the gap junction structures imaged in X-ray and electron diffraction studies.

A further feature of the gap junctional channel protein predicted from functional studies is the high degree of conservation of the extracellular surfaces of the channels. This conservation can be seen by the ability of *Xenopus* fibroblasts in culture to form fully functional gap junctions with hamster fibroblasts (Pitts, 1976). Inspection of the sequences in the presumptive extracellular regions of the 16k proteins from *Nephrops* and mouse and of a bovine cDNA coding for the 16k protein (see below), show complete identity.

The 16k Protein as a Proton Channel Component of Vacuolar H^+-ATPases

A 16k protein has also been identified as the membrane component of vacuolar (V-type) H^+-ATPases from many different sources (fungal, plant and animal). These ATPases function to acidify the insides of vacuoles at the expense of ATP and the resulting proton gradient from inside the vacuole to the cytoplasm is coupled to the inward transport of molecules. (for review see Nelson and Taiz, 1988). They are composed of two separable protein complexes, the cytoplasmic ATPase complex (thought to be the equivalent of the F1 complex of eubacterial F-type H^+-ATPases) and a membrane spanning complex (thought to be equivalent of the F_0 complex) which forms the conduit for protons.

The 16k protein from the V-type H^+-ATPase has been isolated from chromaffin granule membranes of bovine adrenal medulla. Partial sequencing of the N-terminus and of cyanogen bromide fragments shows striking identity with the 16k protein found in gap junctions (Mandel *et al*, 1988). This identity has been confirmed for the whole protein by the deduced sequence from a cDNA isolated from a bovine adrenal medulla library. More recently, a gene has been characterised from yeast which again encodes a 16k protein with a high degree of identity (Nelson & Nelson, 1989). In the case of bovine adrenal medulla and other metazoan sources of the V-type H^+-ATPases, it could be argued the 16k protein in the vacuolar preparations arises from a detergent soluble gap junctional contaminant and is not part of the ATPase complex. Such an argument can not be put forward for yeast because gap junctional structures are not thought to be present. This leads to the possibility that the 16k protein which has the properties of a gap junctional channel, is the membrane component of the V-type H^+-ATPase.

The electron microscopic analyses of isolated gap junctions and the computer model predict a hexamer of the 16k protein with a centrally located pore. Any selectivity of this pore would have to come either from the soluble ATPase complex, or from as yet unidentified membrane components such as the equivalents of subunits 'a' and 'b' of the eubacetrial F_0 complex. Indeed, it has been suggested the 16k protein arose by gene duplication from an ancestral gene coding for an 8k protein which also gave rise to the 8k subunit 'c' proteolipid of the F_0 complex (Mandel *et al*, 1988).

An alternative site of proton translocation exists in the 16k protein and this is between the four helical domains. This might provide the proton selectivity as well as explaining the action of the lipophilic modifying reagant, dichlorohexylcarbodiimide (DCCD). DCCD is an inhibitor of proton translocation of the V-type and F-type H^+-ATPases binding to the 16k and 8k proteins respectively (See Nelson & Taiz, 1989). Recent studies on the 16k protein in its gap junction form show that it too binds DCCD specifically at the glutamate residue in the 4th helix (Thompson *et al*, unpublished results). This residue occupies an analagous site to the reactive acidic residue of the 8k protein.

A proton pathway through the helices makes the central pore not only redundant but also counter-productive because it would dissipate any proton gradient formed if it is not somehow regulated. Moreover, if the 16k protein is serving only a proton channel function, it is difficult to understand why it should form a stable double membrane structure containing an hexagonal array of gap junctional-like channels. This property contrasts with the detergent solubility of the vacuolar H^+-ATPase complex including the membrane component indicating these complexes are not organised into extended arrays.

The possibility exists that the 16k protein may serve different transport functions at different locations in the cell. The double membrane structures formed from the 16k protein channel may provide a role in cell-cell communication and possibly a role in communication between intracellular membrane compartments. It is noteworthy that apparent double membrane channels with gap junction-like properties occur during the early stages of fusion of vacuoles to the plasma membrane forming a 'fusion pore' (Almers *et al*, 1989). The existence of a 'fusion pore' alone without membrane fusion is sufficient to account for the release of neurotransmitters at the chemical synapse (Almers *et al*, 1989). The double membrane form of the 16k protein might provide the structural basis of these hypothetical fusion pores and hence be the pathway for neurotransmitter release as these molecules are small enough to pass through gap junctional channels. The presence of the 16k protein in the synaptic region of neurons seems likely in view of the recent finding of a V-type H^+-ATPase in synaptic vesicles (Cidon & Sihra, 1989).

The channel complex formed by the 16k protein is particularly intriguing. Much work is needed to be done to understand its potentially varied functions and how these might be separated and regulated. However, it is clear that the 16k protein is of importance in eukaryotes and analysis of its structure and properties may provide much information on other membrane transport proteins.

References

Almers W, Breckenridge LJ, Spruce AE (1989) The mechanism of exocytosis during secretion in mast cells 'in' Secretion and its control. The Rockefeller University Press pp 269-282

Baker TS, Sosinsky GE, Caspar DLD, Gall C, Goodenough DA (1985) Gap junction structures VII. Analysis of connexon images obtained with cationic and anionic negative stains. J Mol Biol184:81-98

Benedetti E, Emmelot P (1968) Buultjens TEJ, Finbow ME, Lane NJ, Pitts JD (1988) Tissue and species conservation of the vertebrate and arthropod forms of the low molecular weight (16-18000) proteins of gap junctions. Cell Tiss Res 251:571-580

Caspar DLD, Goodenough DA, Makowski L, Phillips WC (1977) Gap junction structure I. Correlated electron microscopy and x-ray diffraction. J Cell Biol 74:605-628

Cidon S, Sihra TS (1989) Characterization of a H^+-ATPase in rat brain synaptic vesicles. J Biol Chem 264:8281-8288

Finbow ME, Shuttleworth J, Hamilton AE, Pitts JD (1983) Analysis of vertebrate gap junctions. EMBO J 2:1479-1486

Finbow ME, Buultjens TEJ, Lane NJ, Shuttleworth J, Pitts JD (1984) Isolation and characterisation of arthropod gap junctions. EMBO J 3:2271-2278

Findlay, JBC, Pappin DJC (1986) The opsin family of proteins. Biochem J 238:625-642

Huber R (1989) A structural basis of light energy and electron transfer in biology. EMBO J 8:2125-2147

Lear JD, Wasserman W, DeGrado WF (1988) Synthetic amphiphilic peptide models for protein ion channels. Science 240:1177-1181

Makowski L, Caspar DLD, Phillips WC, Goodenough DA (1977) Gap junction structure II. Analysis of of the x-ray diffraction pattern. J Cell Biol 74:629-645

Mandel M, Moriyama Y, Hulmes JD, Pan Y-CE, Nelson H, Nelson N (1988) cDNA sequence encoding the 16-kDa proteolipid of chromaffin granules implies gene duplication in the evolution of H^+ATPases. Proc Natl Acad Sci USA 85:5521-5524

Nelson H, Nelson N (1989) The progenitor of ATP synthase was closely related to the current vacuolar H^+-ATPase. FEBS Letts 247:147-153

Nelson N, Taiz L, (1989) The evolution of H^+-ATPases. TIBS 14:113-116

Pitts JD (1977) Direct communication between animal cells 'in' International Cell Biology 1976-77, eds Brinkley BR, Porter KR, The Rockefeller University Press, New York, pp 43-49

Pitts JD, Finbow ME (1986) The gap junction. J Cell Sci [Suppl] 4:239-266

Unwin, PNT (1987) Gap junction structure and the control of cell-cell communication 'in' Junctional complexes of epithelial cells. eds Bock G, Clark S, CIBA Foundation Symposium 125, J. Wiley & Sons, pp78-103

Unwin PNT, Ennis PD (1984) Two configurations of a channel-forming membrane protein. Nature 307:609-613

Unwin PNT, Zampighi G (1980) Structure of the junction between communicating cells. Nature 283:545-549

MOLECULAR HETEROGENEITY OF GAP JUNCTIONS IN DIFFERENT MAMMALIAN TISSUES

K Willecke, H Hennemann, K Herbers, R Heynkes, G Kozjek,
J Look, R Stutenkemper, O Traub, E Winterhager[1] and
B Nicholson[2]
Institut für Genetik
Abt. Molekulargenetik
Universität Bonn
Römerstr. 164
5300 Bonn 1
Fed Rep of Germany

Introduction

The existence of direct communicating channels between cells was originally postulated from observations of electrical conductance between apposed cells as well as from results of electron microscopy. The latter method revealed clustered arrays of hexagonal particles that were later interpreted to be built up from 6 protein subunits. Two hemichannels each of which in the plasma membrane of apposed cells are thought to dock to each other and form a complete gap junction channel. Ions and metabolites of < 900 Daltons molecular weight can diffuse through gap junctions from one mammalian cell into the next one. Gap junctions are highly conserved from hydra to man where they have been demonstrated in almost all tissues with the exception of skeletal muscle cells and at late stages of hematopoietic differentiation. Although gap junctions from different tissues look quite similar in the electron microscope it has been shown in the last few years that proteins which were isolated with gap junctions from different tissues show different molecular weights. Furthermore antibodies specific to gap junctions clearly differentiate between gap junction proteins from different tissues although other antibodies have been described that recognize different gap junction proteins in rat tissues.

There are several recent reviews on gap junctions that cover most of the aspects of this research field (Loewenstein, 1987; Bennett and Spray, 1987; Bock and Clark, 1987; Warner, 1988; Willecke and Traub, 1989). In addition several monographs on gap junctions have been published or are in press (Hertzberg and Johnson, 1988; De Mello, 1987; De Mello, 1989;

[1] Grünenthal GmbH, Forschungszentrum,. Zieglerstr. 6, 5100 Aachen, Fed. Rep. of Germany
[2] Dpt of Biological Sciences, Suny at Buffalo, Buffalo, NY 14260, USA

NATO ASI Series, Vol. H 46
Parallels in Cell to Cell Junctions in Plants and Animals
Edited by A. W. Robards et al.

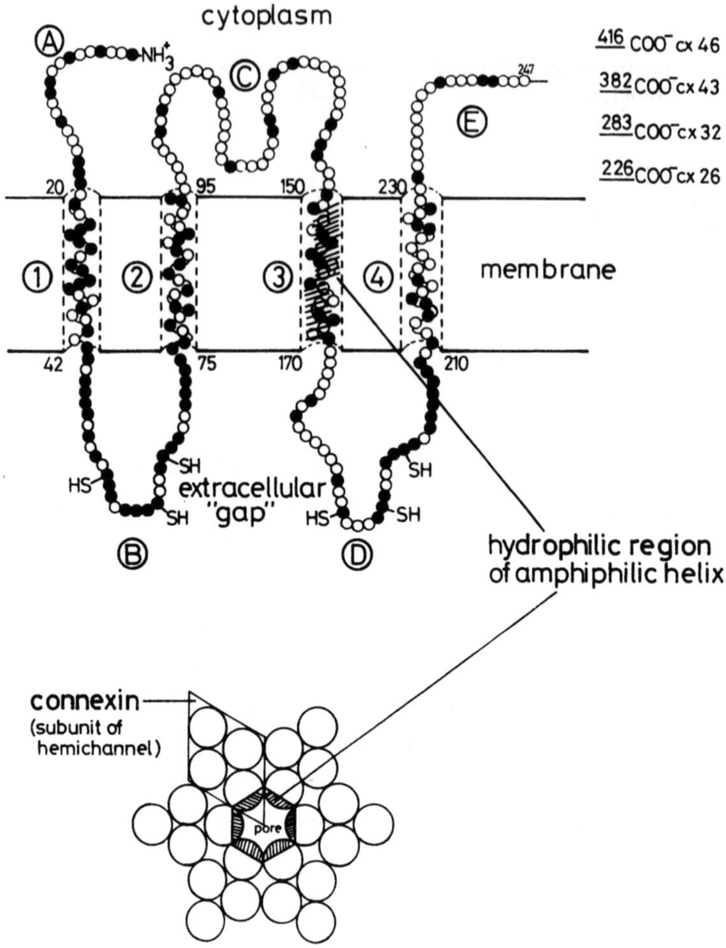

Fig. 1 Hypothetical topological model of connexins (Paul, 1986; Kumar and Gilula, 1986; Zimmer et al. 1987; Beyer et al., 1988)

In the upper part of this figure the partial amino acid sequence of connexin43 (cx43) is schematically shown and compared with that of cx32 (Beyer et al., 1988). Black circles represent positions of identical amino acids, open circles represent positions of non identical or additional amino acids. The positions of amino acid homology are based on an optimized, computer-aided alignment of protein sequences deduced from cDNAs of cx43 and cx32 (Paul, 1986; Beyer et al., 1988). Optimal alignment means that a total of 21 amino acids of cx43 are missing in cx32 whereas two amino residues of cx32 are not represented in cx43. The encircled numbers and letters of the different connexin regions are shown as suggested by Nicholson and

Sperdakis and Cole, 1989). In this chapter we want to discuss some recent results from our laboratory which relate to the heterogeneity of murine gap junctions and their controlled expression in different types of cells.

The Connexin Gene Family

Using rabbit antibodies to the 27 kDa gap junction protein from rat liver the complete cDNA corresponding to this protein was isolated from a lambda gt11 cDNA expression library (Paul, 1986). At the same time partial 27 kDa cDNA sequences from rat liver (Kumar and Gilula, 1986; Heynkes et al., 1986) and the complete cDNA corresponding to the human liver 27 kDa protein (Kumar and Gilula, 1987) were independently described. The amino acid sequence deduced from the cDNA sequence was analyzed for its theoretical hydrophobicity. This analysis resulted in a topological model of the 27 kDa protein spanning the plasma membrane four times (cf. Fig. 1). Each transmembrane region corresponds to about 2O amino acids. Three transmembrane alpha-helical regions exclusively consist of hydrophobic amino acid residues whereas one transmembrane region also contains several hydrophilic amino acids. Thus it has the characteristics of an amphiphilic helix and could possibly contribute to formation of the hydrophilic gap junction channel (cf. Fig. 1, based on data from Paul, 1986; Kumar and Gilula, 1986; Zimmer et al., 1987; and Beyer et al., 1988). Both the N terminal and the C terminal part of the 27 kDa protein were suggested to be on the cytoplasmic side of the membrane. Recent evidence based on reactivity of the 27 kDa protein in liver membranes with antibodies to synthetic oligopeptides (Zimmer et al., 1987; Milks et al., 1988; Goodenough et al., 1988) appears to support this suggestion. However, further experimental proof is needed to verify this topological model.

Zhang, 1988. In E the amino acid sequence ends at amino acid number 247 of cx43 which corresponds to amino acid number 226 of cx32. The total lengths of the different connexin sequences is indicated by the numer of the corresponding carboxy terminal amino acid. Cx26 refers to connexin26, cx46 means connexin46. The transmembrane region 3 has an amphiphilic character.
The lower part of this figure shows the hypothetical structure of the connexin hemichannel viewed perpendicular to the plane of the membrane. The central pore could be formed from 6 amphiphilic regions each of which could be part of another connexin molecule. Each open circle represents a transmembrane region, four of which belong to one connexin molecule. This figure is reprinted from Willecke and Traub, 1989, with permission of CRC Press, Boca Raton, Florida, USA.

When a cDNA library from rat heart was hybridized to rat liver 27 kDa cDNA at low stringency a new cDNA was discovered which showed regions of 58% and 42% homology to the 27 kDa cDNA (Beyer et al., 1987). Thus both cDNAs belong to the same gene family. The corresponding proteins were named connexin32 (cx32, formerly called 26 to 28 kDa gap junction protein) and connexin43 (cx43) according to their theoretical molecular weights (in kDa) calculated from the cDNA-deduced amino acid sequence (Beyer et al., 1987). The hydrophobicity plot of the cx43 amino acid sequence yielded the same hypothetical topological model as derived for cx32 (Beyer et al., 1987). Figure 1 illustrates the homology based on optimal alignment of amino acid sequences deduced from the corresponding cDNAs. The transmembrane region (1, 2, 3, 4) and the extracellular domain (B, D) of the connexin molecules are highly conserved whereas the intracellular domains (A, C, E) are more heterogeneous.

Meanwhile two more connexin cDNAs have been characterized in mammalian tissues. Nicholson and Zhang (1988) sequenced the connexin26 (cx26) cDNA from rat liver and Beyer et al. (1988) described a connexin46 (cx46) cDNA from rat lens fibre cells. The cx26 protein (formerly called 21 kDa gap junction protein) has been characterized in mouse and rat liver using specific polyclonal 21 kDa antibodies (Nicholson et al., 1987; Willecke et al., 1988; Traub et al., 1989). Both the cx26 and the cx46 sequences are homologous to the previously described cx32 and cx43 and fit the topological model of Fig. 1. The pattern of sequence conservation among the known connexins suggests that all the different connexin proteins can recognize each other via highly conserved extracellular sequence motifs. Indeed it has recently been shown that rat cx32 and cx43 proteins can form functional channels with each other when expressed in two different Xenopus oocytes (Swenson et al., 1989).

It is likely that further connexin genes exist in the mammalian genome. Miller et al. (1988) have isolated the rat cx32 gene and have shown that at least 5 more independent DNA sequences homologous to connexins can be cloned from the rat genome. The rat cx32 gene has an uninterrupted coding region and contains a 6.1 kb intron in the 5' untranslated region. Furthermore there are sequences homologous to cAMP response elements in the promotor region and multiple transcription start sites (Miller et al., 1988). Previously it has been shown that the amount of cx32 in mouse hepatocytes is increased in the presence of cyclic AMP (Traub et al., 1987). In our laboratory we have investigated the structure and diversity of connexin sequences in the mouse genome. Table 1 shows that the cx43, cx32, and cx26 connexins coded for by the mouse genome are very similar to the corresponding rat sequences: nucleotide alterations between rat and mouse connexins usually include only the third base of a given codon. The amino acid sequences of rat and mouse cx32 are identical, in spite of 15 base exchanges in the coding region and 26 nucleotide alterations in the 3'nontranslated region. Furthermore similar to the rat cx32 gene the mouse cx32 gene also contains the uninterrupted coding region and an intron in the 5'nontranslated region (G. Kozjek and K. Willecke, unpublished experiments).

Table 1 Comparison of homologies between rat and mouse connexins

rat	cx26 cDNA (liver) (Nicholson et al., 1988)	cx32 gene (Miller et al., 1988)	cx43 cDNA (heart) (Beyer et al., 1987)
mouse	cx26 gene	cx32 gene	cx43 cDNA (brain)
nucleotide-sequence	95.8% (+1 to +36O)	96.4% (coding region)	95.2% (+1O5 to +1O17)
amino acid sequence	99.2% (one exchange)	1OO%	99,7% (one exchange)

Note that the mouse cx26 and mouse cx43 sequences analyzed so far are shorter than the corresponding coding regions. Numbers and parentheses represent the positions of mouse nucleotides relative to the published rat sequences.

The mouse cx43 and cx26 coding sequences are only partially known as yet. It is very likely, however, that the nucleotide sequences are more than 95% homologous to the corresponding rat connexins and there are very few amino acid exchanges in the coding region (K. Herbers, H. Hennemann, B.J. Nicholson, and K. Willecke, unpublished results) (cf. Table 1). Since the corresponding connexins are highly conserved in the mouse and rat genome any new connexin sequence detected in the mouse genome is likely to be derived from an independent connexin gene (or pseudogene). Recently we have discovered two connexin homologous mouse genomic sequences which are only about 3.5 kb apart. One of these regions shows 66 to 7O% homology whereas the second region exhibits between 51 and 55% homology to any of the known connexins. There is about 82% homology between region 1 and region 2. Region 1 is not yet completely sequenced, region 2 shows an uninterrupted reading frame (816 bp) and its deduced amino acid sequence fits the topological model of Fig. 1 (H. Hennemann B.J. Nicholson, and K. Willecke, unpublished results).

Very recently Dermietzel et al.(1989) have found that specific antibodies to cx26, cx32, and cx43 recognize gap junctions in different cells of rodent brain. Cx26 is preferentially expressed before birth (in leptomeningeal cells, ependymal cells and pinealocytes), cx32 preferentially after birth (in oligodendrocytes) similar to cx43 (in leptomeningeal cells, ependymal cells, astrocytes and pinealocytes). Fig. 2 illustrates that the amount of cx43 transcripts in mouse brains is increased after birth when 18 to 2O days old embryos and adult mice are compared (K. Herbers and K. Willecke, unpublished results). This result supports the analysis of the immunoreactivity of the cx43 protein in brain (Dermietzel et al, 1989). Interestingly none of the known connexin antibodies reacts with all gap junctions of neuronal cells. Recently a new connexin cDNA has been isolated from a cDNA library of adult mouse brain using rat cx32 as hybridization probe under conditions of low stringency (R. Heynkes and K. Willecke, unpublished experiments).

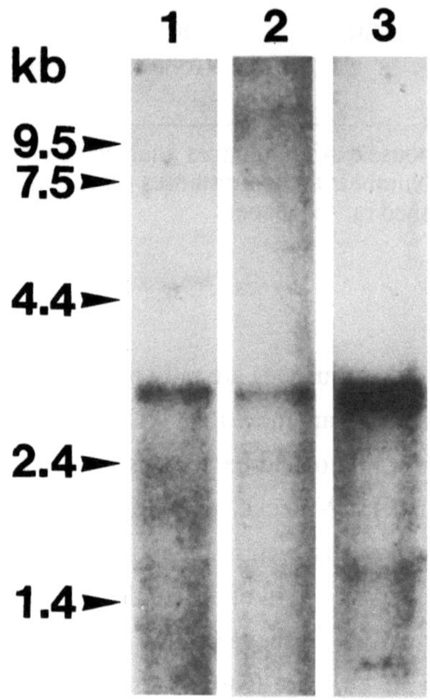

Fig. 2 Northern blot hybridization of poly(A^+) RNA (6,5 μg) from adult mouse brain (lane 1), from mouse brain of 18 to 2O days old embryos (lane 2), and of total cellular RNA (16 μg) from mouse heart (lane 3) to a fragment of cx43 cDNA (cf. Beyer et al., 1987).

The new cDNA contains a region of 93O bp coding for a new connexin protein of about 37 kDa molecular weight. The hydrophobicity plot and the analysis of its connexin homology correspond to the model of Fig. 1 and show more similarities to cx43 than to cx32. The tissue and/or cell specificity of the cx37 remains to be investigated.

We wanted to know whether or not the same connexin43 mRNA is expressed in rat heart, brain and myometrium of pregnant rats. Thus we have analyzed whether two cx43 cDNA fragments (consisting of nucleotides +1 to +560 and of nucleotides +561 to +140O) are protected against degradation by SI nuclease after hybrid formation with RNA from the tissues

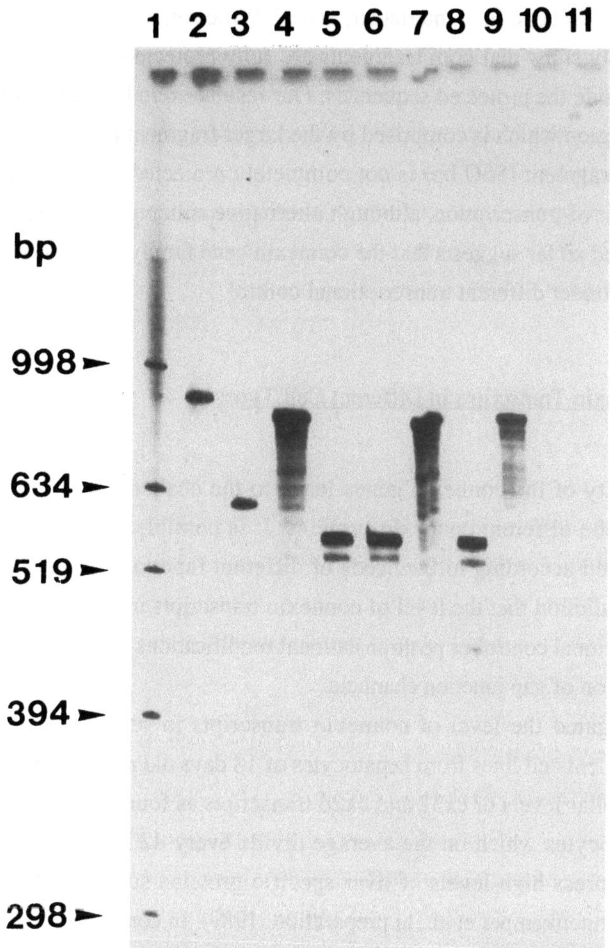

Fig. 3 S1 nuclease analysis of connexin43 mRNA from adult rat brain, rat heart and myometrium of pregnant rats (7 days post coitum, 7dpc). Total cellular RNA (2O µg) from the above rat tissues was hybridized to 5O OOO cpm of ^{32}P-labelled DNA fragments corresponding either to the first 56O nucleotides or to the second fragment (84O nucleotides) of cx43 cDNA (1.4 kb) (Beyer et al., 1987). Both fragments also contain about 4O nucleotides of the M13 vector.

Single-stranded DNA and RNA were degraded by S1 nuclease, the protected fragments were separated on a 4% denaturing polyacrylamide gel and visualized by autoradiography. Lane 1: ^{35}S-labelled pAT153 cut by HinfI and EcoRI. The larger hybridization probe (5OO cpm on lane 2) is completely protected by RNA from brain (lane 4), heart (lane 7) and myometrium (lane 9). The smaller hybridization probe (5OO cpm on lane 3) is also fully protected by RNA from brain (lane 5), heart (lane 6) and myometrium (lane 8). In addition to the major component of 56O bp only one additional minor band of 535 bp is visible. The pattern of protection is the same in all tissues analyzed. Lanes 1O and 11 represent negative controls where the larger (lane 1O) and smaller probe (lane 11) were hybridized to 2O µg of tRNA from yeast.

mentioned above (K. Herbers, E. Winterhager, and K. Willecke, unpublished experiments). Our results (Fig. 3) clearly show that both fragments are fully protected. Thus cx43 mRNAs in the different tissues include the protected sequences. Our results exclude alternative splicing in that part of the coding region which is comprised by the larger fragment (84O bp). In addition some part of the smaller fragment (56O bp) is not completely protected. This is probably due to an alternative start point of transcription, although alternative splicing cannot totally be ruled out. The evidence reported so far suggests that the connexin gene family consists of individual genes whose expression is under different transcriptional control.

Expression of Connexin Transcripts in Different Cell Types

The obvious diversity of the connexin genes leads to the challenging question of functional differences among the different connexin proteins. It is possible that expression of different connexins is regulated according to the needs of different functions of gap junction channels. This leads to the prediction that the level of connexin transcripts may be differently controlled. Alternatively translational control or posttranslational modifications of connexins could also lead to functional regulation of gap junction channels.

We have investigated the level of connexin transcripts in several cell types. Six newly established, independent cell lines from hepatocytes of 18 days old mouse embryos in serum free medium express similar levels of cx32 and cx26 transcripts as found in embryonic liver. These immortalized hepatocytes which on the average divide every 42 h exhibit typical hepatocyte morphology and express high levels of liver specific proteins such as albumin and tyrosine amino transferase (Stutenkemper et al., in preparation, 1989). In contrast to these hepatocyte cell lines, several other cell lines derived from rat and mouse liver (for example BRL, RL, Clone 9) contain very little or no detectable amounts of cx32 and cx26 transcripts but relatively large amounts of cx43 transcripts. Rat liver derived WB-F-344 cells that are classified as "oval" cells, i.e. derived from potential progenitor cells of hepatocytes (Tsao et al., 1984) contain cx26 transcripts, no detectable cx32 transcripts and relatively high level of cx43 transcripts. Thus cx32 and cx26 that are both expressed in liver must be under different transcriptional control. This conclusion can also been drawn from the pattern of expression of cx32 and cx26 in different brain cells (Dermietzel et al., 1989). "Dedifferentiation" of liver derived cells in culture is accompanied by an increased expression of cx43 transcripts. Possibly hepatocyte progenitor cells express cx43. Expression of liver specific phenotypes appears to correlate with expression of relatively high levels of cx32 and cx26 (in mouse liver) and with a strong decrease of cx43 transcripts.

Recently it was reported based on immunoreactivity that cx43 protein is expressed in ovarian granulosa cells, smooth muscle cells in the uterus, fibroblasts in cornea and other tissues, lens and corneal epithelial cells, as well as in renal tubular epithelial cells (Beyer et al., 1989). We have found very high levels of cx43 transcripts in F9 mouse embryonic carcinoma cells and

Table. 2 Expression of connexin transcripts in F9 mouse embryo carcinoma cells before and after differentiation into endoderm cells

		connexins		
	cx43	cx32	cx26	alpha fetal protein
Attached growing cells expressing parietal endoderm				
untreated	++	(+)	+	(+)
db-cAMP	+++	(+)	+	(+)
RA	++++	+	+	(+)
RA+db-cAMP	+++	(+)	(+)	(+)
cells growing in suspension expressing visceral endoderm				
untreated	++	(+)	+	n.s.
db-cAMP	++++	(+)	+	n.s.
RA	++++	(+)	+	++++
RA+db-cAMP	++	(+)	(+)	n.s.

n.s. = not seen

Total RNA (2O μg) was electrophorezed and subjected to Northern blot hybridization using the corresponding rat cDNAs for connexins or alpha fetal protein as control. The amounts of mRNAs detected with the same probe were compared relative to each other on autoradiographs. Samples of total cellular RNA from mouse heart or mouse liver were also analyzed on the same autoradiographs in order to localize positions of connexin mRNAs. F9 mouse embryo carcinoma cells were grown in Dulbecco's modified Eagle medium with 1O% fetal bovine serum and treated with dibutyryl adenosine 3', 5'cyclic monophosphate (db-cAMP, 1 mM), and/or all trans retinoic acid (RA, $1O^{-7}$ M). Attached growing F9 cells were treated for 4 days, F9 cells growing in suspension were treated for 7 days (Strickland and Mahdavi, 1978; Strickland et al., 198O; and Hogan and Taylor, 1981).

lower levels of cx43 transcripts in the blastocyst-derived D3 mouse embryonic stem cell line (J. Look, R. Stutenkemper,and K. Willecke, unpublished experiments). Attached growing F9 cells have been shown to differentiate in culture in the presence of retinoic acid and dibutyryl cyclic adenosine monophosphate to parietal endoderm cells whereas they form visceral endoderm cells when growing in suspension in the presence of retinoic acid (Strickland and Mahdavi, 1987; Strickland et al., 198O; Hogan and Taylor, 1981). Both tissues represent extra-embryonic differentiations of the inner cell mass of mouse embryos. We have investigated by Northern blot hybridization whether these two pathways of differentiation are accompanied by alterations in connexin transcripts. Table 2 lists our results. We find that endoderm differentiation of F9 cells is accompanied by a 3 fold increased level of cx43 transcripts. Little change in the very low amounts of cx32 and cx26 transcripts was seen. Both the F9 parietal endoderm cells (growing attached) and the F9 visceral endoderm cells (growing as aggregates in suspension) show about the same level of cx43 transcripts. It remains to be shown whether these increased levels of cx43 transcripts lead to an increased intercellular conmmunication between differentiated cells.

We have also analyzed the effects of several possible inducers of connexin transcription in fibroblast cells. For our studies we used the mouse fibroblastoid cell line 3T3 A31 1.1, that shows densitiy inhibition of cell proliferation. Selective intercellular communication (transfer of Lucifer yellow) has been demonstrated among chemically transformed, focus forming 3T3 A31 1.1 cells and among surrounding nontransformed cells but not between these two cell types (Yamasaki, 1988; Yamasaki and Katoh, 1988). After treatment of 3T3 A31 1.1 cells with certain inducing compounds (retinoic acid, dibutyryl adenosine 3',5'cyclic monophosphate plus caffeine, and dexamethasone) transfer of Lucifer yellow between focus forming cells and surrounding cells was observed (Yamasaki and Katoh, 1988). We wanted to investigate whether these treatments alter the transcription levels of connexins in these cells. Untreated 3T3 A31 1.1 cells express cx43 transcripts but no cx26 and cx32 transcripts. The results of Northern blot hybridization illustrated in Fig. 4 clearly show that the 3.0 kb cx43 mRNA levels are increased after treatment of these cells with retinoic acid, or dibutyryl cyclic AMP and caffeine but decreased after treatment with dexamethasone (R. Stutenkemper and K. Willecke, unpublished experiments). For comparison the size of cx43 transcripts in the rat liver derived cell line clone 9 and in mouse as well as rat myoblast cell lines, G8 and H9c(c2), respectively, are also shown. Treatment of 3T3 A31 1.1 cells with the tumor promoter phorbol-12,13-didecanoate also leads to an increase of cx43 transcripts whereas treatment with beta estradiol, progesteron or prostaglandine E did not significantly alter the level of cx43 transcripts. Our results suggest that cx43 transcript levels in mouse fibroblasts are up-regulated by retinoic acid as well as by cyclic AMP but down-regulated by glucocorticoids. In spite of the decreased levels of cx43 transcripts after dexamethasone treatment Yamasaki et al. observed increased communication between transformed 3T3 A31 1.1 cells and surrounding non-transformed cells. Thus this effect is not related to the level of cx43 transcripts but must be due to other levels of control (cell adhesion?) that effect intercellular communication.

Table 3 Northern blot analysis of connexin transcripts in rat uterus at two different stages of pregnancy.

	Endometrium	Myometrium	
	7 dpc	7 dpc	21/22 dpc
cx26	+	-	+
cx43	+++++	+	+++++

Total cellular RNA (2O μg) from endometrium 7 dpc (days post coitum), myometrium 7 dpc and myometrium 21/22 dpc was hybridized to rat 43 cDNA (Beyer et al., 1987), and to rat liver cx26 cDNA (Nicholson et al., 1988).

Finally we have investigated the connexin expression in uterus of pregnant and non-pregnant rats (K. Herbers, E. Winterhager, B.J. Nicholson, and K. Willecke, unpublished experiments). Table 3 shows that no transcripts of cx26 and a small amount of cx43 transcripts were found in myometrium at 7 days post coitum. At the time of birth (i.e. 21 to 22 days post coitum) small amounts of cx26 were detected but cx43 transcripts were dramatically increased. The induction of gap junctions analyzed in rat myometrium had previously been studied by morphometry of electron micrographs (Garfield et al., 1978). Expression of cx32 transcripts was not detected in myometrium and endometrium at the level of sensitivity used. In endometrium of pregnant rats low amounts of cx26 but high levels of cx43 transcripts were detected at the time of implantation of the blastocyst (i.e. 7 days post coitum) (Table 3). Apparently during pregnancy the expression of the cx43 gene is differently controlled in different parts of the uterus. Our results show that the increase in gap junctions of myometrium at term is mainly caused by cx43 expression possibly mediated by estrogen and prostaglandins (Garfield et al., 198O).

Summary and Conclusions

The connexin (cx) gene family includes more genes than the previously known cx26, cx32, cx43 and cx46 genes. One additional connexin cDNA tentatively named cx37 has been isolated from a cDNA library of adult mouse brains. Furthermore two additional connexin homologous DNA sequences are localized in the mouse genome about 3.5 kb apart from each other.

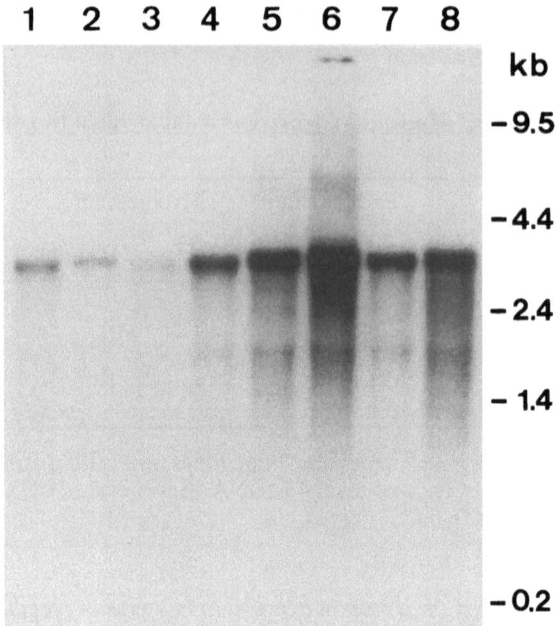

Fig. 4 Northern blot hybridization of connexin43 cDNA to total cellular RNA (2Oμg each) from 3T3 A31 1.1 mouse fibroblastoid cells (lanes 1 to 5), clone 9 (rat liver derived cell line, ATCC CRL 1439, lane 6), G8 mouse myoblast cell line (ATCC CRL 1456, lane 7) (and the H9c2(2.1) rat myoblast cell line, ATCC CRL 1446, lane 8). The 3T3 cells were grown in Dulbeccos modified Eagles medium with 1O% fetal bovine serum (lane 1), and - for 7 days - in the presence of dexamethasone, 1 μM (lane 2) or dexamethasone, 2,5 μM (lane 3), of retinoic acid, 1 μg/ml (lane 4), or of dibutyryl adenosine 3', 5' cyclic monophosphate, 1 mM, plus caffeine, 1 mM (lane 5).

The coding region of cx43 mRNA is likely to be identical in heart, brain and myometrium of pregnant uterus. Thus the same cx43 protein is expressed in these different tissues. Expression of both cx32 and cx26 transcripts coincides with expression of the liver specific phenotype in mouse hepatocytes. In contrast dedifferentiated liver derived cell lines express high levels of cx43. Furthermore cx43 is expressed in mouse embryo carcinoma F9 cells. The level of cx43 transcripts in these cells and in fibroblasts increases after treatment with retinoic acid or cyclic adenosine monophosphate. Strong induction of cx43 transcripts has been detected in myometrium at term compared to myometrium 7 days post coitum. Under these conditions cx26 transcripts are increased to a much lower extent.

Possibly different connexin genes have evolved in the mammalian genome because they allow cells to form gap junction channels according to different functional needs. Some of these functional requirements may be met by transcriptional activation of a single connexin gene (for example cx43) in response to certain inducers. Other functions may require activation of different connexin genes. In order to prove this notion one has to search for functional differences of gap junction channels formed by different connexin subunits. In this way future research may gain answers to the long standing question for the postulated different functions of mammalian gap junctions.

Acknowledgment

Experimental work in the Bonn laboratory is supported by grants of the Deutsche Forschungsgemeinschaft (Wi 27O/13-2), the Dr. Mildred Scheel - Deutsche Krebshilfe and the Fonds der Chemischen Industrie to K. W.

Literature references

Bennett MVL, Spray DC (1987) Intercellular communication mediated by gap junctions can be controlled in many ways, in Edelman, Gall, Cowan (eds) Synaptic function. John Wiley New York: 1O9-135

Bock G, Clark S (eds) (1987) Junctional complexes of epithelial cells, Ciba Foundation Symposium. John Wiley New York

Beyer EC, Paul DL, Goodenough DA (1987) Connexin 43: a protein from rat heart homologous to a gap junction protein from rat liver. J. Cell Biol. 1O5: 2621-2629

Beyer EC, Goodenough DA, Paul DL (1988) The connexins, a family of related gap junction proteins, in Hertzberg EL, Johnson RG (eds) Gap Junctions. Alan Liss New York 165-175

Beyer EC, Kistler J, Paul DL, Goodenough DA (1989) Antisera directed against connexin43 peptides react with a 43-kD protein localized to gap junctions in myocardium and other tissues. J. Cell Biol. 1O8: 595-6O5

Dermietzel R, Traub O, Hwang TK, Beyer E, Spray DC, Willecke K (1989) Differential expression of gap junction proteins in developing and mature brain tissues. Submitted for publication

Garfield RE, Cannan MS, Daniel EE (198O) Gap junction formation in myometrium: control by estrogen, progesteron, and prostaglandins. Am. J. Physiol. 238: C81-89

Garfield RE, Sims SM, Cannan MS, Daniel EE (1978) Possible role of gap junction activation of myometrium during parturition. Am. J. Physiol. 235: C168-178

Goodenough DA, Paul DL, Jesaitis L (1988) Topological distribution of two connexin32 antigenic sites in intact and split rodent hepatocyte gap junctions. J. Cell Biol. 1O7: 1817-1824

Hertzberg EL, Johnson RG (eds) (1988) Gap Junctions. Alan Liss New York

Heynkes R, Kozjek G, Traub O, Willecke K (1986) Identification of rat liver cDNA and mRNA coding for the 28 kDa gap junction protein. FEBS Letters 2O5: 56-6O

Hogan BLM, Taylor A (1981) Cell interactions modulate embryonal carcinoma cell differentiation into parietal or visceral endoderm. Nature 291: 235-237

Kumar N, Gilula NB (1986) Cloning and characterization of the human and rat liver cDNAs for a gap junction protein. J. Cell Biol. 1O3: 767-776

Loewenstein WR (1987) The cell-to-cell channel of gap junctions. Cell 48: 725-726

De Mello WC (ed) (1987) Cell-to-cell communication. Plenum Press New York

De Mello WC (ed) (1989) How cells communicate. CRC Press Boca Raton

Miller T, Dahl G, Werner R (1988) Structure of a gap junction gene: rat connexin-32. Bioscience Reports 8: 455-464
Milks LC, Kumar NM, Hougton R, Unwin N, Gilula NB (1988) Topology of the 32 kD derived gap junction protein determined by site directed antibody location. EMBO J. 7: 2967-2975
Nicholson BJ, Dermietzel R, Teplow D, Traub O, Willecke K, Revel JP (1987) Hepatic gap junctions are comprised of two homologous proteins of MW 28OOO and 21OOO. Nature 329: 732-734
Nicholson BJ, Zhang (1988) Multiple protein components in a single gap junction: cloning of a second hepatic gap junction protein (Mr 21OOO). In Hertzberg EL, Johnson RG (eds) Gap junctions Alan Liss New York 2O7-218
Paul D (1986) Molecular cloning of cDNA for rat liver gap junction protein. J. Cell Biol. 1O3: 123-134
Sperdakis N, Cole WC (eds) (1989) Cell interactions and gap junctions. CRC Press Boca Raton in press
Strickland S, Mahdavi V (1978) The induction of differentiation in teratocarcinoma stem cells by retinoic acid. Cell 15: 393-4O3
Strickland S, Smith K, Marotti K (198O) Hormonal induction of differentiation in teratocarcinoma stem cells: generation of parietal endoderm by retinoic acid and dibutyryl cAMP. Cell 21: 347-355
Swenson KI, Jordan JR, Beyer EC, Paul DC (1989) Formatiaon of gap junctions by expression of connexins in Xenopus oocyte pairs. Cell 57: 145-155
Traub O, Look J, Paul D, Willecke K (1987) Cyclic adenosine monophosphate stimulates biosynthesis and phosphorylation of the 26 kDa gap junction protein in cultured mouse hepatocytes. Europ. J. Cell Biol. 43: 48-54
Traub O, Look J, Dermietzel R, Brümmer F, Hülser D, Willecke K (1989) Comparative characterization of the 21-kDa and 26-kD gap junction proteins in murine liver and cultured hepatocytes. J. Cell Biol. 1O8: 1O39-1O51
Tsao MS, Smith JD, Nelson KG, Grisham JW (1984) A diploid epithelial cell line from normal adult rat liver with phenotypic properties of "oval" cells. Exp. Cell Res. 154: 38-52
Willecke K, Traub O (1989) Molecular biology of mammalian gap junctions. In DeMello WC (ed) How cells communicate. CRC Press Boca Raton
Willecke K, Traub O, Look J, Stutenkemper R, Dermietzel R (1988) Different protein components contribute to the structure and function of hepatic gap junctions. In Hertzberg EL, Johnson RG (eds) Gap Junctions. Alan Liss New York: 41-52
Yamasaki H (1988) Role of gap junctional intercellular communication in malignant cell transformation. In Hertzberg EL, Johnson RG (eds) Gap Junctions. Alan Liss New York: 449-465
Yamasaki H, Katoh F (1988) Further evidence for the involvement of gap-junctional intercellular communication in induction and maintenance of transformed foci in BALB/c 3T3 cells. Cancer Res. 48: 349O-3495
Zimmer DB, Green CR, Evans WH, Gilula NB (1987) Topological analysis of the major protein isolated intact rat liver gap junctions and gap junction-derived single membrane structures. J.Biol. Chem. 262: 7751-7763

BIOCHEMICAL AND STRUCTURAL PROPERTIES OF THE PROTEIN CONSTITUENT OF JUNCTIONAL DOMAINS IN EYE LENS FIBER PLASMA MEMBRANES

E.L. Benedetti, I. Dunia, S. Manenti and H. Bloemendal*

Institut Jacques Monod du C.N.R.S., Université Paris VII, Paris, France
*Department of Biochemistry, University of Nijmegen, Nijmegen, The Netherlands

Eye lens morphogenesis and differentiation are characterized by a unique and common cellular event : the elongation of the fiber cell, derived from the epithelial matrix. During morphogenesis, in embryonic lens a biochemical marker is the gene expression of the crystallin proteins. In particular at 12 days of rat embryo development alpha is the first crystallin to be detected. β and γ crystallins are found later and in parallel with the onset of cell elongation (Yancey, 1988).

The elongation of the epithelial cells that will lead to the formation of the primary lens fibers is also characterized by the expression of a specific transcript and its translational product : a major protein constituent of the plasma membrane indicated with the term of MP26.

The messenger and the protein are predominantly localized in the basal region of the cell cytoplasm and in the plasma membrane, respectively, where the lens vesicle is most closely associated with the developing neural retina and where factors inducing lens formation are delivered. The accumulation of the transcript and of protein synthesis appears to be temporarily closely coupled to each other. As differentiation and elongation of the fibers proceed the entire plasma membrane of the fibers becomes heavily labeled by the MP26 immunoprobe. From these interesting results it is concluded that MP26 expression is a specific marker of fiber cell elongation and membrane assembly during embryonic lens development (Yancey, 1988).

Conversely, the MP26 gene in epithelial cells is transcriptionally inactive at any stage of lens morphogenesis.

In a similar fashion the terminal differentiation of the epithelium into lens fibers which evolves during the cell post-embryonic life of the lens, is characterized by the presence of MP26 restricted to the plasma membrane of the lenticular fiber (Benedetti et al., 1981).

The question remains : what might be the structural and functional role of MP26 that appears as a specific marker of lens fiber elongation.

NATO ASI Series, Vol. H 46
Parallels in Cell to Cell Junctions in Plants and Animals
Edited by A. W. Robards et al.

Biochemical properties of MP26

Experiments in PAGE-SDS have shown that the major component of water and urea-insoluble lens fiber membranes, is a subunit of 26 Kd (MP26). This polypeptide can easily be extracted by chloroform-methanol and butanol (Benedetti et al., 1981). This predominant component which comprises more than 50 % of the protein recovered from the membrane is found in human lenses and in lenses of a great number of animal species (Bloemendal, 1981).

Clearly this protein, during aging of the lens, is affected by a Ca^{++} dependent protease which yields a major component of the nuclear fiber plasma membrane of 22 Kd (MP22). MP22 is also nearly completely extracted by chloroform-methanol (Lien et al., 1985).

The amino-acid analysis and the primary structure of MP26 and its natural or artificial degradation product (MP22) are characterized by a high degree of hydrophobicity (Lien et al., 1985). The intrinsic membrane nature of MP26 is also clearly shown by sequence analysis based on c-DNA cloning, which reveals that MP26 is likely to consist of 6 transmembrane α-helical domains with a short NH_2 terminus and a larger moiety of 43 amino-acid long at the carboxyl terminus, both exposed at the cytoplasmic site of the membrane (Revel et al., 1987). Alternatively, a model derived from the study of MP26 solubilized in octyl glucoside conceives that this polypeptide comprises about 50 % of helical domains and about 20 % of β-sheets (Horwitz and Bok, 1987).

A rather intriguing problem concerns the biosynthetic mechanism of MP26. It is not yet clearly assessed whether or not MP26 possesses the uncleaved signal sequence (Revel et al., 1987). Nevertheless, experiments of Anderson et al. (1983) show that the integration of MP26 into microsomal membranes is cotranslational and requires both Signal Recognition Particles (SRP) and a membrane protein acceptor for integration and translocation. Positively MP26 contains along its chain hydrophobic domains which may have the function of untranslocated insertion sequences. From our *in vitro* translation experiments of different classes of lens polyribosomes in a reticulocyte lysate, it appears that mRNA encoding MP26 was predominantly localized in the class of polyribosomes found in association with the plasma membrane-cytoskeleton complex (Benedetti et al., 1981; Dunia et al., 1985). However, from these results the process of membrane insertion of the newly synthetized MP26 was not fully understood. To investigate this process polyribosomes isolated from the plasma membrane-cytoskeleton complex either by sodium deoxycholate or DNAse I were translated in a reticulocyte lysate. Isolated plasma membranes were added either during translation or after completion of the translation. Thereafter, the membranes were sedimented and repeatedly washed. PAGE-SDS analysis and fluorography of the translation products and the plasma membrane, respectively, revealed that predominantly MP26 is synthetized *de novo* and integrated into membranes only if the isolated fiber plasma membranes are present during translation.

Electron microscopic immunocytochemistry on lens fibers showed that specific antibodies against endoplasmic reticulum and against the "docking" protein labeled a discrete number of membrane compartments either scattered within the fiber cytoplasm or in direct association with the plasma membrane. Therefore, we may assume that a possible mechanism of insertion of MP26 into membranes during biosynthesis might involve either a preceeding cotranslational insertion step into remnants of the endoplasmic reticulum compartment or direct integration into plasma membrane domains where competent acceptors for newly synthetized MP26 are present (Dunia et al., 1985). However, the association of a class of polyribosomes with the cytoskeleton makes it difficult to rule out that MP26 is synthetized far from the membrane and then inserted into the plasma membrane when polypeptide biosynthesis is completed. The insertion step of MP26 should proceed unassisted into the cell membrane cytoplasmic surface as a result of the anchorage of an hairpin loop of the polypeptide chain (Gorin et al., 1984).

The organization of MP26 within the lipid bilayer

We will consider how MP26 and MP22 are organized in the plasma membrane taking into consideration results derived from biochemical analysis, immunocytochemistry, freeze-fracture and reconstitution experiments of MP26 into liposomes.

The maintenance of an electrochemical gradient across the entire lens can be interpreted in the framework of a single cell syncitial model. In this model the source and means of energy production would lie in the epithelium and perhaps in the outer cortex. Through an extensive communicating network the energy would be provided both for ion transport and for keeping the required ion and water balance in the internal "milieu" of each individual cell.

Electrophysiological measurements indicate that the cell membrane of each individual fiber has domains of low electrical resistance (see references in Benedetti et al., 1981). Along this line experimental evidence has been provided showing that the lens cellular elements - epithelial monolayer and fibers - are connected by very extensive junctional complexes of the communicating type (gap junction).

The distribution of MP26 gene transcripts (Yancey, 1988) and immunocytochemistry (Dunia et al., 1985) clearly show that this protein is not expressed in the epithelial monolayer. Hence, it is not yet known which kind of protein forms the epithelial gap junctions. Previous studies on isolated epithelial cell membranes reveal the presence of a 34 Kd protein which might be associated with the gap junctions in the epithelium (Vermorken et al., 1977).

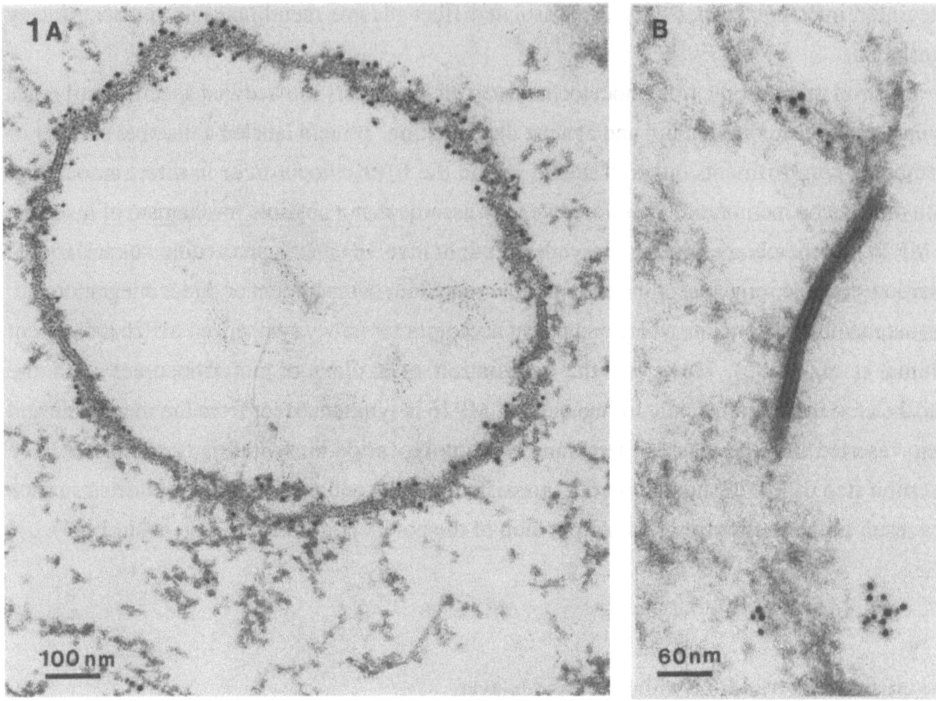

Figure 1 : Immunogold labeling of internalized gap junction using monospecific antibody against MP26 (A). Immunogold labeling of fiber ghost using affinity purified antibody against ankyrin (B)

It is noteworthy that during the process of terminal differentiation the epithelial gap junctions are removed by internalization and shedding whereas in the elongating fibers the process of fiber junctional assembly is simultaneously characterized by the recruitment of identical intramembranous particles, visualized in the cryofracturated membrane faces and by biochemical evidence of the accumulation of the newly synthetized MP26 (Benedetti et al., 1981; Dunia et al., 1985).

Fluorescent , gold and peroxidase immunolabeling of fiber cells, isolated plasma membranes and junctions, using mono- or polyclonal antibodies raised against MP26, purified by chloroform-methanol from urea-alkaline extracted fiber plasma membranes, lead to the conclusion that MP26 is associated with the junctional domains (Fig. 1A). The ubiquitous nature of MP26 is not in contradiction with the latter conclusion; it probably indicates that we are simultaneously scanning several steps of junction assembly which takes place during differentiation of the lens fibers and along the total life-span of the lens (Vallon et al., 1985). Internalization and shedding of plasma membranes and junctions even in the deepest lens fibers demonstrate the presence of a slow but constant assembly and removal of membrane constituents. The junctional nature of MP26 is also consistent with the experiments of Johnson et al. (1988), using a variety of monoclonal and polyclonal antibodies raised against MP26 (MP28) and applying either the "tube method" or the "ELISA method". This study showed that MP26 is indeed a junction associated protein. However, the MP26 conformation and the presentation of specific epitops is affected by the various methods of membrane preparation and immunocytochemical techniques.

In agreement with the assumption of the junctional nature of MP26 are also several experiments indicating that intracellular injection of antibodies raised against MP26 may drastically reduce dye transfer and electrical coupling between differentiated lens fibers *in vitro* (Johnson et al., 1988).

Other lines of evidence demonstrate that lens fiber plasma membranes display pleomorphic features seen by ultrathin sectioning and freeze-fracture. Different sub-classes of "pentalaminar profiles" according to their thickness and/or symmetry have been visualized by Costello et al. (1989) and by Zampighi et al. (1989).

In our hands two types of membrane domain are identified by freeze fracture in sites where the intercellular space is abruptly reduced and the close apposition of two adjoining plasma membranes is visible. One type consists of membrane domains where an assembly of identical particulate entities is visualized on PF and where EF is characterized by complementary pits. The other type is characterized by the presence of square arrays of tetragonal subunits on the fracture faces. This latter type of particle arrangement is more frequently demonstrated in plasma membranes of the nuclear fibers where also the thin pentalaminar profiles are found (Dunia et al., 1985). Likely, the non-geometrical particulate pattern characterizes the true communicating junctional domains that usually connect the cortical lens fibers. In some membrane domains of nuclear lens fibers the rectangularly packed rows of particles on the PF fracture face and the geometrical linear series of grooves in the EF fracture face appear even not restricted to true junctional regions (Fig. 2A). Furthermore, the square array organization can be generated when outer cortical plasma membranes are isolated following homogeneization and extraction by urea and sodium hydroxide at pH 11 (Fig. 2 B and C). The presence of identical geometrical packing can be even induced by incubation of isolated lens cortical fiber plasma membranes with V8 *S.aureus* protease (Lien et al., 1985; Dunia et al., 1985).

In a previous study we demonstrate that the square array organization is retained by lens fiber plasma membranes following homogeneization, purification and solubilization with detergents (Benedetti et al, 1981). This type of geometrically packed intramembranous particle

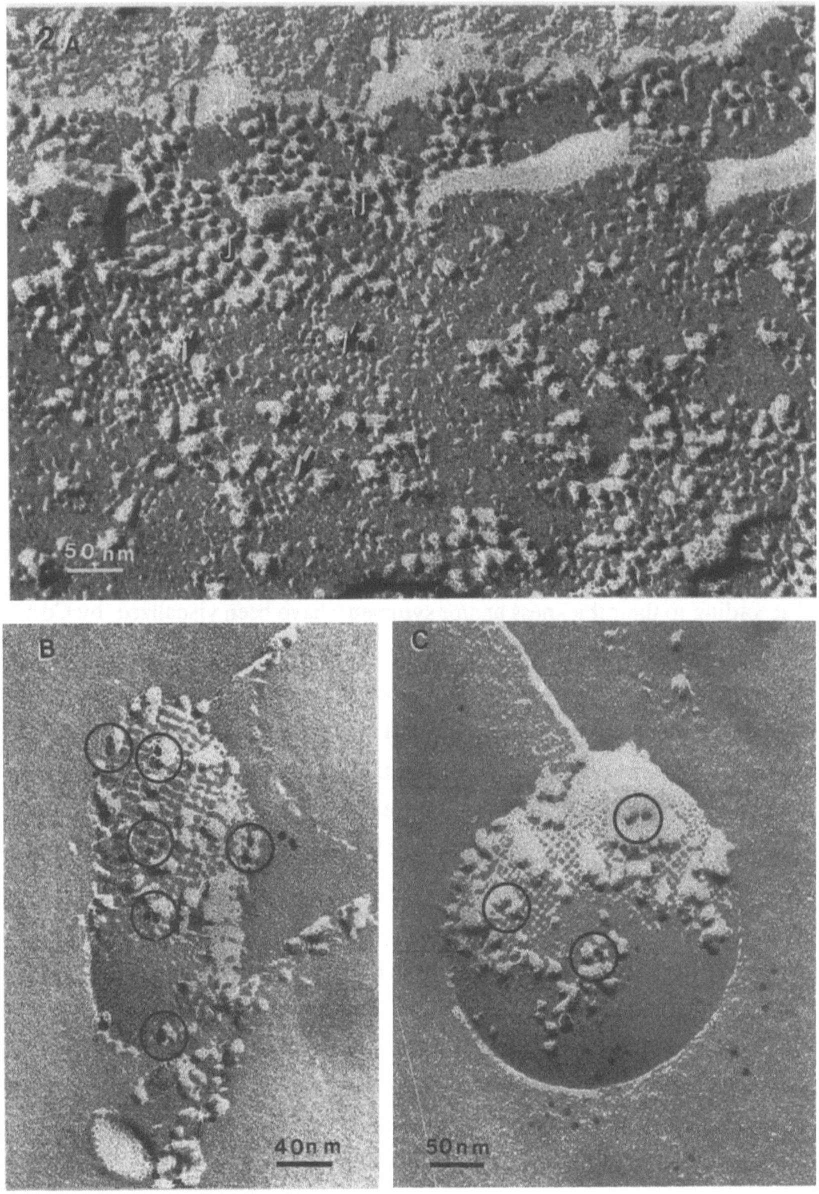

Figure 2 :

A : Freeze-fracture replica of embryonic lens nuclear fibers. Orthogonal array is visible in close proximity to a junctional domain (J) (arrow).

B and C : Freeze-fracture immunogold labeling of urea-alkaline extracted sheep cortical plasma membranes with the monospecific antibody against MP26. Note that many intramembranous particles are labeled by the immunogold probe (circles) array is found as local differentiation in plasma membranes of several types of tissus. In some cases a correlation between water mouvement across plasma membranes and number and extension of the square array has been experimentally prooved. Increase in sodium transepithelial transport and increased water permeability were proportional to the differentiation of the square particulated array found in plasma membranes (see references in Benedetti et al. 1981).

Therefore we anticipated that the square array may represent membrane domains with increased permeability. In contrast to the gap junctions, which ensure rapid horizontal communication between one fiber and another one, the square array of transmembrane particulate entities could be the site of preferential and controlled permeability between the intracellular milieu and the narrow extracellular domain (Benedetti et al, 1981).

Hence the existence of a structural transition between the two intramembranous particle arrangements seems to be dependent upon regional membrane specializations, chemical extraction and enzymatic cleavage of the membrane component which in natural membranes may occur concomitantly with the aging process of the lens.

It is noteworthy that SDS-PAGE of the urea-alkaline plasma membranes is characterized by the sole presence of MP26 and eventually of MP22 which becomes the major component of the protease-treated membranes (Lien et al., 1985).

Application of the "fracture label method" developed by Pinto da Silva and Torrisi (1982) to isolated urea-alkaline extracted cortical plasma membranes labeled with the monospecific antibody against MP26 and with gold-labeled secondary antibody prior to freeze fracture, shows indeed that the particulate entities associated to the orthogonal pattern comprises MP26 (Fig. 2 B and C). These results have been obtained both on calf and sheep lens cortical fiber membranes.

Further support to the assumption that MP26 is involved in the junctional assembly was obtained by the study of reconstituted proteoliposomes, containing MP26, integrated after solubilization in octyl glucoside. The study of solubilized MP26 in octyl glucoside by gel filtration in high performance liquid chromatography (HPLC) and analytical centrifugation indicates that more than 90 % of the protein is monomeric (Manenti et al., 1988). From this study, we could establish that the solubilized fraction from cortical fibers contains a main species with a sedimentation coefficient of 4.6 S and a molecular mass of 27 Kd for the protein, without the bound detergent, which has a molecular weight closer to that calculated from c-DNA cloning (28.2 Kd). It is remarkable that the partial removal of the NH_2 and COOH termini by V_8 *S. aureus* protease increases both the insolubility of the polypeptide in the detergent and its tendency to aggregate into oligomers (Manenti et al., 1988).

Fatty acid analysis of the octyl-glucoside soluble and insoluble fractions, obtained from the cortical and nuclear fiber plasma membranes, respectively, shows that the nuclear insoluble fraction is characterized by a higher concentration of the fully saturated behenic acid (C22:0) than that found in the cortical insoluble fraction (Manenti et al., 1988).

From these results we may assume that the polar character of MP26 and the presence of specific classes of lipids governs the protein-protein and protein-lipid interactions responsible for the various structural features of MP26 and MP22 in the lipid bilayer.

A number of studies have already shown that MP26, reconstituted into liposomes, is a channel forming protein and that MP26 channels in one single vesicle are likely regulated by voltage and by calmodulin dependent Ca^{++} concentration (Girsh and Peracchia, 1985; Zamphigi et al., 1985). A sensitive fluorometric assay has been applied for measuring permeability of liposomes containing or not bovine lens MP26. Positively the fluorescence-quenching assay clearly shows that permeability was correlated with the amount of MP26 present in the liposomes and with increasing purity of the protein (Scaglione and Rintoul, 1989).

In our experiments the structural polymorphism which can be observed on the fracture faces of the reconstituted MP26 proteoliposomes (Fig. 3A,B and E), seems quite comparable to that seen in natural fiber plasma membrane domains (Dunia et al., 1987). Since SDS-PAGE patterns of proteoliposomes show that MP26 is the major protein component of the reconstituted material we concluded that the particles found on the fracture faces are mainly composed of MP26.

When in some experiments the SDS-PAGE pattern of the isolated membranes prior to octyl-glucoside solubilization shows a relevant amount of MP22 and in parallel the SDS-PAGE of the reconstituted material is characterized both by MP26 and by MP22 we assume that the proteolytic derivative of the MP26 may also contribute to the formation of intramembranous particles.

It is likely that individual intramembranous particles visualized on the fracture faces of reconstituted proteoliposomes contain more than one copy of MP26. If we consider MP26 as the only protein component of the reconstituted liposomes then each individual 8 nm intramembranous particle would contain no more than four copies of the polypeptide. One may expect that MP22 displaying the same transmembrane segments will also form tetrameric oligomers (Fig. 3C).

The junctional oligomers formed by MP26 copies could possess, among other helical domains, an amphipathic helix that could easily accomodate in tetramers or hexamers with a circular symmetry to form a pore-like structure (Revel et al., 1987). There has been controversy about whether highly charged amphipatic helices or uncharged ones provide the lining. As far as the ionic channel of the nicotinic-acetyl choline receptor hetero-oligomers is concerned, results from various recent studies favor a pore lined by the uncharged helix (Dani, 1989).

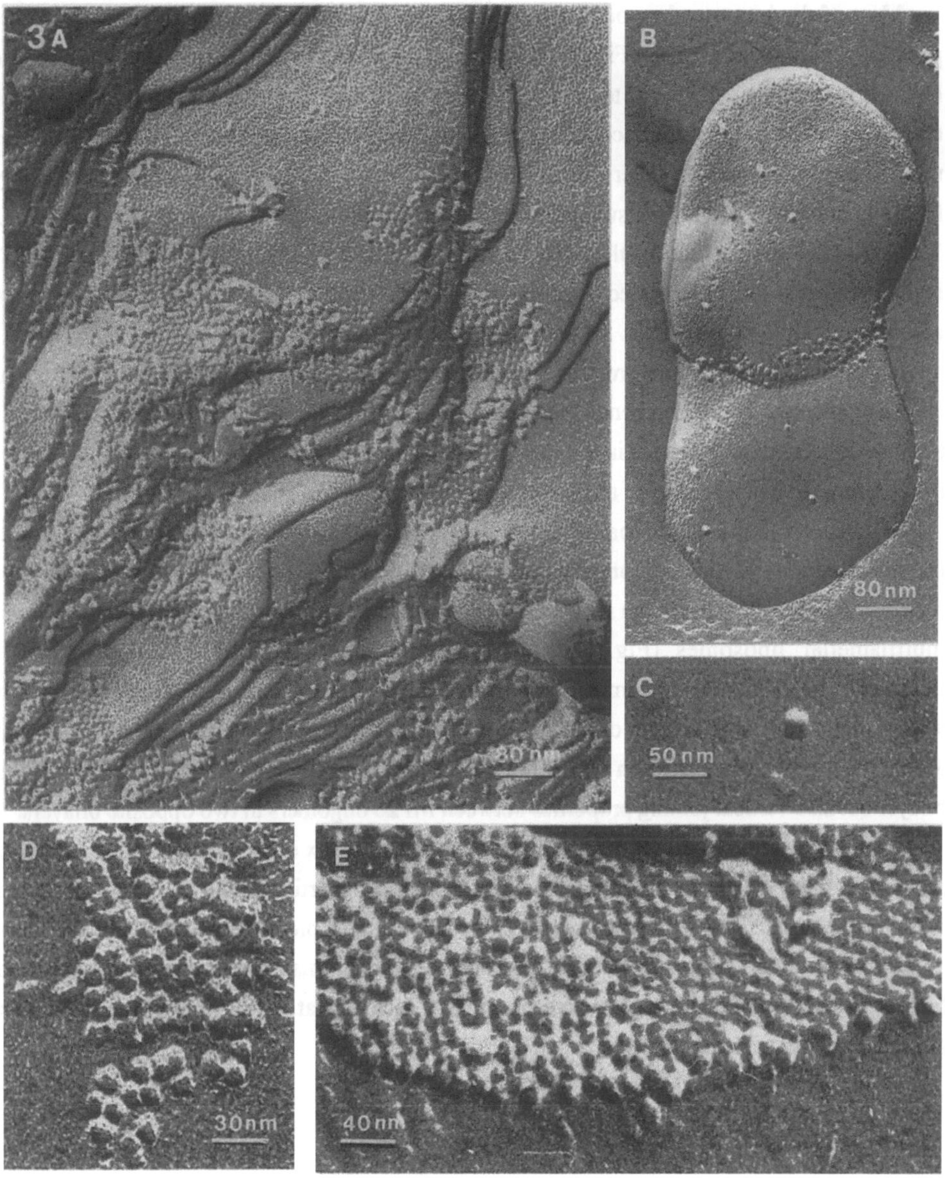

Figure 3: Freeze-fracture replica of MP26 reconstituted proteoliposomes (A,B,C and E) and of urea-alkaline extracted bovine cortical plasma membranes (D).
Note the pleomorphic arrangement of MP26 distribution within the lipid bilayer and the tetragonal aspect of the particulate entities

Most of the intramembranous particles found on the fracture faces of proteoliposomes possess a tetragonal profile (Fig. 3C and D). We are aware of the fact that asymmetric hexameric structure can appear to approach a tetrameric symmetry when the particle image is tilted about 30°, a value not uncommon in freeze fracture replicas (Hanna et al., 1985). It is noteworthly, that there are other instances in which junctional particles have been positively found to exhibit less than six subunits, for instance, pentamers or tetramers (Hanna et al., 1985). A more direct proof for the existence of a tetragonal arrangement of MP26 and MP22 is their tendency to be associated in orthogonal and rhombic packing in large domains of the reconstituted lipid bilayers (Fig. 3A and E).

For other transmembrane proteins such as bacteriorhodopsin and CHL a/b LHC-D6, the choice of appropriate environmental parameters during reconstitution, in particular of the lipids and the ionic conditions, results in the transition of random assembly to a bidimensional lattice of repeating subunits (Michel et al., 1980; Sprague et al., 1985).

In our reconstituted preparations we observed that the accumulation of MP26 intramembranous particles is preferentially found in association with local membrane-to-membrane interactions between reconstituted vesicles and in differentiated domains of large multilamellar liposomes (Fig. 3A and B). It is remarkable that in the region of intramembranous particles accumulation the interbilayer space is abruptly reduced in a similar way as in the junctional domain of lens fiber plasma membranes. Hence we may conclude that MP26 possesses a conformation which not only favors lateral interaction in the plane of the bilayer but also vertical joining properties between MP26 oligomers in two opposite bilayers.

The intramembranous particle distribution on either the convex or the concave fracture face of the reconstituted vesicles is random, without any significant partition coefficient as it occurs in natural membranes. This fact indicates that in the reconstituted proteoliposomes there is little or no preferential orientation of MP26 and that the assembly of this polypeptide, within the lipid bilayer, occurs non assisted by the morphopoïetic factors that during junctional formation govern the correct sideness of the transmembrane polypeptide.

Junctional assembly and the plasma membrane-cytoskeleton complex

Junctional assembly can be regarded as a process involving serial steps : a) the reciprocal and stereospecifical binding of cell surface receptors (Edelman, 1985) followed by b) conformational and/or topological changes of the receptor proteins which in turn generates c) the recruitment of the transmembrane junctional protein oligomers. But another crucial aspect of the junctional assembly could be the interaction of the plasma membranes with the cytoskeletal constituents.

Post-translational processing of the junctional proteins such as phosphorylation mediated by a cAMP-dependent protein kinase described for MP26 (Johnson et al., 1986) may be an essential step anchoring this transmembrane protein to cytoskeletal amphitropic polypeptides (Bum, 1988). In fact post-translational phosphorylation of membrane and cytoskeletal constituents has been described as an important step for the assembly of other types of junctions, in particular for cell adhesion plaques (Burridge et al., 1988; Niggli and Burger, 1987).

Recently observations of Franke et al. (1987) have revealed that plakoglobin is a component of the filamentous sub-plasmalemma coat of lens cells. This protein characterizes the intramembrane adhering junction in close connection with other cytoskeletal proteins. Another amphitropic protein, vinculin, has also been revealed by fluorescence and immunogold labeling techniques both on intact lens cryostat sections and in lens fiber ghosts (Fig. 4A and B).

Beside the role of vinculin for triggering actin polymerization and for anchoring the actin filaments to the cytoplasmic inner surface of the plasma membrane (Burridge et al., 1988), we demonstrate, by immunogold labeling experiments, that vinculin appears not to be directly associated to the inner cytoplasmic surface of the junctional domain. This protein is concentrated in the general plasma membrane inner cytoplasmic surface flanking the gap junction pentalaminar profile. Hence vinculin could generate in the plane of the lipid bilayer a tangential seggregation of protein oligomers specifically involved in cell to cell interaction and communication. In this view, vinculin would not stabilize directly the junctional domain but indirectly by the formation of a boundary surrounding the junctional areas and hence could have an important role in the formation of the specialized membrane domains.

α-Actinin is also present as a constituent of the filamentous network found in close association to the cytoplasmic surface of the lens fiber plasma membrane (Fig. 5A) This polypeptide is co-distributed with vinculin, spectrin (Fig. 5B) and actin in the general plasma membrane and like vinculin, α-actinin flanks and seggregates the junctional domain.

Another protein, amphitropic in nature, named ankyrin, has been studied by immunofluorescence and immunogold labeling techniques (Fig. 1B). The firm attachment of ankyrin to the plasma membrane is known to be mediated by two types of molecular interactions. The amphitropic ankyrin subunit (Bum, 1988) interacts with the cytoplasmic domain of band-3, anion carrier of the red blood cell plasma membrane and is acylated (Niggli and Burger, 1988; Geiger, 1985). Likely the same situation occurs in lens fiber plasma membrane where the presence of band 3 can be demonstrated by immunofluorescence, using a monospecific antibody raised against the cytoplasmic domain of this transmembrane protein .

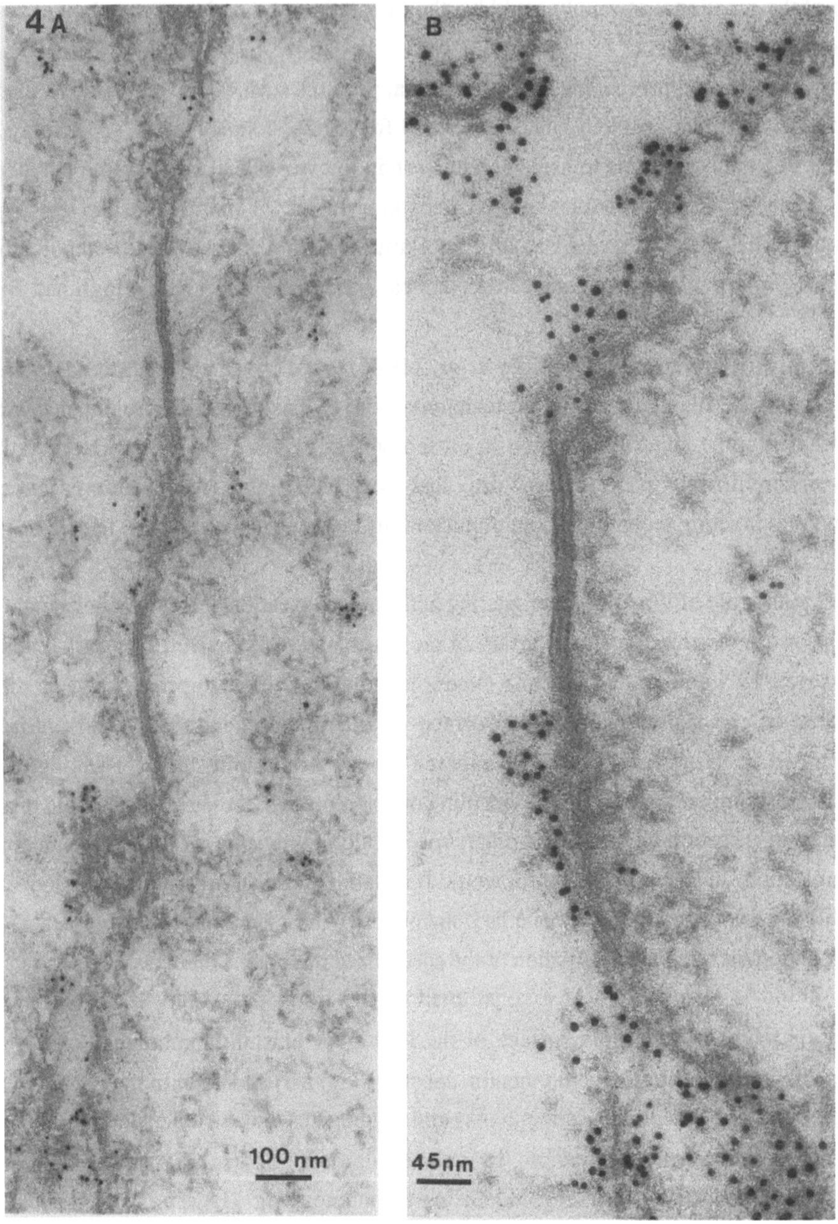

Figure 4 A and B : Immunogold labeling of cortical fiber ghosts with monoclonal antibody against vinculin. The junctional domains are not labeled

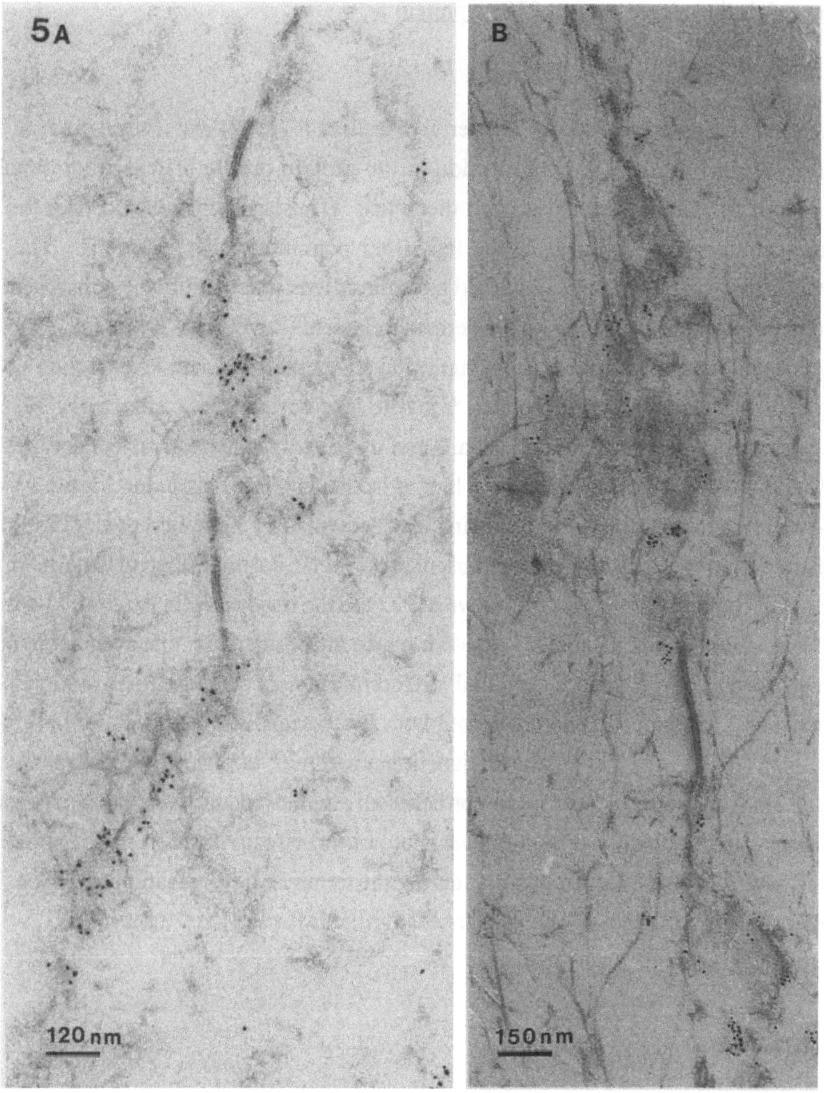

Figure 5 : Immunogold labeling of isolated fiber ghosts with affinity purified antibodies directed against alpha-actinin (A). and against spectrin (B). The immunogold labeling is associated with the general plasma membrane flanking the junctional domain

Dilemma of the junctional nature of 70Kd protein

Experimental data provided by Kistler and Bullivant (1989) and Zampighi et al (1989) show the presence of a 70 Kd polypeptide in the protein profile of lens fiber membranes isolated from the outer cortex of calf and sheep lens. These authors conclude that the 70 Kd protein is a putative constituent of the lens fiber communicating junctions. The 70 Kd polypeptide has been previously described as a minor lens fiber membrane constituent easily removed from the lipid bilayer by mild detergent treatment (Revel et al, 1987).

Experiments in our laboratory show that exhaustive urea and alkaline extraction of calf or bovine fiber plasma membranes isolated from the lens outer cortex, remove several protein constituents including the 70 Kd polypeptide and its proteolytic derivatives but not MP26 (Fig 6A). The SDS-PAGE profile and western blot of isolated plasma membrane fractions from the cortical region of both calf and adult bovine lens presented by Zampighi et al. (1989) clearly demonstrate that the 70 Kd protein is absent from outer cortical fiber plasma membranes whereas MP26 and its proteolytic derivative MP22 are the major protein bands of this fraction. It is surprising that Zampighi et al (1989) do not take in considaration that even in bovine lens, the outer cortex is the site of terminal differentiation of young fibers which possess communicating gap junctions demonstrated both by freeze-fracture and ultrathin sectioning observations (Fig 6B and C). Hence the junctional nature of the 70 Kd protein remains to be clarified. This polypeptide could represent either an amphitropic protein which may establish association with the junctional lipid bilayer, but not directly involved in the communicating junctional path, or participate temporarily, during the terminal differentiation of the lens fibers, to the junctional assembly in a way which needs further investigation.

Concluding remarks

The data discussed in this paper imply that the expression of MP26 is one of the key steps of cell surface domain formation during morphogenesis and terminal differentiation of the lens fibers.

We have demonstrated that MP26 is a major constituent of lens fiber plasma membrane domains involved in cell interaction and cell communication. The formation of various membrane domains can be interpreted as a self-assembly of identical or quasi-equivalently related protein oligomers, affected by the lipid environment and likely dependent on the

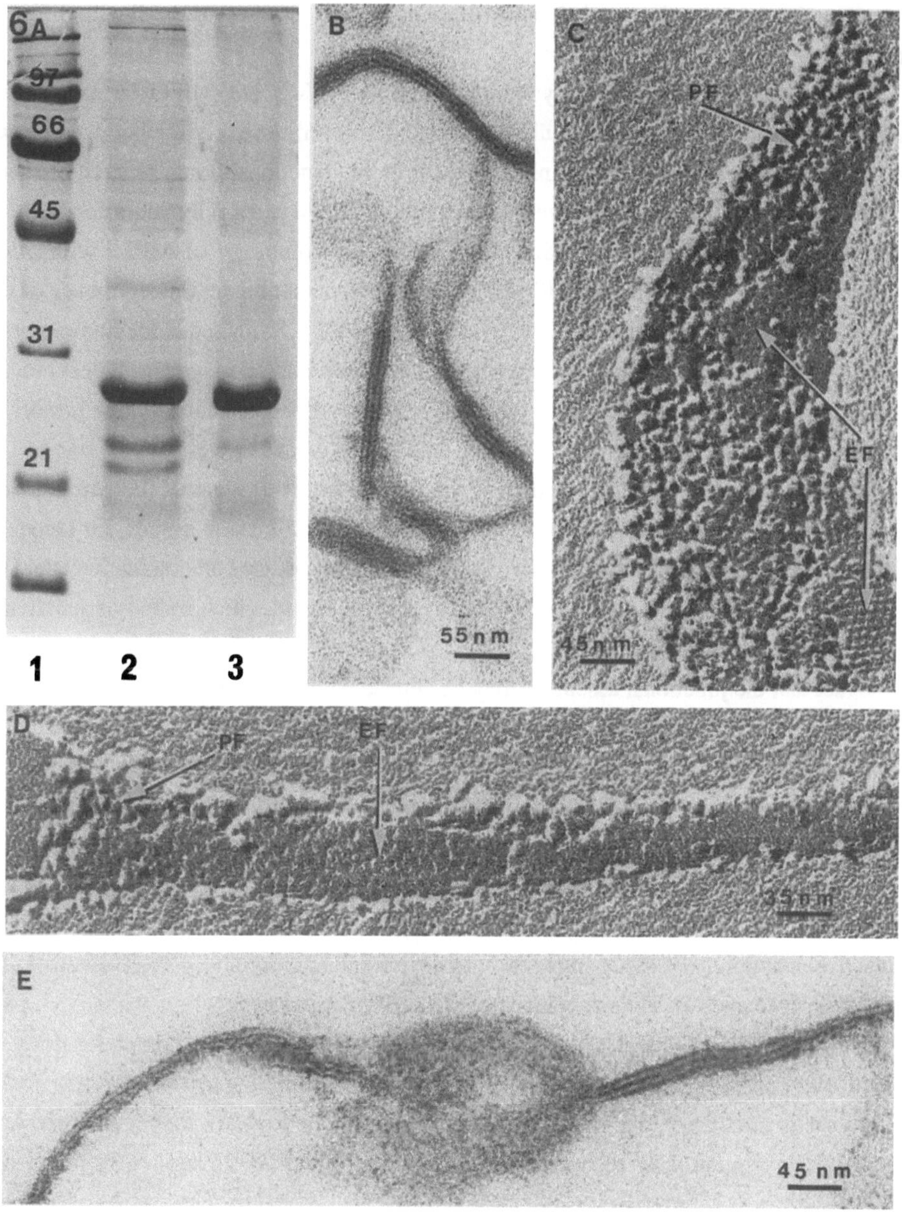

Figure 6 : A) SDS-PAGE of markers (lane 1), 7M urea (lane 2) and urea-alkaline extracted calf lens outer cortex fiber plasma membranes (lane 3). MP26 is the major protein constituent. B) Thin section of urea-alkaline extracted plasma membranes from bovine lens outer cortex where "thick" pentalaminar profils are shown. C and D) Freeze-fracture replica of gap junction domains of urea-alkaline extracted plasma membranes from calf lens outer cortex. The clivage exposes simultaneously EF and PF fracture faces. E) Thin section of same material as C and D. "Thick" and "thin" pentalaminar profiles are present interaction with amphitropic cytoskeletal constituents, which may vary according to different species and aging process.

Other membrane associated polypeptides such as 70 Kd, are purported as putative junctional constituents (Kistler and Bullivant, 1989). However, the experimental data that we have collected show that MP26 certainly is present in the junctional domains even of sheep lens fiber plasma membranes. Therefore the exact role of MP70 is not fully understood yet.

The fact that sequence information does not reveal homology of MP26 with other junctional proteins - connexins - identified in other tissues, do not hinder the legitimacy of the MP26 as a major lens junctional protein (Beyer et al., 1988). Strict sequence homology of protein(s) forming the permeable transmembrane junctional channels is probably less crucial than conformational equivalence which may characterize non-identical amino-acid sequences of connexins spanning the lipid bilayer.

To date most of the functional studies of the regulatory mechanism of the gap junction channel show that the control may not be the same in all cells. In some tissues, for instance, c-AMP depressed communication. Conversely, in other cell types the communication could be up-regulated by c-AMP and a c-AMP-dependent protein kinase which increases junctional permeability (Loewenstein, 1987).

The fact that the junctional assembly may be dependent upon a variety of morphopoïetic factors which may vary for different classes of connexins, is demonstrated by the experiments of Swenson et al. (1989) describing the formation of gap junctions by expression of connexins in Xenopus oocyte pairs. Unexpectedly the injection of RNA-coding for MP26 triggers the biosynthesis of this polypeptide but it is not followed by the correct assembly of junctional domains, though MP26 is incorporated into oocyte plasma membranes. Conversely, in the same system, connexins 32 and 43 induced large increase of junctional conductance. Likely this inconvenient failure of MP26 may reside in the intriguing biosynthetic mechanism of this polypeptide which evolves while elongation of the lens fibers proceeds.

We have anticipated that a stringent step of the biosynthesis and membrane insertion of MP26 involves activation of its specific cytoskeleton-bound messenger RNA and transfer of the biosynthetic machinery from the cytoskeleton domain to the plasma membrane target. Any variation of this program may generate a misleading assembly of MP26 and hinder junctional formation.

Acknowledgments

The authors want to thank Mrs G. Raposo-Benedetti for valuable advice and suggestions during this research . This work was supported by Institut National de la Santé et de la Recherche Médicale (INSERM), grant CRE 880002 and by "Fondation pour la Recherche Médicale", France.

References

Anderson DJ, Mostov KE and Blobel G (1983) Mechanisms of integration of *de novo*-synthesized polypeptides into membranes : signal recognition particle is required for integration into microsomal membranes of calcium ATPase and of lens MP26 but not of cytochrome b5. Proc. Natl. Acad. Sci. USA 80:7249-7253.

Benedetti EL, Dunia I, Ramaekers FCS and Kibbelaar MA (1981) Lenticular plasma membranes and cytoskeleton, *in:* Molecular and Cellular Biology of the Eye Lens (Bloemendal H, ed.), John Wiley and Sons New York, pp. 137-184.

Beyer EC, Goodenough DA and Paul DL (1988) The connexins, a family of related gap junction proteins, *in:* Modern Cell Biol. (Hertzberg EL and Johnson RG, eds) Alan R. Liss, Vol. 7, pp. 167-175.

Bloemendal H (1982) Lens proteins. CRC Critical Rev. Biochem. 1-38.

Bum P (1988) Amphitropic proteins: a new class of membrane proteins. TIBS 13:79-83.

Burridge K, Fath K, Kelly T, Nuckolls G and Turner C (1988) Focal adhesions: Transmembrane junctions between the extracellular matrix and the cytoskeleton. Ann. Rev. Cell Biol. 4:487-525.

Costello MJ, Mc Intosh TJ and Robertson JD (1989) Distribution of gap junctions and square array junctions in the mammalian lens. Invest. Ophthal. Visual Sci. 30:975-989.

Dani JA (1989) Site-directed mutagenesis and single-channel currents define the ionic channel of the nicotinic acetylcholine receptor. TINS 12:125-128.

Dunia I, Lien DN, Manenti S and Benedetti EL (1985) Dilemnas of the structural and biochemical organization of lens membranes during differentiation and aging. Cur. Eye Res. 4(11):1219-1234.

Dunia I, Manenti S, Rousselet A and Benedetti EL (1987) Electron microscopic observations of reconstituted proteoliposomes with the purified major intrinsic membrane protein of eye lens fibers. J. Cell Biol. 105:1679-1689.

Edelman GM (1985) Specific cell adhesion in histogenesis and morphogenesis, *in:* The Cell in Contact (Edelman G and Thiery JP, eds), John Wiley and Sons New York, pp. 139-168.

Franke WW, Kapprell HP and Cowin P (1987) Plakoglobin is a component of the filamentous subplasmalemmal coat of lens cells. J. Cell Biol. 43:301-315.

Geiger B (1985) Microfilament-membrane interaction. TIBS 10:456-461.

Girsh SJ and Peracchia C (1985) Lens cell-to-cell channel protein. I - Self-assembly into liposomes and permeability regulation by calmodulin. J. Membr. Biol. 83:217-225.

Gorin MB, Yancey B, Cline J, Revel JP and Horwitz J (1984) The major intrinsic protein (MIP) of the bovine lens fiber membrane: characterization and structure based on c-DNA cloning. Cell 39:49-59.

Hanna RB, Ornberg RL and Reese TS (1985) Structural details of rapidly frozen gap junction, *in:* Gap Junctions (Bennet M and Spray D, eds) Cold Spring Haln. Let., pp. 23-30.

Horwitz J and Bok D (1987) Conformational properties of the main intrinsic polypeptide (MIP 26) isolated from lens plasma membranes. Biochemistry 26:8092-8098.

Johnson KR, Lampe PD, Hur KC, Louis CF and Johnson RG (1986) A lens intercellular junction protein, MP26, is a phosphoprotein. J. Cell Biol. 102:1334-1343.

Johnson RG, Klukas K, Lu Tze-Hong and Pray DS (1988) Antibodies to MP28 are localized to lens junctions, alter intercellular permeability and demonstrate increased expression during development, *in:* Modern Cell Biol. (Hertzberg EL and Johnson RG, eds) Alan R. Liss, Vol. 7, pp. 81-98.

Kistler J and Bullivant S (1989) Structural and molecular biology of the eye lens membranes. Crit. Rev. Biochem. Mol. Biol. 24(2):151-181.

Lien ND, Paroutaud P, Dunia I, Benedetti EL and Hoebeke J (1985) Sequence analysis of peptide fragments from the intrinsic membrane protein of calf lens fibers MP26 and its natural maturation product MP22. FEBS Lett. 181(1):74-78.

Loewenstein WR (1987) The cell-to-cell channel of gap junctions. Cell 48:725-726.

Manenti S, Dunia I, Le Maire M and Benedetti EL (1988) High-performance liquid chromatography of the main polypeptide (MP26) of lens fiber plasma membranes solubilized with n-octyl b-D-glycopyranoside. FEBS Lett. 233(1):148-152.

Michel HD, Hoestenhelt H and Henderson R (1980) Formation of a new 2-D-crystalline form of purple membrane with orthorombic lattice, *in:* Electron Microscopy at Molecular Dimensions (Baumeinster W and Vogell W, eds) Springer Verlag Berlin Heidelberg New York, pp. 61-70.

Niggli V and Burger MM (1987) Interaction of the cytoskeleton with the plasma membrane. J. Membrane Biol. 100:97-121.

Pinto da Silva P and Torrisi MR (1982) Freeze-fracture cytochemistry: partition of glycophorin in freeze-fractured human erythrocyte membranes. J. Cell Biol. 93:463-469.

Revel JP, Yancey SB, Nicholson BJ and Hoh J (1987) Sequence diversity of gap junction proteins, *in:* Junctional Complexes of Epithelial Cells. Ciba Foundation Symp. 125 (Bock G and Clark S, eds) John Wiley and Sons United Kingdom, pp. 108-127.

Scaglione BA and Rintoul DA (1989) Fluorescence-quenching assay for measuring permeability of reconstituted lens MIP26. Invest. Ophthal. Vis. Sci. 30:961-966.

Sprague SG, Camm EL, Green BR and Staehelin LA (1985) Reconstitution of light-harvesting complexes and photosystem II cores into galactolipid and phospholipid liposomes. J. Cell Biol. 100:552-557.

Swenson KI, Jordan JR, Beyer EC, Paul DL (1989) Formation of gap junction expression of connexins in Xenopus oocyte pairs. Cell 57:145-155.

Vallon O, Dunia I, Favard-Sereno C, Hoebeke J and Benedetti EL (1985) MP26 in the bovine lens: a post-embedding immunocytochemical study. Biol. Cell. 53:85-88.

Vermorken AJH, Hildering JHHC, Dunia I, Benedetti EL and Bloemendal H (1977) Changes in membrane protein patterns in relation to lens cell differentiation. FEBS Lett. 83:301-306.

Yancey B (1988) MIP gene expression during development, *in:* Modern Cell Biol. (Hertzberg EL and Johnson RG, eds) Alan R. Liss, Vol. 7, pp. 199-206.

Zampighi GA, Hall JE and Kreman M (1985) Purified lens junctional protein forms channels in planar lipid films. Proc. Natl. Acad. Sci. USA 82:8468-8472.

Zampighi GA, Hall JE, Ehring GR and Simon SA (1989) The structural organisation and protein composition of lens fiber junctions J. Cell. Biol. 108:2255-2275.

JUNCTIONAL COMMUNICATION: THE ROLE OF COMMUNICATION COMPARTMENTS IN COMPLEX MULTICELLULAR ORGANISMS

John D. Pitts
Beatson Institute for Cancer Research Garscube Estate
Bearsden
GLASGOW G61 1BD
Scotland

Introduction

The concept of junctional communication first arose more than a hundred years ago. In 1879 Eduard Tangl, a plant physiologist working in Cerowitz, suggested that his observations on intercellular bridges left "only one correct explanation of the matter; the protoplasmic bodies of the cells are united by thin strands passing through connecting ducts in the walls which put the cells in connection with each other and so unite them as an entity of higher order" (translated by Carr, 1976).

The general idea of direct intercellular communication via some form of cytoplasmic continuity was widely adopted by both plant and animal physiologists of the time but then fell into relative oblivion until it was revived in the 1930s with the concept of the symplast (Münch, 1930). Intercellular bridges in plants, it was suggested, produced a continuous cytoplasmic network (symplast) separated by a continuous plasma membrane from the extracellular space (apoplast). The cellular state within the symplast was seen as intermediate between the independence of complete isolation and the suppresion of individuality seen in a true syncytium. This partial syncytial state was thought to provide a mechanism for intercellular transport of materials and synchronization of cellular activities at the tissue level.

These ideas were not considered seriously by animal biologists until much more recently but it is now realized that the general concepts of Tangl and Münch are directly applicable to animal tissues.

NATO ASI Series, Vol. H 46
Parallels in Cell to Cell Junctions in Plants and Animals
Edited by A. W. Robards et al.

Plasmodesmata and Gap Junctions

Plasmodesmata in plants and gap junctions in animals provide a good example of convergent evolution. Functionally they are very similar but stucturally they are unrelated, presumably because the problems of producing cytoplasmic continuity are quite different in plants and animals. They both act like intercellular sieves, allowing the free exchange of small ions and molecules between coupled cells. The permeability limit is based primarily (or solely) on molecular size and the cut-off, which is normally expressed for convenience in terms of molecular weight rather than molecular dimensions, is eqivalent to about Mr 800 – Mr 1000 (Finbow and Pitts, 1981; Terry and Robards, 1987).

Plant cells are in general separated by a much wider intercellular space than animal cells. This space contains the cellulose and other components which provide the structural rigidity of the plant. Plasmodesmata bridge this space as stabilized plasma membrane bounded cell-cell channels which form (i.e. are left behind) during what is in effect incomplete cytokinesis. The maximum curvature of a phospholipid bilayer would result in a minimum tube diamter which is too large for the necessary permeability limit but this has been achieved through the formation of a partial blockage in the form of an annulus or plug. A combination of ultrastructural observations and permeability measurements suggest the annulus is penetrated by 10-20 discrete cylindrical channels, about 1.5 nm in diameter around the neck regions and possibly a little wider internally (Terry and Robards, 1987).

Most plant cells are joined by plasmodesmata though the number per cell-cell interface varies considerably. Similarly most animal cells are joined by gap junctions, again with wide variations in density.

Gap junctions form in regions of close cell-cell contact. They are plaque like structures containing regular arrays of intercellular channels which cross both plasma membranes and the much reduced extracellular space. Each channel has two identical halves, one contributed by each cell. The half channels are made of six protein subunits arranged around an axially located aqueous pore which has an internal diameter of 1.5 – 2 nm. As with the channels in plasmodesmata, the gap junctional channels appear to be somewhat constricted at the (cytoplasmic) ends. There is also evidence that gap junctions can switch between open and closed conformations, possibly through changes in the constriction at these sites (Unwin and Ennis, 1984).

Unlike plasmodesmata, gap junctions form *de novo* when cells come into contact. It is generally believed that precursor half channels, present in a closed state in the membranes of the approaching cells, interact end-to-end when contact occurs. The complete channels can then switch to an open state. The aggregation of channels into the close-packed, paracrystalline arrays which are termed gap junctions, is possibly the consequence of minimizing electrostatic replusion between the closely apposed phospholipid bilayers although it may be the consequence of the mechanism of formation.

Specificity of Gap Junction Formation

Gap junctions are found in nearly all animal tissues. Junctional communication is lost in only a few terminally differentiated cell types (e.g. skeletal muscle, some circulating blood cells, some nerve cells).

Animal cells also form gap junctions in culture and communication is normally established very rapidly after the cells make contact. Cells from widely different species (of *Mammalia, Aves, Amphibia, Osteichthes*) form functionally normal junctions in culture but cells from different tissues can show specificity. For example, epithelial cells and fibroblasts which rapidly form (homologous) gap junctions among themselves (>99% coupled in 30 min), form heterologous junctions much more slowly (<5% coupled in 3 h). Cells which show specificity also tend to sort out in culture, producing clearly segregated groups of the different cell types (Pitts & Bürk, 1976).

The difference in rate of junction formation between cells of the same type within a group and that between different cell types in adjacent groups results in a well defined steady state situation in mixed cultures. Cells within the same group form the partial syncytial state (i.e. form a communication compartment) and share common pools of small cytoplasmic ions and molecules (metabolites, inorganic ions, cofactors, second messengers). The boundaries between adjacent compartments in these mixed cultures are functionally emphasized because relatively slow junctional transfer via the much reduced number of channels joining cells on opposite sides of the boundary is followed by more rapid dispersion once inside the next compartment.

Mechanism of Specificity

Several different proteins are required for the formation of functional gap junctions. The identity of the structural protein of the junctional channels has been the subject of debate for some time but current evidence points to a highly conserved hydrophobic protein, Mr 16,000 (Finbow *et al*, 1983, 1984). The same protein is used for junctional channels in different tissues and very similar proteins are found in different species (there is >85% identity between the amino acid sequences of the 16k proteins from mammalian and arthropod sources). The protein appears to have evolved from a more primative Mr 8,000 channel protein by a gene duplication event. A relative of this 8k protein provides the membrane channel for proton translocation by the eubacterial F1F0 ATPases (see chapter by Finbow *et al*).

Current models for the organization of the gap junction channel protein in the membrane (4 trans-membrane alpha helices, 70% mass in membrane, 25% extracellular and 5% cytoplasmic) and the hexameric structure of the half-channels agree very well with the three-dimensional images produced from electron and X-ray diffraction data.

Proteins of the connexin family, are also found in gap junction preparations (Hertzberg, 1984) if these are made by less stingent procedures (e.g. dissociation of plasma membranes with increased pH rather than using detergents, 6M urea and proteases). These proteins, which vary in size from Mr 26,000 to Mr 47,000 (and perhaps to Mr 75,000), have about 50% homology in their Mr 20,000 N-terminal regions and are expressed in a tissue specific manner (see chapters by Willecke *et al* and Dermietzel *et al*).

Antibodies to the connexins and the the 16k channel protein disrupt junctional communication when injected into cells, showing both types of protein are required for the maintenance of functional coupling (Hertzberg *et al*, 1985; Serras *et al*, 1988).

There is also some functional data implicating other components in the processes of junction formation and/or stabilization. Proteoglycans appear to be necessary for the normal patterns of junctional communication in Drosophila (Bargiello *et al*, 1987) and addition of various extracellular matrix components, particularly the heparins which have a tissue specific distribution, to the medium of cultured primary hepatocytes stimulates coupling (Spray *et al*, 1987). Junctional communication is also stimulated by the expression of tissue specific cell adhesion molecules (CAMs; Keane *et al*, 1988; Mege *et al*, 1988).

It seems likely therefore that junction formation is regulated by a variety of factors which influence the frequency with which cells make close contact. The adhesion systems are likely to play a primary role by stabilizing membrane-membrane alignment and providing favourable conditions for the next stage in the process. This step, which may invlove the connexins, must bring the membranes into the thermodynamically unstable, very close aposition required for the precursor half channels to make contact. The tissue specific expression of the connexins suggests they may be responsible for the observed specificity of junction formation.

Patterns of Junctional Communication *in vivo*

Complex patterns of junctional communication are produced in mixed cultures due to specificity of junction formation (Pitts and Kam, 1985). Communication compartments form while the cells are still intermingled due to preferential coupling between cells of the same type often via processes under or over intervening cells of the other type. After two or three days, cell types which show specificity normally sort into separate domains. The size and shape of the more physically separate communication compartments produced by this process is a function of the proportion and distribution of the cells at the time they were originally plated.

Specificity of junction formation also operates *in vivo* producing complex patterns of compartmentation. The most detailed study of the patterns of junctional communication in an intact tissue has been made in skin (Kam *et al*, 1986; Salomon *et al*, 1988). The analytical method uses the fluorescent dye Lucifer Yellow which is injected through a micro-elecrode inserted into a piece of freshly excised tissue. As skin is opaque it is not possible to micro-manipulate the tip of the electrode into a pre-selected cell. However intracellular and extracellular injections can be distinguished subsequently by the distribution of fluorescence in serial sections. Intracellular injections (which are more common because most space inside most tissues is intracellular) produce cytoplasmic and nuclear fluorescence while extracellular injections produce fluorescent extracellular matrix with contrasting dark cells. After analysing the dye spread in sections of many injections it is possible to construct a detailed map of the pathways of junctional communication in the different parts of the tissue.

This type of analysis shows that the dermis is an apparently limitless communication compartment. Dye injected for five minutes into any one dermal cell spreads detectably into 500 – 2,500 other dermal cells. It does not spread into epidermal cells or other epithelial tissues (e.g. sebacious glands) but it does diffuse into the endothelial cells of the small blood vessels. This endothelial-stromal cell coupling may provide a direct entry route for nutrient supply.

Dye injected into an epidermal cell, in contrast, diffuses into only 15 – 25 cells. The coupled cells form a well delineated column made up of 5 – 6 basal cells and the overlying differentiating cells. Sharp boundaries to the dye spreads are seen between the basal cells in each column and the underlying dermal cells. Boundaries are also seen between apparently contiguous epidermal cells in adjacent columns. The dermal-epidermal boundary follows the basement membrane although this in itself is probably insufficient to prevent cell contact and coupling as the similar structure between the dermal cells and the endothelial cells of the small blood vessels is not associated with a communication boundary.

More complex patterns of communication compartmentation are found in hair folicles where more complex controls of growth and differentiation are required.

The Role of Junctional Communication in Epidermal Growth Control

The epidermis is subject to continuous regulative growth control and division occurs in the basal cells at a rate which balances the loss of terminally differentiated cells from the skin surface. The basic unit of growth control in the epidermis appears to be the epidermal proliferative unit (EPU; Potten, 1981). These units were identified by autoradiographic analysis after ^{3}H-thymidine pulse-chase labelling. Each EPU is composed of 6 or 7 basal cells and the overlying column of differentiating cells. One of the basal cells is thought to be a stem cell while the others, which are clonally derived from it, are committed to the differentiation programme but remain in the basal layer where they continue to divide for several generations. After each division, one daughter cell normally remains in the basal layer to divide again while the other moves up into the next layer. Once above the basal layer the cells stop dividing and start to differentiate as they begin to transit towards the surface of the skin, driven by further division below and the loss of terminally differentiated cells above.

The size, shape and organisation of the EPUs are similar to those of the epidermal communication compartments. Growth control appears to operate therefore within groups of cells joined by gap junctions. This observation prompted the suggestion that junctional communication may play some role in epidermal proliferative control.

It is generally believed that extracellular growth factors are required to stimulate cell division. Many growth factors, including those known to operate on epidermal cells, affect target cells by interacting with appropriate cell surface receptors. Subsequent signal transduction then results in an alteration in the effective level of second messenger. Second messengers are small cytoplasmic molecules which pass freely through gap junctions.

A Model for Epidermal Proliferative Control

It has been proposed (Pitts *et al* 1988; Kam and Pitts, 1988) that second messengers produced in the committed basal cells of the epidermis will diffuse through gap junctions into the overlying, non-responding, differentiating cells in each epidermal communication compartment (proliferative unit). The equilibrium concentration of second messenger in the basal cells will depend on the number (total cytoplasmic volume) of overlying cells. When the cell number increases sufficiently, through continuing division of the basal cells, the second messenger concentration will fall below the threshold value and further division will be arrested. Loss of cells from the compartment by terminal differentiation and subsequent sloughing will lead to a decrease in total cytoplasmic volume, an increase in second messenger concentration in the basal cells and consequent stochastic onset of cell division. Junctional communication, by simply providing a pathway for intercellular cytoplasmic homeostasis, results in a skin thickness dependent modulation of growth factor stimulated cell division.

A problem with this model is the length of time it takes for the first stimulated cell to pass through the cell cycle and divide. During this time, before an extra unresponsive cell is produced to again dilute the second messenger concentration below the theshold value, other division competent cells will be stimulated and observations show that only one or at the most two cells in each unit divide at any one time.

Other observations have provided a solution to this problem (Kam and Pitts, 1988). Topical application of the phorbol ester TPA to mouse skin induces hyperproliferation and within four hours of treatment the communication compartment boundary between the epidermis and the dermis breaks down. Similar breakdown is observed in other hyperproliferative situations (follicular invagination, tumour growth, homozygous *Er* (repeated epilation) mice – see below). So, returning to the epidermal growth control model, if the first stimulated cell breaks the boundary between the EPU and the dermis, second messenger concentration should fall rapidly below the threshold value due to loss via gap junctions into the extensively coupled dermal "sink". After completion of division and release of one daughter cell into the higher layer, the basal layer should reorganize and reform the dermal-epidermal boundary to allow the whole process to be repeated.

Again it is important to note that control is operating by intercellular cytoplasmic homeostasis. Proliferation is controled by homeostasis within the epidermal communication compartment and by regulation of the boundary condition between the epidermis and the underlying dermis.

The Er Mutation, Tissue Specific Regulation of Junctional Communication and a Test for the Model

Mouse embryos homozygous for the repeated epilation (*Er*) mutation become "pupoid" due to excessive proliferation and abnormal differentiation of the epidermis (Guénet *et al*, 1979). If allowed to go to term the new-born mice suffocate as all external orifices are blocked by epidermal overproduction. Analysis of the patterns of junctional communication in the skin of these homozygotes (Kam and Pitts, 1989) shows normal dye spread in the epidermis, widespread epidermal-dermal coupling and, unexpectedly, a marked down-regulation of junctional communication in the dermis. The injected dye spreads into only one or a very few cells instead of spreading into thousands of cells as it does in the wild-type skin.

This observation that the primary lesion is in the dermis and not the epidermis is consistent with recent work which shows that grafting mutant epidermis into a wild-type background results in the restoration of normal epidermal growth and differentiation (Fisher, 1987). It also provides a serendipitous test for the growth control model described above.

From the model it can be predicted that removal of the dermal "sink" will result in the maintenance of high second messenger concentrations and the continued stimulation of basal cells into the cell cycle; namely the hyperproliferation which is characteristic of this mutant.

Control of Junctional Communication

Gap junctional communication can be controlled in different ways. The channel can adopt open and closed conformations (Unwin and Ennis, 1984) and the gating mechanism appears to respond to H^+ and Ca^{++} concentrations. This provides a rapid and reversible response which may be particularly important when gap junctions act as electrotonic synapses in the central nervous system. It may also provide for emergency closure when calls are damaged. The increase in cytoplasmic Ca++ concentration in the damaged cell which results from equilibration with the extracellular fluid should close the channels and prevent metabolites, second messengers etc leaking out of adjacent, otherwise intact cells.

Specificity of junction formation is also a form of control and the observations in skin show that junctional communicaiton across the boundaries which form as a result of specificity are also subject to dynamic regulation.

The tissue specific loss of junctional communication seen in the *Er* mice may result from reduced channel permeability, reduced junction formation or decreased stability.

Compartmentation and the Expression of Difference

The concentrations of cytoplasmic ions and small molecules play a central role in the regulation of cell division and gene expression. It may therefore be necessary for cells to separate themselves by compartmentation from the rest of the population in order to successfully express different activities. Such separation has the further advantage that different cellular functions can be controllled by the same ions or small molecules if the cells are in different compartments.

Junctional communication is first expressed at the eight cell stage in mouse embryogenesis (Lo and Gilula, 1979). All the cells form a single compartment to the blastocyst stage but as morphological differentiation becomes apparent different cells types are found to be in different compartments (Kalimi and Lo, 1988). It is possible that the observed compartmentation is a necessary prerequisite for the expression of these differences. If this is so, the positional expression of the molecules which control specificity of junction formation and hence the location of the compartment boundaries is a key process in morphogenesis.

Differences between Plants and Animals

The general principles of Tangl and Münch apply in both plants and animal. The main difference appears to lie in the extent of compartmentation. In plants, the compartments appear to be large and a single compartment may, as Münch suggested, account for most of the organism. In animals the patterns of communication and compartmentation appear to be more complex and individual compartments vary markedly in size. Such complexity may be possible in animals because junction formation is a dynamic process and new communication pathways can be established as cells move and form new contacts.

This difference in compartmentation may represent a fundamental distinction between the oportunities for developmental programming in plants and animals and may in part account for the extra complexity which has evolved in the animal kingdom.

References

Bargiello AB, Saez L, Baylies MK, Gasic G, Young MW, Spray DC (1987) The Drosophila clock gene *per* affects intercellular junctional communication. Nature 328:686-689

Carr DJ (1976) Historical perspectives on plasmodesmata. In "Intercellular communication in plants" (Eds Gunning BES & Robards AW) Springer Berlin, 291-295

Finbow ME, Shuttleworth J, Hamilton AE, Pitts JD (1983) Analysis of vertebrate gap junction protein. EMBO J 2:1479-1486

Finbow ME, Buultjens TEJ, Lane NJ, Shuttleworth J, Pitts JD (1984) Isolation and characterization of arthropod gap junctions. EMBO J 3:2271-2278

Finbow ME, Pitts JD (1981) Permeability of junctions between animal cells. Exp Cell Res 131:1-13

Fisher C (1987) Abnormal development in the skin of the pupoid fetus (*pf/pf*) mutant mouse; abnormal keratinization, recovery of normal phenotype and relationship to repeated epilation (*Er/Er*) mutant mouse. Curr Topics Devel Biol 22:209-221

Hertzberg EL (1984) A detergent independent method for the isolation of gap junctions from rat liver. J Biol Chem 259:9936-9943

Hertzberg EL, Spray DC, Bennett MVL (1985) Blockade of gap junctional conductance by an antibody prepared against liver gap junction membranes. Proc Natl Acad Sci USA 82:2412-2416

Kalimi GH, Lo CW (1988) Communication compartments in the gastrulating mouse embryo. J Biol Chem 107:241-255

Kam E, Melville L, Pitts JD (1986) Patterns of junctional communication in skin. J Invest Dermatol 87:748-753

Kam E, Pitts JD (1988) Affects of the tumour promotor 12-0-tetradecanoylphorbol-13-acetate on junctional communicaiton in intact mouse skin: persistance of homologous communication and increase in dermal-epidermal coupling. Carcinogenesis 9:1389-1394

Kam E, Pitts, JD (1989) Tissue-specific regulation of junctional communication in the skin of foetuses homozygous for the repeated epilation (*Er*) mutation. Submitted

Keane RW, Mehta PP, Rose B, Honig LS, Loewenstein WR, Rutishauser U (1988) Neural differentiation, N-CAM mediated adhesion and gap junctional communication in neuroectoderm: a study *in vitro*. J Cell Biol 106:1307-1309

Lo CW, Gilula NB (1979) Gap junctional communication in the pre-implantation mouse embryo. Cell 18:411-410

Mege R, Matsuzaki F, Gallin WJ, Goldberg JI, Cunningham BA, Edelman GM (1988) Construction of epithelioid sheets by transfection of mouse sarcoma cells with cDNAs for chicken cell adhesion molecules. Proc Natl Acad Sci USA 85:7274-7278

Münch E (1930) Die Stoffbewegung in der Pflantz. Gustav Fischer Jera

Pitts JD, Bürk RR (1976) Specificity of junctional communication between animal cells. Nature 264:762-764

Pitts JD, Kam E (1985) Communication compartments in mixed cell cultures. Exp Cell Res 156:439-449

Pitts JD, Finbow ME, Kam E (1988) Junctional communicaiton and cellular differentiation. Br J Cancer 58 Suppl IX:52-57

Potten CS (1981) Cell replacement in epidermis (keratopoesis) via discrete units of proliferation. Int Rev Cytol 69:271-291

Salomon D, Saurat J, Meda P (1988) Cell-to-cell communicaiton within intact human skin. J Clin Invest 82:248-257

Serras F, Buultjens TEJ, Finbow ME (1988) Inhibition of dye-coupling in *Patella* (*Mollusca*) embryos by microinjection of antiserum against *Nephrops* (*Arthropoda*) gap junctions. Exp Cell Res 179:282-288

Spray DC, Fujita M, Saez JC, Choi H, Watanabe T, Hertzberg E, Rosenberg LC, Reid LM (1987) Proteoglycans and glycosaminoglycans induce gap junction synthesis and function in primary liver cultures. J Cell Biol 105:541-551

Tangl E (1879) Ueber offene Communicationen zwishen den Zellen des Endosperms einiger Samen. Jb Wiss Bot 12:170-190

Terry BR, Robards AW (1987) Hydrodynamic radius alone governs the mobility of molecules through plasamodesmata. Planta 171:145-157

Unwin PNT, Ennis PD (1984) Two configurations of a channel forming membrane protein. Nature 307:609-613

ELECTROPHYSIOLOGICAL PROPERTIES OF GAP JUNCTION CHANNELS

David C Spray
Department of Neuroscience
Albert Einstein College of Medicine
Bronx NY 10461

Cells communicate with one another and with the extracellular environment through channels in their membranes. Characteristics that distinguish among the many types of channels in biological membranes include location and tissue distribution (whether channels are concentrated at membrane specializations), morphological features of the channels (as visualized with ultrastructural techniques), biochemical and molecular composition of proteins that form the channels and functional features (which ions and molecules pass through the channels and how this permeation is regulated).

Our interest has centered on these latter properties of gap junction channels: size (in terms of the diameters of permeant species and in terms of unitary conductances), selectivity (the degree to which charge governs the permeability), and gating properties (mechanisms that affect the fraction of time that the channels exist in the open configuration). These studies have defined properties for gap junction channels that distinguish them from nonjunctional channels and from other forms of intercellular communication.

The purpose of this Chapter is to present an overview of the functional properties of gap junction channels in various tissues. Because gap junctions comprise a family of channel forming proteins, there is similarity and diversity in properties. Diverse functional properties within this class of channels gives rise to tissue - and species -specific regulation of gap junction channels that may be relevant to the control of exchange of ions and signalling molecules in normal physiological as well as in pathological conditions.

Distribution of Gap Junctions

Cellular distribution of gap junction channels is one of the characteristics that is most dichotomous: Gap junction channels are only found at intercellular interfaces and form plaques interconnecting cells. A classical example of this polarized distribution is in ventricles of the vertebrate heart, where gap junctions are a prominent component of the intercalated disk, connecting myocytes in an end-to-end orientation. Immunofluorescent localization in a large

NATO ASI Series, Vol. H 46
Parallels in Cell to Cell Junctions in Plants and Animals
Edited by A. W. Robards et al.

variety of tissues has indicated that the general case is staining of punctate regions at intercellular contacts, which are often multiple at lateral membranes between the cells (Fig. 1a). Freeze fracture electron micrographs of gap junctions of vertebrates shows that the thin section (Fig. 1c) and immunofluorescence images correspond to aggregates of 8-10 nm particles that can be as large as a few um in diameter (Fig. 1c) or as small as a few particles (e.g., Ginzberg et al., 1985). Each particle seen in freeze fracture is believed to form a channel composed of 6 identical protein subunits surrounding a central pore.

Gap junction distribution is thus distinct from that of other membrane channels. The clustering of particles is, however, similar to the higher order arrangement of the ACh receptor, which in mature cholinergic synapses is found localized beneath the presynaptic terminal. The gross morphology of gap junctions is also distinct from cytoplasmic bridges, the latter being larger in diameter, much longer and are often filled with microtubules and other cytoskeletal elements (Fig. 1d).

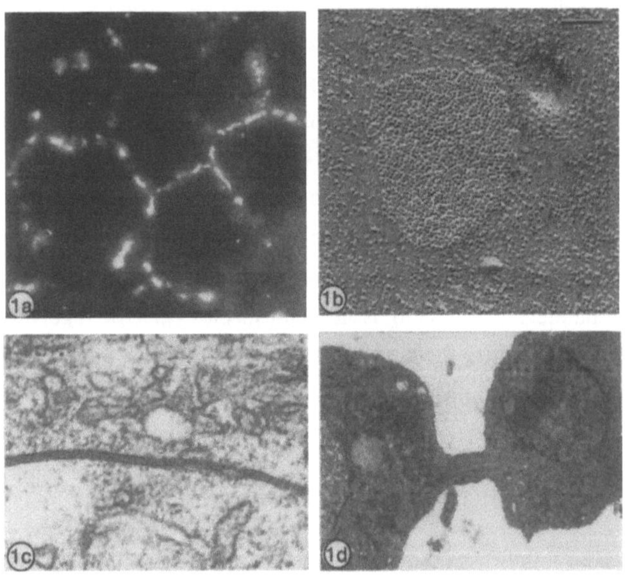

Figure 1. Distribution of intercellular channels. a. Antibodies to the liver gap junction protein label punctate regions of cell contact in frozen sections of rat liver viewed with immunofluorescence using a fluoresceinated secondary antibody (Saez et al., 1989b). b,c. Freeze fracture and thin sections of groups of freshly dissociated rat hepatocytes reveal characteristic particle aggregates (b) and linear membrane appositions (c) of gap junctions (Spray et al., 1986b). d. Thin section of cytoplasmic bridge between sibling cells of a *Drosophila* cell line (K_c cells) (Spray et al., 1989c). Bar in upper righthand corner of b represents 10 μm in a, 0.1 μm in b,c and 1 μm in d

The Gap Junction Gene Family

It has become increasingly clear that gap junctions are formed by a closely related group of membrane proteins encoded by a gene family (Paul, 1986; Kumar & Gilula, 1986; Beyer et al., 1987; Nicholson and Zhang, 1988; Ebihara et al., 1989; Gimlich and Gilula, 1988, Miller et al., 1988). Initially, cDNAs encoding the 27 kDa gap junction proteins from rat and human liver were identified and sequenced using oligonucleotides corresponding to partial amino acid sequences (Paul, 1986; Kumar and Gilula, 1986). The molecular weight of the protein predicted from the coding sequence was about 32,000, and in the new terminology proposed for these molecules (Beyer et al., 1987) the rat protein has been referred to as rat connexin 32. Low stringency hybridization with connexin 32 probes has also led to identification and sequencing of cDNAs encoding the heart gap junction protein, connexin 43, and the 21 kDa rodent liver gap junction protein, connexin 26; hybridization with other connexin probes has detected two gap junction cDNAs in early amphibian embryonic cells (*Xenopus* connexin 38: Ebihara et al., 1989; and another that is homologous to the mammalian connexins: Gimlich et al., 1988).

Hydropathicity analysis of the predicted amino acid sequences for these proteins suggests the presence of four transmembrane segments (and possibly a fifth in *Xenopus* connexin 38), three of which (and a fourth in *Xenopus* connexin 38) are predominantly hydrophobic. The third predicted transmembrane segment is the most amphipathic of the helices, suggesting that it may provide a hydrophilic face lining the intercellular pore. For connexin 32, protease sensitivity and antibody binding to isolated junctions indicates that both amino- and carboxyl-termini are on the cytoplasmic face of the channel (Zimmer et al., 1987; Hertzberg et al., 1988; Milks et al., 1988); homology of the other connexins suggests that their topologies may be similar.

Comparison of the sequences of these three gap junction proteins indicate regional homologies and differences. The four putative transmembrane segments are highly conserved, including the third, amphipathic segment. The two extracellular segments, connecting transmembrane segments 1 and 2 and 3 and 4, are also highly homologous, including positions of three extracellular cysteine residues per loop. These homologies are likely to account for the similar channel properties, such as size and selectivity, among these proteins, and the fact that gap junctions can form between cell types which express different gap junction proteins. Differences among these proteins are observed predominantly in the cytoplasmic segments of the protein, notably the carboxyl-terminal "tail" and a loop connecting transmembrane segments 2 and 3. These differences may be responsible for different mechanisms by which channel activity may be regulated, e.g. differential sensitivity to pH and voltage and to protein phosphorylation.

Size and selectivity of Gap Junction Channels

The size of the channels formed by the family of gap junction proteins (connexins) is larger than those of most channels of the nonjunctional membrane. Gap junction channels are permeable to ions and molecules with molecular weights below about 1 kDa (corresponding to an ionic radius of 0.8-1.0 nm: see Simpson et al., 1977). Permeant species thus include most ions (Fig. 2b) and small molecules (including the brilliantly fluorescent dye Lucifer Yellow CH: Fig. 2a), second messenger molecules and amino acids. Excluded on the basis of size are larger molecules including proteins, and cellular organelles and nucleic acids. Because the junctional permeability to Lucifer Yellow (1.4 nm diameter) distinguishes junctional from nonjunctional channels, it is the basis for the most commonly employed screening methods for the presence of gap junctions between cells (dye injection, GAP-FRAP, scrape loading). The few cases in which gap junctions are present, as judged morphologically or by other criteria, but dye permeability is not observed (e.g., *Aplysia* neurons: Bodmer et al., 1988; developmental boundaries in insects: Blennerhassett and Caveney, 1984) have generally been attributed to gap junctions of unusually small size, although there are other possible explanations including mismatch in size of pre- and postsynaptic elements (as in *Aplysia*) or an asymmetry in the distribution of gap junctions that a cell makes with its neighbors (as in *Tenebrio*) so that dye spread is more likely in one direction than in the other (Safranyos and Caveney, 1985; Safranyos et al., 1986).

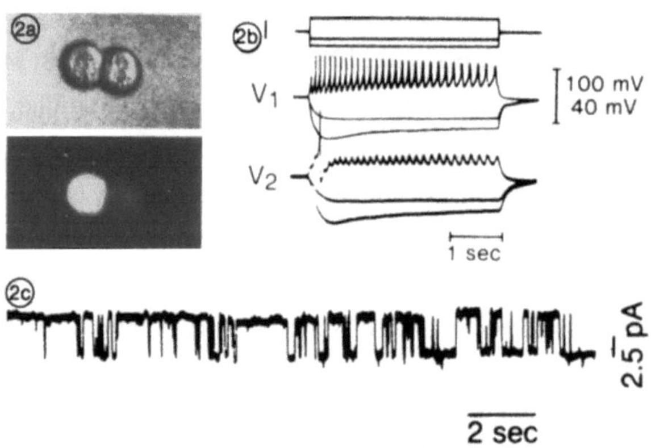

Figure 2. Permeability of gap junction channels. a. Lucifer Yellow CH (M_r457) injected into the lefthand hepatocyte rapidly diffuses into its coupled neighbor, as shown in the lower panel with FITC illumination 1 min after injection (Hertzberg et al., 1985). b. One function of gap junctions in the nervous system is presumably in synchronizing neuronal activity. Depolarization of one *Aplysia* neuron (upward-going current I represents depolarization of cell 1; downward currents represent hyperpolarizing currents in this cell). The voltage in cell 1 reaches threshold, causing a barrage of impulses (V_1) and electrotonic potentials in the second cell that lead to its firing (V_2) (Bodmer et al., 1988). c. Events corresponding to currents through single gap junction channels can be recorded from isolated liver gap junction membranes incorporated into the tips of patch pipettes. Here the transpipette voltage was about 40 mV and the single channel conductances are about 150 pS (Spray et al., 1986c)

Gap junction channels are not particularly selective, except on the basis of size, as mentioned above. The published selectivity series of the gap junction channel are as predicted for a large channel lacking a selectivity filter (e.g., Neyton and Trautmann, 1985).

Single channel recording techniques applied to cardiac and other gap junctions have recently allowed another type of glimpse into the size of the gap junction channel, and are indicating that the diversity of biochemically distinct channel proteins in various tissues is reflected in differences in unitary conductance. In acinar cells from salivary gland and in isolated gap junction membranes incorporated into bilayers or patch pipette tips (Fig. 2c), the amplitude of the dominant unitary event has been found to be about 120-150 pS (Neyton and Trautmann, 1985; Spray et al., 1986c; Young et al., 1987). For these tissues the principal gap junction protein is connexin 32 {Because these cells also express the putative gap junction protein connexin 26, the possibility had remained that this other protein formed the identified channels. Single channel studies of a communication-incompetent cell line stably transfected with connexin 32 has verified this estimate of unitary conductance: Eghbali et al., 1989}. In cardiac myocytes, unitary conductances are substantially smaller [about 50 pS in neonatal and adult mammalian ventricular cells (Burt and Spray, 1988b; Rook et al, 1988). (As noted above, the cardiac gap junction protein, connexin 43, is homologous to, but distinct from, the junctional proteins present in liver (Beyer et al., 1987; Kumar and Gilula, 1986; Paul, 1986); presumably the differences in unitary conductance are related to differences in primary sequences of the proteins. Two other gap junction channel sizes (20 and about 90-100pS) have been detected in other tissues (leptomeningeal cells: Spray et al., 1989c; clonal "WB" cells derived from liver: Spray et al., 1989b; embryonic frog muscle cells: Chow and Young, 1987). Channels of even larger sizes have been reported in chick embryonic heart cells (Veenstra and DeHaan, 1986, 1988). Other gap junction proteins may form the channels identified by these unitary conductance measurements. It is useful to point out that these unitary conductances are all less than those reported for the large Ca-activated K^+ channels in biological membranes and are in the range of conductances generally found for ligand-gated channels (eg. NMDA, ACh and glycine receptors). A parsimonious explanation for the gap junction's modest unitary conductance in the face of permeability to large ions and molecules is that the channel (or a large portion of it) may be a right cylindrical' pore spanning a large distance (two bilayers instead of one); for other membrane channels, conductance and selectivity have apparently been optimized by combining a large funnel-shaped mouth with a narrow selectivity filter.

Unitary conductance of gap junction channels can be predicted from constraints imposed by structural and permeability features (see Hille, 1984). Voltage imposed across the channel drops along the pathway from one cytoplasmic face to another through access resistances (r_a) symmetrically placed on each cytoplasmic surface with a pore resistance (r_p) in the center of the channel. The resistance of a single channel (r_j) is thus:

$$rj = 2p/4r + p\,(1/\pi r^2)$$

Where r is channel radius (about 0.7-0.8 nm, based on the channel's permeability to Lucifer Yellow), l is channel length (about 15-20 nm, although possibly longer in connexins where carboxyl-termini contribute to effective channel length) and p is resistivity of the pore fluid (about 100-200 Ohm-cm, depending on whether values from axoplasm or myoplasm are chosen). Therefore, conductance of a single gap junction channel would be predicted to lie between about 50 and 100 pS, agreeing closely with the values obtained experimentally.

Conductance of cytoplasmic bridges can also be predicted using the equations above. For a bridge with a diameter of 0.5 um and a length of 1 um (Fig. 1d), unitary conductance would be expected to be about 100-150 nS. The lack of observable flux of fluoresceinated dextrans through cytoplasmic bridges of a *Drosophila* cell line (Kc cells) is presumably due to the tortuosity of the pathway, which is filled with microtubules. If flux of large molecules were significantly retarded, photobleaching and cytoplasmic binding could interfere with detection of FITC-dextran transfer. An estimate of the tortuosity can be made by comparing the measured intercellular conductance with that calculated from the equations above. Taking into account the pore's access resistance (see above) and assuming cytoplasmic sensitivity of 100 ohm-cm, pore conductance would be expected to be about 2.5 times as high as those conductances that were measured (Spray et al., 1989c).

Physiological variability within the gap junction family

The degree to which cells communicate through gap junction channels is dictated by the number of these channels in the membranes, the size of the channels, and the fraction of time that the channels are open. As noted above, channel size is a property that differs with connexin type; although early reports suggested the possibility that channels might constrict during closure, more recent experiments indicate that closure of gap junction channels is all-or-none rather than graded (Verselis et al., 1986b; Zimmerman et al., 1985). The other determinants of coupling strength can be distinguished on the basis of kinetics: The fraction of time that the channels are open is influenced by factors operating over the short term (seconds to minutes), and the number of channels available on which the short-term regulatory agents may act is influenced over a longer time course (hours to days). The rapid action of various agents we have termed gating (with the analogy used by workers on other membrane channels of a door's opening and closing: Spray et al., 1984); the long-term regulation is measureable in terms of protein and mRNA synthesis and seems to involve expression of the junctional molecules (e.g., Spray et al, 1988).

The high degree of homology among the gap junction family members gives rise to similar properties but also allows considerable divergence, especially in terms of mechanisms of gating and regulation of expression. For gap junction channels in various tissues, treatments that gate the channel include transjunctional voltage, where the sensed field is across the junctional membranes, or inside-outside, where the relevant field is from cell interior to extracellular space, intracellular pH (pH_i), certain anesthetics (e.g. octanol, heptanol and halothane), and unsaturated fatty acids (e.g., oleic, arachidonate and myristolate), second messengers (including calcium, cyclic nucleotides, diacylglycerol and arachidonic acid) and halomethanes.

Voltage dependence. Transjunctional voltage dependence is most strongly displayed in the crayfish rectifying synapse (Furshpan and Potter, 1966) and in early embryonic stages of invertebrate and vertebrate embryos (Fig. 3; Spray et al., 1979, 1981a). In these systems voltage imposed across the junctional membrane rapidly (tens of microseconds, in crayfish) or more slowly (hundreds of msec, in amphibian and tunicate embryo) closes gap junction channels. In crayfish this effect is unidirectional: Depolarizing pulses spread from giant fiber to motoneuron to allow orthodromic conduction; hyperpolarizing pulses in the giant axon or depolarizing pulses in the motoneuron are not spread, so that antidromic conduction is prohibited. In embryonic cells, the g_j-voltage relation displays symmetric closure of the channels for either positive or negative potentials applied to either cell. Recent cloning of an

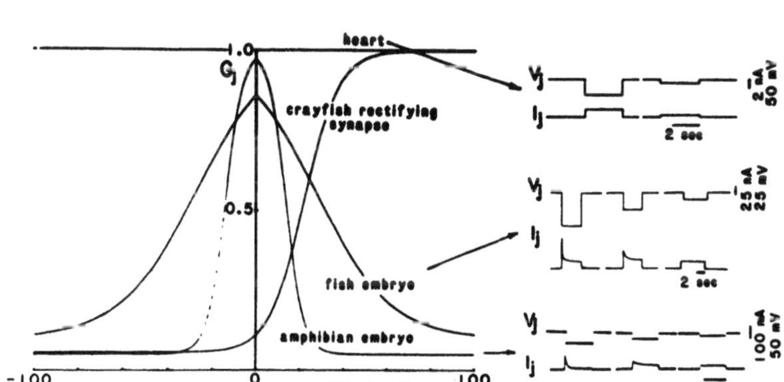

Figure 3. Voltage dependence of selected gap junctions. Normalized junctional conductance (G_j) is plotted as a function of transjunctional voltage (V_j) for several voltage dependent junctions. For comparison, neonatal and adult cardiac cells show little voltage sensitivity (Spray et al., 1985)

embryonic amphibian gap junction protein (*Xenopus* connexin 38) reveals differences from other voltage sensitive channels (it lacks the transmembrane S_4 segment) and from other gap junction proteins (hydropathicity plots of connexin 38 predict the possibility of a fifth hydrophobic domain, with the C-terminus located extracellularly: Ebihara et al., 1989). Experiments comparing wild type and mutant channels expressed in *Xenopus* oocytes or mammalian systems should resolve whether this novel hydrophobic region is involved in sensing the transjunctional voltage.

It was believed until very recently that voltage dependence was a feature unique to early embryonic cells and certain invertebrate tissues. However, later embryonic cells of chick heart have now been reported to display decreased junctional conductance at high transjunctional voltages (R. DeHaan and RD Veenstra, personal communications), and gap junctions between liver cells have been found to be even more sensitive (Young et al., 1987; Spray et al., 1989e). In the latter case, pairs of hepatocytes and isolated junctional membranes from rat liver incorporated in lipid bilayers have similar dependence of junctional conductance on transjunctional voltage (the voltage at which conductance is reduced by half is about 25-35 mV), and experiments on communication-deficient cells stably transfected with connexin 32 cDNA indicate that the voltage dependent gap junction protein in this case is connexin 32 (Eghbali et al., 1989).

Nevertheless, not all gap junctions display voltage dependent conductance. Most neonatal and adult tissues do not ordinarily show it (e.g., White et al., 1985; Burt and Spray, 1988; Somogyi and Kolb, 1988; Perez-Armandariz, et al., 1988), although at high voltages conductance is reduced in rat neonatal cardiocyte pairs (Rook et al., 1988).

Inside-outside voltage dependence, where the resting potential of the cell influences strength of coupling, is a property of some gap junctions, including those in *Chironomus* and *Drosophila* salivary gland cells, in salamander rod photoreceptors and after pharmacological treatments in squid blastomeres (Obaid et al., 1983; Verselis and Bargiello, 1989; Spray et al., 1984; Stern and MacLeish, 1985). It appears that in some of these tissues inside-outside voltage dependence coexists with transjunctional voltage dependence to a variable extent (Spray et al., 1984; Verselis and Bargiello, 1989; Bennett et al., 1988). In these systems depolarization of either cell, or both cells together, reduces junctional conductance and hyperpolarization increases g_j. This property is presumably due to a voltage sensor located in the wall of the gap junction channel.

pH, Ca and Calmodulin. Studies supporting the concept of gating of gap junction channels can be traced back a century, to Engelmann's demonstration (1877) that cardiac muscle fibers adjacent to an injured area first depolarized in response to the injury and then repolarized as the uninjured cells recovered. Although the experiment is complicated by the possible resealing of ruptured nonjunctional membrane, this phenomenon, termed "healing over", has been interpreted as indicating the closure of gap junction channels, and the time course has been shown to be dependent on the extracellular Ca concentration (Deleze, 1965; Nishiye, 1977). Ca as a general effector of gap junction channel closure has been championed by a number of laboratories, beginning with the work of Loewenstein (1966). Raising intracellular free Ca^{++} to high levels, by injecting Ca^{++}, by treating with the Ca ionophore A23187 or by various manipulations that elevate free cytoplasmic Ca^{++} levels by increasing entry from extracellular medium or by evoking release from intracellular stores, decreases electrical coupling (see DeMello, 1984; Loewenstein, 1981), and can decrease junctional conductance in the few cases where g_j has been measured (e.g., Spray et al., 1982).

The experiments in which Ca^{++} sensitivity of the gap junction has been evaluated are complicated by the secondary effects that Ca^{++} would be expected to have on levels of other intracellular ions and also by changes in levels of enzymes, peroxidation products and other second messengers that could result from Ca^{++} loading of the cell. Furthermore, coupling strength depends on conductance of the nonjunctional membrane as well as that of the junction (for a cell pair, the coupling coefficient, $k = g_j/(g_j + g_{nj})$, where g_j and g_{nj} are junctional and nonjunctional conductances: Bennett, 1966), so that cells might uncouple without changing g_j. The basic issue in deciding whether and to what extent Ca^{++} is involved in the regulation of intercellular communication via gap junctions is the sensitivity of the junctions to Ca^{++} Free intracellular Ca^{++} levels under normal conditions are usually below 100 nM, and can even be much lower than this. Measurements of intracellular Ca^{++} levels obtained by uncoupling treatments where the effect was attributed to Ca^{++} ions have all been orders of magnitude above these normal intracellular levels; in *Chironomus* salivary gland and early embryonic cells of the fish *Fundulus* , for example, the uncoupling threshold is apparently about 0.1 mM (Oliveira-Castro and Loewenstein, 1972; Spray et al., 1982). These high concentrations could be significant in allowing channel closure following cell injury, when intracellular Ca^{++} would approach extracellular levels, but would be unlikely to mediate changes in intercellular communication under normal physiological conditions.

In 1977 intracellular pH was found to uncouple embryonic cells of the amphibian *Xenopus* (Turin and Warner, 1977), and we subsequently showed that the relation between intracellular pH and junctional conductance was described by a form of the Hill equation, suggesting that the interaction between H^+ ions and the channel protein might be direct (Spray et al., 1981a). According to this interpretation, H^+ ions would titrate acidic amino acid residues on the gap junction protein, leading to conformational change and channel closure. Gap junction channels of many types have now been shown to be closed by intracellular acidification, although the apparent pK of the g_j-pH_i relation and its steepness (the Hill coefficient, n) are distinctive features of gap junctions in various tissues. For example, in embryos of fish and amphibian, the apparent pK has been found to be about 7.3 or 6.4-6.5 and n is 3-4 or 2 respectively (compare Spray et al., 1981a to Bennett et al., 1988); in heart, apparent pK is variable (6.6 or lower) and n is 2 (see Spray et al., 1985); in liver the apparent pK is about 6.5 and n is 7-8 (Spray et al.,1986b). In the latter tissues, the Hill coefficient correlates closely with the number of histidine residues in the cytoplasmic loop of the connexins forming the channels, suggesting that titration of these groups (the pK value of ionizable HIS residues in proteins is around 6.5) might lead to cooperative conformational change. In the amphibian protein there is at most one

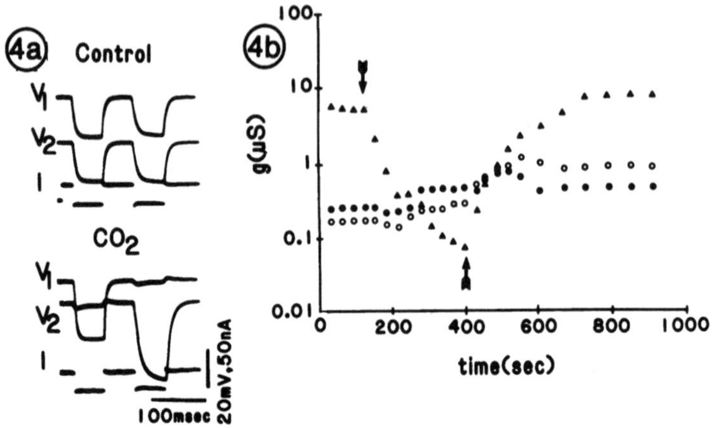

Figure 4. pH sensitivity of junctional conductance. Exposure to solution equilibrated with CO_2 (a) or saline in which acetate is substituted for Cl (between the arrows in b) rapidly and reversibly uncouples the cells. Currents (I) in a are alternately applied to each cell, allowing the calculation of junctional (filled triangles) and nonjunctional conductances (circles) shown in b

histidine in this segment (Ebihara et al., 1989);together with the higher pK this may indicate titration of other groups. In summary, although all systems studied are sensitive to intracellular pH; sensitivity varies from system to system which might be partially explained by differential expression of connexins in different tissues. Although most of the findings mentioned above have been interpreted as if H^+ and Ca^{++} ions interacted directly with molecular components of gap junction channels to affect their permeability, it has also been suggested that a calmodulin-like cytoplasmic factor may participate in crayfish axon as well as in other systems (Johnston and Ramón, 1981; Arellano et al., 1986). Evidence in support of the "calmodulin hypothesis" is of several types: Calmodulin (CaM) inhibitors have been reported to reduce pH sensitivity of a variety of preparations; CaM perfusion restores pH sensitivity to perfused crayfish septate axons, where the responsiveness is lost by perfusion and has been attributed to a soluble intermediary molecule; and there is interaction between the effects of H^+ and Ca^{++} ions in certain systems. Furthermore, pH sensitivity is hormonally altered in a diurnal manner in lateral axons of the crayfish nerve cord (Moreno et al., 1987), which might be explained either by alteration in levels or affinity of an intermediate molecule or by altered properties of the sensor within the gap junction protein.

CaM antagonists (W7, calmidazolium, trifluoperazine, chlorporomazine) were reported to reduce pH sensitivity of several gap junction types (Peracchia and Bernardini, 1984; Peracchia,

1987), leading to the hypothesis that pH sensitivity is mediated by a calmodulin-like molecule (Perrachia, 1987). However, changes in pH sensitivity by CaM inhibitors has not been confirmed in experiments in which both junctional conductance and intracellular pH were measured simultaneously (Verselis et al., 1986a). In the latter studies, a long-term increase in junctional conductance was found in response to the CaM antagonists which might be consistent with the notion that CaM or a CaM-like molecule exerts an action on the expression, rather than on gating of gap junction channels. Lack of responsiveness of internally perfused crayfish axons to Ca^{++}(which is consistent with the lack of responsiveness seen in perfused *Fundulus* blastomeres: Spray et al., 1982, and in isolated liver gap junctions incorporated into lipid bilayers: Young et al., 1987; Spray et al., 1986), has been attributed to washout of a cytoplasmic intermediary molecule (Johnston and Ramón, 1981). CaM replacement restored Ca^{2+} sensitivity (Arellano et al., 1988; albeit to levels that were still not physiologically relevant: 5 μM).

CaM binds to connexin 32 in isolated junctions with high affinity (Van Eldik et al., 1985; Zimmer et al., 1987), although there is as yet no evidence that this binding occurs *in situ* The results cited above indicating that CaM inhibitors may increase g_j as a consequence of channel accrual, thus increasing the availability of channels on which gating stimuli may act, would be consistent with a role for CaM in reducing channel number, and perhaps CaM is involved in the process of retrieval of gap junctions from the surface membrane which is the first step in degradation. Whatever the outcome of this controversy, it should be re-emphasized that there are as yet no compelling data to support the hypothesis that Ca^{++} affects gap junctions directly (or through an intermediary) at physiological concentrations. In fact, the relative insensitivity of the gap junction channel to Ca^{++} ions permits the intercellular diffusion of this important second messenger ion (Saez et al., 1989a).

A number of reports indicate that the gating treatments considered above may interact with each other, so that the combination of moderate acidification and application of moderate transjunctional voltage, for example, could effectively close junctional channels. Ca^{++}may modify voltage dependence in hepatocytes (Young et al., 1987) and in *Drosophila* and *Chironomus* salivary gland cells (Verselis and Bargiello, 1989; Obaid et al, 1982); Ca^{++} apparently alters or interacts synergistically with pH dependence in cardiac myocytes (Burt, 1987; White et al., 1986; Noma and Tsuboi, 1987).

In crayfish lateral septate axons and in blastomeres of Rana pipens, pH dependence is abolished by glutaraldehyde and EEDQ, reagents that crosslink amino acid residues (lysine or carboxylic and lysine residues, respectively) and by retinoic acid through a mechanism not yet identified (Campos de Carvalho et al., 1986; Spray et al., 1986a). In frog blastomeres moderate intracellular acidification reduces junctional conductance but does not affect voltage dependence of the residual conductance evaluated by large pulses (Spray et al., 1986a); with stronger acidification, the voltage dependence is reduced slightly (Verselis et al., 1987). These data may indicate that the sensors for H^+ and voltage are located in different regions of the channel or even on different molecules and that the charges of the channel that confer the voltage sensitivity can be to some extent neutralized with H^+, reducing sensitivity to the applied voltage field.

Lipophilic molecules. Lipid soluble molecules are a group of agents that reduce gap junctional conductance rapidly and reversibly. The first of these agents discovered were the alcohols octanol and heptanol (Johnston et al., 1980), which uncouple within seconds to minutes in every system tested except *Aplysia* neurons (Bodmer et al., 1988). The uncoupling effect can be rapidly reversed by washing and occurs independently of changes in intracellular Ca^{2+}(Meda et al., 1986; Sáez et al., 1989a) or pH_i (White et al., 1986). Halothane and related volatile anesthetics uncouple all tested cell types also (e.g., Burt and Spray, 1989b), which is of special interest for heart their arrhythmogenicity is pronounced. These agents are more rapidly

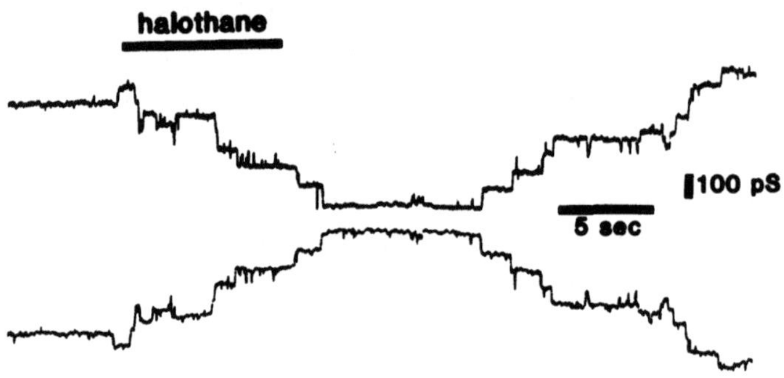

Figure 5. Halothane reversibly uncouples cell pairs. Here leptomeningeal cells with low junctional conductance display unitary channel closures as halothane reduces conductance (seen as convergence of the upper and lower traces). When halothane is rinsed from the dish cells recouple in a step-like manner, representing discrete openings of individual channels (Spray et al., 1989d)

reversible than the alcohols, and are thus useful in revealing single channel events in cases where junctional conductance is initially high. New agents with presumably similar effects are unsaturated lipids (Aylsworth et al., 1986; Burt and Minnich, 1989) and the doxyl stearic acid spin probes (Burt, 1989), all of which may interact at the protein-lipid interface of the channel to perturb its function.

Other lipophilic molecules are also potent blockers of intercellular communication via gap junctions, including halomethanes (Sáez et al., 1987b) and arachidonic acid (Sáez et al., 1987a), but metabolic products of these agents, rather than the agents themselves, may be the active molecular species. For example, the effect of halomethanes in pairs of rat hepatocytes is partially prevented by ß mercaptoethanol and by SK&F 525A, suggesting that action is via cytochrome P_{450} metabolism (Sáez et al., 1987b). In the case of arachidonic acid, action on heart cells is apparently largely direct, as inhibitors of lipoxygenase and cylooxygenase pathways may delay but do not prevent uncoupling (Burt and Spray, 1989a); action on hepatocytes, however, can be prevented by lipoxygenase inhibitors, indicating that direct action is unlikely (Sáez et al., 1987a).

Protein kinases. Cyclic nucleotides affect junctional conductance in most tissues and cell lines which have been studied to date, and these agents appear to be the most likely candidates for physiological regulation under most conditions (Spray and Saez, 1988). Membrane permeant cAMP derivatives increase junctional conductance in pairs of cells derived from heart (DeMello, 1988; Burt and Spray, 1988a) and liver (Sáez et al., 1986) and decrease junctional conductance between pairs of fish retinal horizontal cells (Lasater and Dowling, 1985). These changes are all consistent in direction and magnitude with those obtained in response to hormones and neurotransmitters that elevate intracellular cAMP (glucagon in liver, isoproterenol in heart, dopamine in horizontal cells). In heart cells the effect of cAMP derivatives can be reversed to an uncoupling action under conditions of C^{++} overload (Burt and Spray, 1988a). In liver and heart cells the physiological action of the agents that increase cAMP is apparently through the cAMP-dependent protein kinase (kinase a) as indicated by the decreased response when recording pipettes are filled with kinase a inhibitor (Sáez et al., 1986; DeMello, 1988).

Membrane permeant cGMP derivatives decrease g_j in cardiac myocytes (Burt and Spray, 1988a) but these agents have no detectable effect on hepatocytes (Sáez et al., 1986). In heart cells, the action of cGMP derivatives is similar in magnitude and time course to that of the cholinergic agonist carbachol, which would be consistent with acetylcholine action through a cGMP second messenger pathway (Burt and Spray, 1988a).

We have shown in both isolated liver junctional membranes and dissociated groups of liver cells that the 27 kDa liver gap junction protein (connexin 32) is directly phosphorylated at a serine residue by cAMP-dependent protein kinase (cAMP-dPk), with kinetics similar to those associated with increased conductance in response to glucagon or 8 Br-cAMP (Sáez et al., 1986). Subsequent experiments have localized the phosphorylated SER residue and have shown that a synthetic peptide corresponding to this region of connexin 32 is an adequate physiological substrate for phosphorylation by kinase a (Sáez et al., 1989c). Presumably the increased conductance of junctional membranes in liver cells is due to increased open time of the gap junction channels caused by phosphorylation of the junctional protein; proof will require measurements of single channel currents under conditions of low intrinsic g_j, which has been elusive thus far but may be obtainable in exogenous expression systems. In the primary sequence of the cardiac gap junction protein, connexin 43, there are a number of possible kinase a phosphorylation sites as well as a site that may be a substrate for phosphorylation by a cGMP dependent kinase. The availability of high affinity antibodies against this gap junction protein will soon allow immunoprecipitation experiments to ascertain whether any of these putative sites is actually phosphorylated and to compare the kinetics of phosphorylation with kinetics of conductance change.

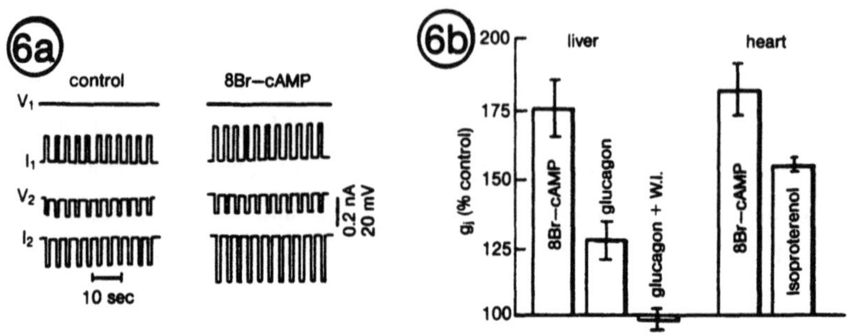

Figure 6. Effects of cyclic nucleotides on junctional conductance. a. Exposure of neonatal cardiac myocytes to 1 mM 8 Br-cAMP for 3 min leads to an increase in junctional current, indicated by the amplitude of the upward going deflections in I_1 in response to a voltage command in the other cell (V_2). b. In both heart and liver cell pairs the membrane permeant cAMP derivative increases junctional conductance, as do agents that increase cAMP as a second messenger (glucagon in liver, isoproterenol in heart). In liver, the response to glucagon is abolished when the patch pipettes contain an inhibitor of cAMP-dependent protein kinase (Burt and Spray, 1988a; Sáez et al., 1986)

The effects of protein kinase c (PKC) on gap junctional conductance have been explored widely since the initial reports that treatment of cells with tumor promoting phorbol esters leads to uncoupling (Murray and Fitzgerald, 1979; Yotti et al., 1979; see also Trosko et al., 1988). Since diacylglycerol, a physiological activator of protein kinase C (PKC), causes the same effect as phorbol esters, it has been proposed that the uncoupling effect is mediated by PKC (Gainer and Murray, 1985; Enamoto and Yamasaki, 1985; Yada et al., 1985). In several cell lines the phorbol ester blockade has been shown to be rapid (within 10-20 min: Gainer and Murray, 1985; Yada et al., 1985; Chanson et al., 1988; Oh et al., 1988), which correlates temporally with kinase c activation, assayed as translocation of PKC from cytosolic to membrane compartments (Gainer and Murray, 1985; Chanson et al., 1988; Oh et al., 1988). In other systems, phorbol esters increase electrical coupling (Grassi et al., 1986; JM Burt & DC Spray, unpublished observations). In pancreatic acinar cells, which express the same gap junction protein as hepatocytes (Traub et al., 1989), treatment with PDBu causes rapid translocation of PKC, but coupling is not reduced and may even be slightly increased over time (Chanson et al., 1988).

Thus, depending on the cell type, phorbol esters may lead to rapid increases or decreases in junctional conductance that are temporally associated with kinase c activation. In addition, phorbol ester treatment can lead to longer term changes, including decreased occurrence of gap junctions at the interfaces of treated cells (Yancey et al., 1982). The rapid effects have been attributed to increased intracellular free Ca^{++} or to phosphorylation of the gap junction protein, which might be associated with a conformational change leading to channel closure (Yada et al., 1985). It should be noted that other secondary effects, for example through the arachidonic acid cascade, have for the most part been ignored (see Sáez et al., 1989d).

We have analyzed the effect of phorbol esters on the phosphorylation state of gap junctions in adult rat hepatocytes (Sáez et al., 1989c), which represents a differentiated cell system that in primary culture displays abundant gap junctions (Spray et al., 1986b). Phosphorylation of connexin 32 was found to be stimulated by the tumor promoting phorbol ester PDBu, but not by its inactive analogue, 4∝PB (Sáez et al., 1989c). The surprise came when we analyzed the two dimensional tryptic phosphopeptide maps of connexin 32 immunoprecipitated from hepatocytes radiolabeled with ^{32}P and treated with PDBu or 4∝PB for 15 min. These maps revealed that the pattern of phosphopeptides were practically identical to those from cells treated with cAMP. Consistent with this finding and with the effects of cAMP on hepatocytes, concentrations of PDBu that stimulate the phosphorylation of connexin 32 did not reduce the incidence of dye coupling or electrical coupling after treatment for as long as 6 hrs (Sáez et al., 1989c). The possiblity of cross-talk between the signal transduction pathways using diacylglycerol and cAMP (e.g., Byus et al., 1983) was ruled out, indicating that PKC stimulates phosphorylation of connexin 32 independently and at the same site as does cAMP-dPk. The convergence on the same SER residue of connexin 32 presumably causes similar functional consequences of phosphorylation by the two classes of kinases. This finding raises the intriguing possibility that it might not be the activation of kinase c that leads to the reduction in junctional conductance, but the inactivation that characteristically follows. If PKC activity is necessary to maintain communication between cells, its disappearance should reduce coupling; and after the cells recouple, the gap junctions should be refractory to further application of phorbol esters; such an effect has been reported (Enamoto and Yamasaki, 1985). Moreover, if these protein kinases converge, activation of cAMP-dPk might reverse the effect of inactivation of PKC by phorbol esters; this could be the basis of the spontaneous reversibility seen in some cases (Yada et al., 1985) and of the finding that elevation of the intracellular cAMP protects against uncoupling by tumor promoters (Kanno et al., 1984).

Long-term changes in gap junction expression

For the most part, the changes in gap junction - mediated intercellular communication that have been considered above occur within minutes after the effective treatment is applied. As was pointed out above, gap junctions are composed of proteins and as such are subject to the same sets of regulatory mechanisms that act to control expression of other membrane proteins (Fig. 7).

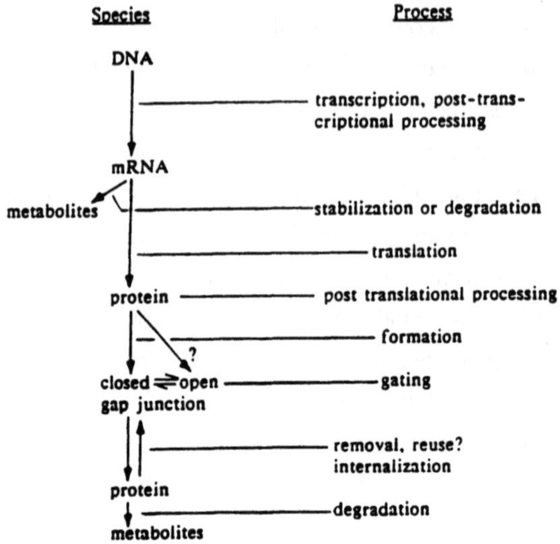

Figure 7 Control of intercellular communication via gap junctions (Spray et al., 1988).

Studies on these regulatory steps are only beginning, but they already reveal that multiple points in the pathway may be facilitated or interrupted by various treatments. cAMP is one example, where so-called "upregulation" of gap junctions has been reported in a variety of cell types (e.g., Flagg-Newton et al, 1981; Arzania et al., 1981; "downregulation" has been reported in uterine smooth muscle cells treated with cAMP derivatives for comparable periods: Cole and Garfield, 1986). Within 6 hours, cAMP exposure leads to increase in coupling between sympathetic neurons which is blocked by protein and mRNA synthesis inhibitors and the effect is therefore presumably transcriptional in neurons (Kessler et al., 1984, 1985). In hepatocytes, cAMP extends the lifetime of gap junctions between dissociated cell groups, which is largely due to an increase in stability of the mRNA encoding connexin 32 (Sáez et al., 1989b). Transcriptional effects are expected as well, based on the plausible cAMP response element sequence in the genomic clone of connexin 32 (Miller et al., 1988).

Components of extracellular matrix also exert strong effects on expression of gap junctions. In hepatocytes, there is an interaction of matrix components and certain hormones such that treatment of cultured cells with glucagon together with proteoglycans leads to a marked increase in connexin 32 transcription as well as increased stability of its mRNA (Watanabe et al., 1987). A possibly related case is that of the *Drosophila* clock mutant *per*, where the gene product is a proteoglycan-like molecule. Here the periodicity of the biological clock is directly related to gene dosage and the strength of electrical coupling is much lower in the mutant lacking the *per* locus than in wild type (Bargiello et al., 1988). Regulation of connexin expression by matrix is an exciting finding worth pursuing in regard to pathological conditions, where in various tissues including liver specific changes occur under chronic pathological conditions (e.g., Rojkind, 1988).

Conclusions

Considered *in toto* , these short- and long-term regulatory mechanisms define the gap junction channel as one with enormous plasticity in response to intra- and extracellular signals. The range of permeant signalling molecules, while not as large as that passed between cells connected by cytoplasmic bridges, allows gap junctions to serve as major conduits for exchange of second messenger molecules, metabolites and substances that might serve as either triggers or brakes for various enzymatic reactions. The unitary conductance of the channels, while not as large as for some channels of nonjunctional membranes, can effectively propagate impulses and synchronize electrical activity.

Tools are now available with which to pursue the nature of the regulation of channel conductance in the context of short-term gating and longterm expression. At the level of the gap junction channel itself, alterations in structure of the connexins and the use of exogenous expression systems (e.g., Dahl et al., 1987; Eghbali et al., 1989; Young et al., 1987; Spray et al., 1986c) should allow identification of the molecular domains responsible for sensitivities to various gating stimuli. At the other levels of regulation, including precursor synthesis, membrane trafficking and degradation, the necessary probes are also now available and in the immediate future should lead to elucidation of other cell biological aspects of the channel-forming proteins. These approaches should also illuminate the relationships, both analogous and homologous, among the various types of membrane channels. Cells use both gap junctions and cytoplasmic bridges as conduits for exchange of ions, metabolites and signalling molecules. The properties of the gap junction channel outlined here may serve as a key to identifying one class of intercellular channel and in distinguishing it from others in the surface membranes of biological systems.

Acknowledgements

The contributions of past and present colleagues (especially JC Sáez, AL Harris, MVL Bennett, EL Hertzberg, AC Campos de Carvalho) to the ideas, the labor, and the interpretation of the material presented here is gratefully acknowledged, as is grant support: NIH NS07512, NIH NS16524, NIH HL38449.

Literature Cited

Arellano RO, Ramón F, Rivera A, Zampighi GA (1986). Lowering of pH does not directly affect the junctional resistance of crayfish lateral axons. J Membr Biol 94:293-299.

Arellano RO, Ramón F, Rivera A, Zampighi GA (1988). Calmodulin acts as intermediary for the effect of calcium on gap junction from crayfish lateral axons. J Membr Biol 101:119-131.

Arzania R, Dahl G. Loewenstein WR (1981). Cell junction and cyclic AMP. III. Promotion of membrane permeability and junctional membrane particles in a junction-deficient cell type. J Membr Biol 63:133-146.

Aylsworth CF, Trosko JE and Walsh CW (1986). Influence of gap junction-mediated intracellular communication between Chinese hamster cells in vitro. Cancer Res 46:4527-4533.

Bargiello TA, Sáez L, Baylies MK, Gasic G, Young MW, Spray DC (1987) The *Drosophila* clock gene *per* affects intercellular junctional communication. Nature 328:686-691.

Bennett MVL (1966). Physiology of electrotonic junctions. Ann NY Acad Sci 37:509-539.

Bennett MVL, Verselis V, White RL and Spray DC (1988). Gap junctional conductance: Gating. In: Gap Junctions (EL Hertzberg & RG Johnson eds) Alan R. Liss. Inc. New York pp. 287-304.

Beyer EC, Paul D and Goodenough DA (1987). Connexin 43: a protein from rat heart homologous to a gap junction protein from liver. J Cell Biol 105:2621-2629.

Blennerhassett MG and Caveney S (1984). Separation of developmental compartments by a cell type with reduced junctional permeability. Nature 309:361-364.

Bodmer R, Verselis V, Levitan IB and Spray DC (1988). Electronic synapses between Aplysia neurons in situ and in culture: Aspects of regulation and measurements of permeability. J Neurosci 8:1656-1670.

Burt JM (1987) Block of intercellular communication: Interaction of intracellular H^+ and Ca^{++} Amer J Physiol 253:C607-C612.

Burt JM (1989) Uncoupling of neonatal rat cardiocytes by doxylstearic acids: specificity and mechanisms of action. Amer J Physiol, in press.

Burt JM, Minnich B (1989) Uncoupling of cardiac cells by fatty acids: Structure-activity relations. Amer J Physiol, in press.

Burt JM, Spray DC (1988a) Inotropic agents modulate junctional conductance between cardiac myocytes. Amer J Physiol 254:H1206-H1210.

Burt JM, Spray DC (1988b) Single channel events and gating behavior of the cardiac gap junction channel. Proc Natl Acad Sci USA 85:3431-3434.

Burt JM, Spray DC (1989a) Arachidonic acid uncouples cardiac myocytes. Biophys J 55:217a.

Burt JM, Spray DC (1989b) Volatile anesthetics block intercellular communication between neonatal rat myocardial cells. Circ Research, in press.

Byus C, Trevillyan JM, Cavit LJ and Fletcher WH (1983). Activation of cyclic adenosine 3':5'-monophosphate protein kinase in H35 hepatoma and Chinese hamster ovary cells by a phorbol ester tumor promoter. Cancer Res 43:3321-3326.

Campos de Carvalho AC, Ramon F, Spray DC (1986). Effects of group specific protein reagents on electrotonic coupling in crayfish septate axons. Amer J Physiol 251:C99-C103. Chanson M, Bruzzone R, Spray DC, Regazzi R Meda P (1988). Cell uncoupling and protein kinase C: Correlation in a cell line but not in a differentiated tissue. Amer J Physiol 255:C699-C704.

Chow, I., and Young, S.H. (1987) Opening of single gap junction channels during formation of electrical coupling between embryonic muscle cells. Devel Biol 122: 332-337.

Cole WC, Garfield RE (1986). Evidence for physiological regulation of myometrical gap junction permeability. Amer J Physiol 251:C411-420.

Dahl G, Miller T, Paul D, Voellmy R, Werner R (1987) Expression of functional cell-cell channels from cloned rat liver gap junction complementary DNA. Science 236:1290-1293.

Deleze J (1965) Calcium ions and the healing-over of heart fibers. In: Electrophysiology of the Heart (B Taccardi & G Marchetti, eds), pp. 147-148. Pergamon Press, N.Y.
DeMello WC (1984) Modulation of junctional permeability. Fed Proc 43:2692-2696.
De Mello WC (1988). Increase in junctional conductance caused by isoproterenol in heart cell pairs is suppressed by cAMP-dependent protein-kinase inhibitor. Biochem Biophys Res Commun 154:509-514.
Dermietzel R, Yancey SB, Traub O, Willecke K, Revel J-P (1987) Major loss of the 28-kD protein of gap-junction in proliferating hepatocytes. J Cell Biol 105:1925-1934.
Ebihara L, Beyer EC, Swenson KI, Paul DL, Goodenough DA (1989). Cloning and expression of a Xenopus embryonic gap junction protein. Science 243:1194-1195.
Eghbali B, Roy C, Kessler JA, Spray DC (1989) Transfection of a hetatoma cell line lacking functional gap junctions with cDNA encoding connexin 32, the major gap junction protein of rat liver. Fed Proc
Enamoto T, Yamasaki H (1985). Rapid inhibition of intercellular communication between BALB/c 3T3 cells by diacylglycerol, a possible endogenous functional analogue of phorbol esters. Cancer Res 45:3706-3709.
Enamoto T, Martel N, Kanno Y, Yamasaki H (1984). Inhibition of cell communication between Balb/c 3T3 cells by tumor promoters and protection by cAMP. J Cell Physiol 121:323-333.
Engelmann TW (1977). Vergleichende Untersuchungen zur Lehre von der Muskel- und Nervenelektricitat. Pflug Arch 15:116-148.
Flagg-Newton JL, Dahl G, Loewenstein WR (1981). Cell junction and cyclic AMP: I. Upregulation of junctional membrane permeability and junctional membrane particles by administration of cyclic nucleotide or phosphodiesterase inhibitor. J Membr Biol 63:105-121.
Furshpan EJ, Potter DD (1959). Transmission at the giant motor synapses of the crayfish. J Physiol 145:289-325.
Gainer H St C, Murray AW (1985). Diacylglycerol inhibits gap junctional communication in cultured epidermal cells: Evidence for a role of protein kinase C. Biochem Biophys Res Commun 126:1109-1113.
Gimlich RL, Kumar NM, Gilula NB (1988). Sequence and developmental expression of mRNA coding for a gap junction protein in Xenopus. J Cell Biol 107:1065-1073.
Ginzberg RD, Morales EA, Bennett MVL, Spray DC (1985) Cell junctions in early embryos of squid (Loligo pealei) Cell Tissue Res 239:477-484.
Grassi F, Monaco L, Fratamico G, Dolci S, Iannini E, Conti M, Eusebi F, Stefanini M (1986). Putative second messengers affect cell coupling in the seminiferous tubules. Cell Biol Intern Reports 10:631-639.
Harris AL, Spray DC, Bennett MVL (1981). Kinetic properties of a voltage dependent junctional conductance.J Gen Physiol 77:95-120.
Harris AL, Spray DC, Bennett MVL (1983). Control of intercellular communication by a voltage dependent junctional conductance. J Neuroscience 3:79-100.
Hertzberg EL, Disher RM, Zhou Y, Cook RG (1989). Topology of the 27,000 M_r liver gap junction protein: Cytoplasmic localization of amino- and carboxy-termini and a hydrophilic domain which is protease-hypersensitive. J Biol Chem
Hertzberg EL, Spray DC, Bennett MVL (1985). Blockade of gap junctional conductance by an antibody prepared against liver gap junction membranes. Proc Natl Acad Sci USA 82:2412-2416.
Hille B (1982). Ionic Channels of Excitable Membranes. Sinuaer Associates, Sunderland, Mass. Johnston MF, Ramón F (1981). Electrotonic coupling in internally perfused crayfish segmented axons. J Physiol 317:509-518.
Johnston MF, Simon SA, Ramón F (1980). Interaction of anesthetics with electrical synapses. Nature 286:498-500.
Kanno Y, Enamoto T, Shiba Y, Yamasaki H (1984). Protective effect of cyclic AMP on tumor promoter-mediated Inhibition of cell-cell communication (electrical coupling). Exp Cell Res 152:31-37.
Kessler JA, Spray DC, Sáez JC, Bennett MVL (1984). Determination of synaptic phenotype: Insulin and cAMP independently initiate development of electrotonic coupling between cultured sympathetic neurons. Proc Natl Acad Sci USA 81:6235-6239.

Kessler JA, Spray DC, Sáez JC, Bennett MVL (1985). Development and regulation of electrotonic coupling between culutred sympathetic neurons. In: Gap Junctions (MVL Bennett & DC Spray, eds) Cold Spring Harbor Laboratory, Cold Spring Harbopr, Ny. Pp 231-240.
Klaunig J, Ruch R (1987). Role of cyclic AMP in the inhibition of mouse hepatocyte intercellular communication by liver tumor promoters. Toxicol Appl Pharmacol 91:159-170.
Kumar NM, Gilula NB (1986). Cloning and characterization of human and rat liver cDNA for liver gap junction protein. J Cell Biol 103:123-134.
Lasater EM, Dowling JE (1985). Electrical coupling between pairs of isolated fish horizontal cells is modulated by dopamine and cAMP. In: Gap Junctions (MVL Bennett & DC Spray, eds) Cold Spring Harbor Laboratory, NY, pp 393-404.
Loewenstein WR (1966). Permeability of membrane junctions. Ann NY Acad Sci 137:441-472.
Loewenstein WR (1981). Junctional intercellular communication. The cell top cell membranbe channel. Physiol Rev 61:809-913.
Meda P, Bruzzone R, Knodel S, Orci L (1986). Blockage of cell-to-cell communication within pancreatic acini is associated with increased basal release of amylase. J Cell Biol. 103:475-483.
Milks LC, Kumar NM, Houghten R, Unwin N, Gilula NB (1988). Topology of the 32-kd liver gap junction protein determined by site-directed antibody localizations. EMBO J 7:2967-2975.
Miller T, Dahl G, Werner R (1988). Structure of a gap junction gene: Rat connexin-32. Biosc. Reports 8:455-464. Moreno AP, Ramón F, Spray DC (1987). Variation of gap junction sensitivity to H ions with time of day. Brain Res 400:181-184.
Murray AW, Fitzgerald DJ (1979). Tumor promoters inhibit metabolic cooperation in cocultures of epidermal and 3T3 cells. Biochem Biophys Res Commun 91:395-401.
Neyton J, Trautmann A (1985). Single channel currents of an intercellular junction. Nature 317:331-335.
Nicholson BJ, Zhang J-T (1988). Multiple protein components in a single gap junction: Cloning of a second hepatic protein (M_r 21,000). In: Gap Junctions (EL Hertzberg & RG Johnson eds) Alan R. Liss, Inc N.Y. pp.207-218.
Nishiye H (1977). The mechanism of Ca action on the healing-over process in mammalian cardiac muscles: A kinetic analysis. Japn J Physiol 27:451-466.
Noma A, Tsuboi N (1987). Dependence of junctional conductance on proton, calcium and magnesium ions in cardiac paired cells of guinea-pig. J Physiol 382:193-211.
Obaid AJ, Socolar SA, Rose B (1983). Cell-to-cell channels with two independent gates in series: Analysis of junctional conductance modulated by membrane potential, calcium and pH. J Membr Biol 73:69-89.
Oh SY, Madhukar BV, Trosko JE (1988). Inhibition of gap junctional blockage by palmitoyl carnitine and TMB-8 in a rat liver epithelial cell line. Carcinogenesis 9:135-139.
Oliviera-Castro GM., Loewenstein WR (1971) Junctional membrane permeability: Effects of divalent cations.J Membr Biol 5:51-77.
Paul DL (1986). Molecular cloning of cDNA for rat liver gap junction protein. J Cell Biol 103:123-134.
Peracchia C (1987). Calmodulin-like proteins and communicating junctions: Electrical uncoupling of crayfish septate axons is inhibited by the calmodulin inhibitor W7 and is not affected by cyclic nucleotides. Pflugers Arch 408:379-385.
Peracchia C, Bernardini G (1984). Gap junction structure and cell-to-cell coupling. Is there a calmodulin involvement? Fed Proc 43:2681-2691.
Perez-Armendariz ME, Spray DC, Bennett MVL (1988). Properties of gap junctions between pairs of pancreatic beta cells of mice. Biophys J 53:53a.
Rojkind M (1988) Extracellular matrix. In: Liver Biology and Pathology (IM Arias, WB Jakoby, H Popper, D Schachter & D Sharfritz, eds) Raven Press, NY pp 707-717.
Rook MB, Jongsma HJ, van Ginneken AC (1988). Properties of single channels between isolated neonatal rat heart cells. Amer J Physiol 255:H770-H782.

Sáez JC, Bennett MVL, Spray DC (1987a). Oxidant stress blocks hepatocyte gap junctions independently of changes in intracellular pH, Ca^{2+} or MP27 phosphorylation state. Biophys J 51:39a.
Sáez JC, Bennett MVL, Spray DC (1987b). Carbon tetrachloride at hepatotoxic levels reversibly blocks gap junctional communication between rat liver cells. Science 236:967-969.
Sáez JC, Connor JA, Spray DC, Bennett MVL (1989a). Hepatocyte gap junctions are permeable to the second messengers inositol 1,4,5-triphosphate and calcium ions. Proc Natl Acad Sci USA 86:2708-2712.
Sáez JC, Gregory WA, Dermietzel R, Hertzberg EL, Watanabe T, Reid L, Bennett MVL, Spray DC (1989b). Cyclic AMP extends the lifetime of gap junctions between pairs of rat hepatocytes. Amer J Physiol in press.
Sáez JC, Nairn AC, Czenick AJ, Hertzberg EL, Spray DC, Greengard P, Bennett MVL (1989c). Phosphorylation of the 27 kDa hepatocyte gap junction protein (connexin 32) by cAMP-dependent protein kinase, protein kinase C and Ca^{2+}-calmodulin protein kinase II. (Submitted for publication).
Sáez JC, Spray DC, Hertzberg EL (1989d). Gap junctions: Biochemical properties and functional regulation under physiological and toxicolcogial conditiuons. Toxicology, in press.
Sáez JC, Spray DC, Nairn AC, Hertzberg EL, Greengard P, Bennett MVL (1986). cAMP increases junctional conductance and stimulates phosphorylation of the 37 kDa principal gap junction polypeptide. Proc Natl Acad Sci USA 83:2473-2477.
Safranyos RA, Caveney S (1985). Rates of diffusion of fluorescent molecules via cell-to-cell membrane channels in a developing tissue. J Cell Biol 100:736-747.
Safranyos R, Spray DC, Bennett MVL (1986). Electrical and dye coupling at a develoipmental compartment boundary. J Cell Biol 103:363a.
Simpson, I., Rose, B. and Loewentstien, W.R. (1987). Size limit of molecules permeating the junctional membranechannels. Science 197: 294-296.
Somolgyi R, Kolb HA (1988). Cell-to-cell channel conductance during loss of gap junctional coupling in pairs of pancreatic acinar and Chinese hamster ovary cells. Pflüger Arch 412:43-65.
Spray DC, Bennett MVL (1985). Physiology and pharmacology of gap junctions. Ann Rev Physiol 47: 281-303.
Spray DC, Burt JM, White RL, Wittenberg BA (1989a). Cardiac gap junctions: Status and dynamics. In: Conduction of the Cardiac Impulse (W Giles, Ed). Alan R. Liss, NY., in press.
Spray DC, Campos de Carvalho AC, Bennett MVL (1986a). Sensitivity of gap junctional conductance to H ions in amphibian embryonic cells is independent of voltage sensitivity. Proc Natl Acad Sci USA 83:3533-3536.
Spray DC, Chanson M, Moreno AP, Dermietzel R, Meda P (1989b). Multiple types of gap junction channels between individual pairs of WB cells. Submitted for publication.
Spray DC, Cherbas L, Cherbas P, Morales EA, Carow G (1989c). Ionic and metabolic coupling and synchronous sibling mitsis in a clonal cell line of Drosophila Exp Cell Research, in press.
Spray DC, Dermietzel R, Kessler JA (1989d). Unitary conductances and physiological responses of leptomeningeal cells in culture. Soc Neursci Abstr, in press.
Spray DC, Fujita M, Sáez JC, Choi H, Watanabe T, Hertzberg EL, Rosenberg LC, Reid L.M. (1987) Glycosaminoglycans and proteoglycans induce gap junction synthesis and function in primary liver cultures. J Cell Biol 105:541-551.
Spray DC, Ginzberg RD, Morales EA, Gatmaitan Z, Arias IM (1986b). Electrophysiological properties of gap junctions between dissociated pairs of rat hepatocytes. J Cell Biol 101:135-144.
Spray DC, Harris AL, Bennett MVL (1979). Voltage dependence of junctional conductance in early amphibian embryos. Science 204:432-434.
Spray DC, Harris AL, Bennett MVL (1981a). Gap junctional conductance is a simple and sensitive function of intracellular pH. Science 211:712-715.

Spray DC, Harris AL, Bennett MVL (1981b). Equilibrium properties of a voltage dependent junctional conductance. J Gen Physiol 77:75-94.

Spray DC, Sáez JC (1988). Agents that regulate gap junctional conductance: Sites of action and specificities. In: Biochemical Regulation of Intercellular Communication Advances (MA Mehlman, ed) Alan Liss N.Y. pp. 1-26.

Spray DC, Sáez JC, Brosius D, Bennett MVL, Hertzberg EL (1986c). Isolated liver gap junctions: Gating of transjunctional currents is similar to that in intact pairs of rat hepatocytes. Proc Natl Acad Sci USA 83:5494-5497.

Spray DC, Sáez JC, Burt JM, Hertzberg EL, Watanabe T, Reid LM, Bennett MVL (1988). Gap junctional conductance: Multiple sites of regulation. In: Gap Junctions (EL Hertzberg & RG Johnson, eds) Alan Liss, N.Y. in press.

Spray DC, Stern JH, Harris AL, Bennett MVL (1982). Comparison of sensitivities of gap junctional conductance to H and Ca ions. Proc Natl Acad Sci USA 79:441-445.

Spray DC, Verselis V, Moreno AP, Eghbali B, Bennett MVL, Campos de Carvalho AC (1989e). Voltage dependent gap junctional conductance. In: Biophysics of Gap Junction Channels (C. Peracchia, ed) CRC Press Baton Rouge, in press.

Spray DC, White RL, Campos de Carvalho AC, Harris AL, Bennett MVL (1984). Gating of gap junction channels.Biophys J 45:219-230.

Spray DC, White RL, Verselis V, Bennett MVL (1985). General and comparative physiology of gap junction channels. In: Gap Junctions (MVL Bennett & DC Spray, eds). Cold Spring Harbor: NY. pp. 139-153.

Stern JH, MacLeish PR (1985). Isolated pairs of rod photoreceptors are electrically coupled by a large voltage-dependent conductance. Invest Ophthal Visual Sci 26:193.

Traub O, Look J, Dermietzel R, Brümmer F, Hülser D, Willecke K (1989). Comparative characterization of the 21-kD and 26-kD gap junction proteins in murine liver and cultured hepatocytes. J Cell Biol 108:1039-1051.

Trosko JE, Chang C-C, Madhukar BV, Oh SY, Bombick D, El-Fouly MH (1988). Modulation of gap junction intercellular communication by tumor promoting chemicals, oncogenes and growth factors duringmcarcinogenesis. In: Gap Junctions (EL Hertzberg & RG Johnson), Alan R. Liss, Inc., N.Y. pp. 435-448.

Turin L, Warner AE (1978). Carbon dioxide reversibly abolishes ionic communication between cells of early ¡amphibian embryo. Nature 270:56-57.

Van Eldik LJ, Hertzberg EL, Berdan RC, Gilula NB (1985). Interaction of calmodulin and other calcium-modulated proteins with mammalian and arthropod junctional membrane proteins. Biochem Biophys Res Commun 126:825-832.

Veenstra RD, DeHaan RL (1986). Measurement of single currents from cardiac gap junctions. Science 233:972-974. Veenstra, R.L. and DeHaan R. (1988). Cardiac gap junction channel activity in embryonic chick ventricle cells Amer. J. Physiol. 363: H170-H180.

Verselis VK, Bargiello TA, Spray DC, Bennett MVL (1988). Conductance of gap junctions in Drosophila salivary gland depends on both transjunctional and inside-outside voltages. Biophys J 45:219-230.

Verselis VK, Bargiello TA (1989). Dual voltage control in gap junctions. In: Biophysics of Gap JunctionChannels. (C Perracchia, ed) CRC Press, Inc. (in press).

Verselis VK, Campos de Carvalho AC, White RL, Bennett MVL, Spray DC (1986a). Calmodulin and gap junction regulation. J Cell Biol 103: 73a.

Verselis, V., White, R.L., Spray, D.C. and Bennett, M.V.L. (1986b) Gap junctional conductance and permeability are linearly related. Science 234:462-464.

Verselis VK, White RL, Spray DC, Zavilowitz J, Bennett MVL (1987). Induced asymetry of gating of gap junctions in amphibian blastomeres. J Cell Biol 105:307a.

Watanabe T, Sáez JC, Spray DC, Reid LM (1987). Heparins potentiate the regulation by hormones and growth factors of liver-specific mRNA expression in cultured hepatocytes. J Cell Biol

White RL, Verselis VK, Spray DC, Bennett MVL, Wittenberg BA (1986). Acidification-resistant junctional conductance between some pairs of ventricular myocytes from rat. J Cell Biol 103:363a.

White RL, Spray DC, Campos de Carvalho AC, Wittemberg BA, Bennett MVL (1985). Some electrical and pharmacological properties of gap junctions between adult ventricular myocytes. Amer J Physiol 249:C447-C455.

Yada T, Rose B, Loewenstein WR (1985). Diacylglycerol down regulates junctional membrane permeability,TMB blocks this effect. J Membr Biol 88:217-232.

Yancey SB, Edens JE, Trosko JE, Chang C-C, Revel J-P (1982). Decreased incidence of gap junctions between Chinese hamster V-79 cells upon exposure to the tumor promoter 12-O-tetradecanoyl phorbol-13-acetate.Exp Cell Res 139:329-340.

Yee A, Revel J-P (1978). Loss and reappearance of gap junctions in regenerating liver. J Cell Biol 78:554-564.

Yotti LP, Chang CC, Trosko JE (1979). Elimination of metabolic cooperation in Chinese hamster cells by a tumor promoter. Science 206:1089-1091. Young JD-E, Cohn ZA, Gilula NB (1987). Functional assembly of gap junction conductance in lipid bilayers: Demonstration that the major 27 kD protein forms the junctional channel. Cell 48:733-743.

Zimmer DB, Green CR, Evans WH Gilula NB (1987). Topological analysis of the major protein in isolated intact rat liver gap junctions and gap junction-derived single membrane structures. J Biol Chem 262:7751-7763.

Zimmerman AL, Rose B (1985). Permeability properties of cell-to-cell channels: Kinetics of fluorescent tracer diffusion through a cell junction. J Membr Biol 84:269-283.

CARDIAC GAP JUNCTIONS: GATING PROPERTIES OF SINGLE CHANNELS

H J Jongsma and M B Rook
Department of Physiology
University of Amsterdam Academic Medical Centre
Meibergdreef 15
1105 AZ Amsterdam
The Netherlands

Sjöstrand and Anderson (1954) demonstrated that the heart is composed of single cells contradictory to the till then prevailing thought that the coordinated activation and contraction of the heart was made possible by its syncytial nature. Since that time the nature of the cell contacts that allow the heart to behave like a functional syncytium have been investigated intensively.

It is well established now that between all myocardial cells membrance specialisations known as gap junctions are present which have a low electrical resistance with regard to the plasma membrane and which therefore allow the conduction of action potentials from cell to cell. The specific resistance of gap junctions has been estimated to range from 0.01 to 10 Ω cm^2 compared to 1 to 10 KΩ cm^2 for cardiac plasma membranes (Fozzard 1979). It has become clear also that gap junctions consist of arrays of intercellular channels allowing the easy passage of ions (see Loewenstein, 1981 for a review). Already in 1977 Simpson *et al.* estimated the conductance of individual gap junctional channels to be around 100 pS based on theoretical considerations. Recent measurements have shown that the actual conductance of gap junctional channels varies from organ to organ and from species to species. Neyton and Trautmann (1985) reported a single channel conductance of 70 to 180 pS for rat lacrimal gland cells. For embryonic chick heart a value of about 165 pS was reported (Veenstra and DeHaan, 1986) while for neonatal rat heart the single channel conductance was estimated to be between 40 and 50 pS (Burt and Spray, 1988, Rook *et al.* 1988). Somogyi and Kolb (1988) measured a range of gap junctional conductances in chinese hamster ovary cells from 22 to 120 pS and a conductance of 130 pS in murine pancreatic acinar cells. In fibroblasts a gap junctional channel conductance of 22 pS was found (Rook *et al.* 1989). In many cases more than one peak in the frequency distribution of the conductances was found. At present it is not clear whether these subsidiary conductances, which are always lower than the main conductance, represent subconductance states of the same channel or separate channels.

There is ample evidence now that the junctional conductance may be controlled by a variety of chemical agents (see Herzberg and Johnson, 1988) although precise mechanisms are not yet clear. Whether gap junctional channels are controlled by transcellular voltage (i.e. whether they

NATO ASI Series, Vol. H 46
Parallels in Cell to Cell Junctions in Plants and Animals
Edited by A. W. Robards et al.

are voltage dependent) also is a matter of controversy. The steep voltage dependence reported by Spray *et al.* (1979) for gap junctions in early amphibian embryos is not seen in most other preparations. In general it appears that gap junctions between mammalian cell pairs isolated as such do not show any voltage dependence (e.g. White *et al.* 1985, Noma and Tsuboi 1987, Weingart 1986, Reverdin and Weingart 1988). Gap junctions consisting of relatively few channels however do exhibit voltage dependence (Neyton and Trautmann 1985, Rook *et al* 1988, Veenstra, 1989). In the present paper we will contribute to this issue by presenting data on the voltage dependence of gap junctional conductance in different cell systems derived from the neonatal rat heart.

Methods

Neonatal rat heart cells were obtained by collagenase dissociation as described previously (de Bruijne and Jongsma 1980). The cells were used either directly after the dissociation or after culturing them for 24-48 h in 35 mm petri dishes (Falcon) in a humidity and 0_2 controlled incubator at 37° C. To obtain cultures the cells were plated at a final concentration of 0.5 x 10^6 cells/ml in Ham's Fl0 (Flow laboratories) modified to contain 2.2 mM ca^{2+}, 5% horse serum (Flow) and 5% of fetal calf serum (Flow). For experimentation with freshly isolated cells 1 ml of cell suspension in Ham's Fl0 was pipetted in 35 mm petri dishes containing already 1 ml of Ham's Fl0. When cultured cells were used the culture medium was changed to 2 ml Ham's Fl0 prior to experimentation. The dishes were placed on the stage of an inverted microscope (Nikon Diaphot TMD) and viewed with phase contrast optics at a total magnification of 400x. After 15 min necessary for the settling down of the suspended cells or for the adaptation of cell cultures to the ambient temperature suitable pairs of cells were selected for impalement with suction electrodes for whole cell recording according to Hamill *et al.*, 1981. In the case of suspensions we choose cells lying less than 5 µm apart. These cells almost always turned out to be myocytes. The majority of the fibrocytes, which originally were abundantly present, remained stuck to the glass of the container in which the suspension was kept. In cell cultures we selected cell pairs consisting of two fibrocytes or of a fibrocyte and myocyte. The fibrocytes were identifiable by the fact that they were very thin and translucent while myocytes were much thicker and more opaque. Moreover the fibrocytes never contracted while the myocytes did. In mixed pairs this was sometimes difficult to assess by eye because of the mechanical interaction of the myocyte with the fibrocyte. The electrical measurements however provided unambiguous identification (see results). The cells were impaled with suction electrodes pulled from 1 mm borosilicate capillaries containing a glass fiber. They were heat polished and backfilled with a filtered solution containing (in mM) 10 KCl, 130 K-gluconate, 10 Na_2ATP, 10 $MgCl_2$, 5

HEPES, 10 EGTA and 1 $CaCl_2$. The solution was brought to pH 7.1 with KOH. The free calcium concentration was buffered with the EGTA at 5×10^{-8} M. This low calcium concentration was used to avoid possible interference of Ca ions with the junctional conductance and to prevent contractile activity. All experiments were performed at room temperature (21° C).

The electrodes were connected to custom built amplifiers allowing both current clamp or voltage clamp recordings. The electrodes typically had resistances of 5 to 15MΩ. The access resistance between pipette tip and cell interior which was estimated in current clamp from the fast rising step in voltage displacement caused by application of 50 pA pulses to both cells directly after making contact between cell interior and pipette interior by breaking the membrane under the tip by applying some extra suction, ranged between 20 and 60 MΩ. In experiments with isolated cells the next step was to gently push the cells together. The evolvement of coupling was initiated by adding a 1:1 mixture of fetal calf serum and horse serum to the dish to a final concentration of 10 %. The development of coupling was monitored by applying rectangular current pulses (50 pA,5-10 ms) to one of the cells. Once the coupling had started indicated by weak electronic interaction between the cells, both cells were voltage clamped at identical holding potentials as near as possible to the zero current potential which was generally around - 60 mV. The potential in one cell was then changed in stepwise fashion to varying levels of de- and hyperpolarisation resulting in a constant voltage drop across the intercellular junction during the step.

In experiments with cell pairs from cultures the presence of electrical coupling was assessed in current clamp by injecting current pulses of 50 - 100 pA, 10 ms in either of the two cells and recording of the electrotonic effect in the other cell prior to switching to voltage clamp mode.

In principle the determination of the junctional resistance is very simple because the current recorded in the cell held at constant potential is solely determined by the voltage drop across the junction and the junctional resistance. In practice however one should take into account the access resistances of the electrode tips because they are in series with the junctional resistance (see Rook *et al* 1988, for an extensive discussion of this point). As long as the junctional resistance is much larger than the access resistances these can be neglected and in that case the junctional resistance R_j is given by $R_j = -\Delta V_j/i_j$ in which ΔV_j is the difference between the voltage in the cell held at constant potential and the voltage in the de- or hyperpolarised cell and i_j is the current measured in the cell held at constant potential. The current measured in the polarized cell is the sum of the membrane current of that cell and the junctional current. The junctional current component in this cell has the opposite sign to the junctional current measured in the other cell. Channel openings and closings in the junction should therefore appear as current displacements in both tracings simultaneously with equal amplitude but opposite sign. The current and voltage signals were FM recorded on tape (NAGRA TI). After low pass filtering at 100 Hz to reduce noise from nonjunctional membranes and digitising the signals were analysed on a DEC VAX-750 computer using fortran written programs for e.g. current amplitude measurements and histogram construction.

Results

Figure 1 shows voltage tracings from 2 myocytes (A and B) in suspension pushed together for 10 min. Stimulation of cell A elicited an action potential in this cell and a subthreshold depolarisation in cell B. A few minutes later the procedure was repeated. Apparently the coupling between the cells had improved so much that the action potential in cell A now induced a suprathreshold depolarisation in cell B which therefore generated in action potential also, be it with a delay of 35 ms (Fig. 1, middle panel).

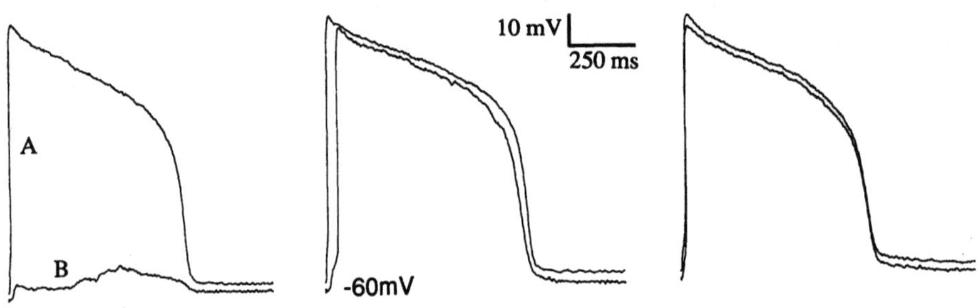

Fig. 1. Voltage tracings from a pair of myocytes in the process of coupling; left panel: action potential elicited in cell A 10 min after contact was made; middle panel: 12 min after contact; right panel: 25 min after contact

Further improvement of coupling is evidenced by the decrease in delay between the two action potentials in the right panel of Figure 1 which was taken 25 min after the start of the experiment.

When the two cells of a pair are voltage-clamped after the first signs of electrotonic interaction are seen, recordings as shown in Figure 2 can be obtained. The cells were clamped at a holding potential, of -58 mV. A depolarising voltage step to +42 mV was applied for 2 s to cell B. The upper trace shows the junctional current, the lower trace the sum of junctional current and membrane current. In the upper trace steps of current presumably representing opening and closing of junctional channels are seen. The amplitude of the steps is 4.5 pA at a transjunctional voltage difference of 100 mV. The single channel resistance would thus amount to 22 GΩ or the single channel conductance to 45 pS, which agrees well with our previously published figures

(Rook *et al.* 1988). Steps in current in the lower trace are seen simultaneously with those in the upper trace. They have the same amplitude but opposite polarity as expected.

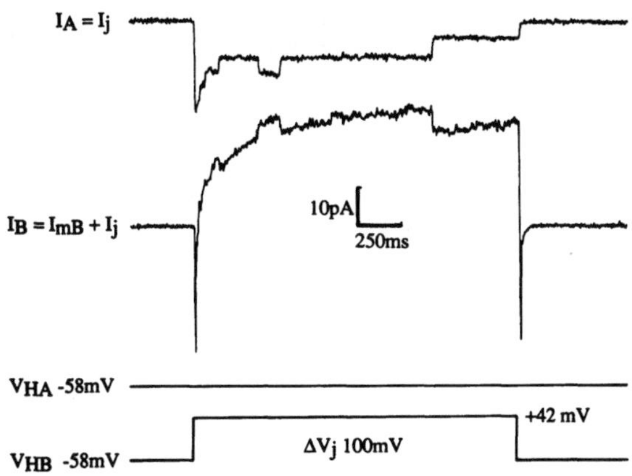

Fig. 2. Current records from a pair of myocytes held together for 20 min. Cell A (upper trace) was kept at a holding potential of -58 mV. Cell B (lower trace) was clamped from a holding potential of -58mV to 42mV for 2 s as indicated

At the beginning of the voltage step the amplitude of the junctional current (Fig. 2 upper trace) is about 22.5 pA corresponding to 5 open channels while at the end the amplitude has decreased to less than 5 pA corresponding to 1 open channel. It seems therefore that channels are closing during the voltage step. This is clear also from Figure 3. Here the relation between junctional current and transcellular voltage difference is plotted at the beginning (□) and end (Δ) of a 2 s voltage step applied to a pair of myocytes in the process of coupling. It can be seen that between +50 and -50 mV the two lines superimpose with a slope of 1.6 GΩ(600 pS ;13 junctional channels). At transjunctional voltage differences of more than 50 mV the slope of the line measured at the beginning of the pulse remains unchanged but that of the line measured at the end of the pulse is decreased, i.e. the junctional conductance has decreased. In all experiments with weakly coupled myocytes we have found this voltage dependence. In gap junctions consisting of a large number of channels voltage dependence is not observed as is illustrated in Figure 4. Here the current-voltage relations of the junction measured at the beginning and the end of the pulse superimpose over the whole range of voltage differences. The slope of both lines is 135 MΩ. Taking into account the access resistances of the electrodes, a R_j of 76 MΩ can be calculated which means that 290 open channels are present in this junction.

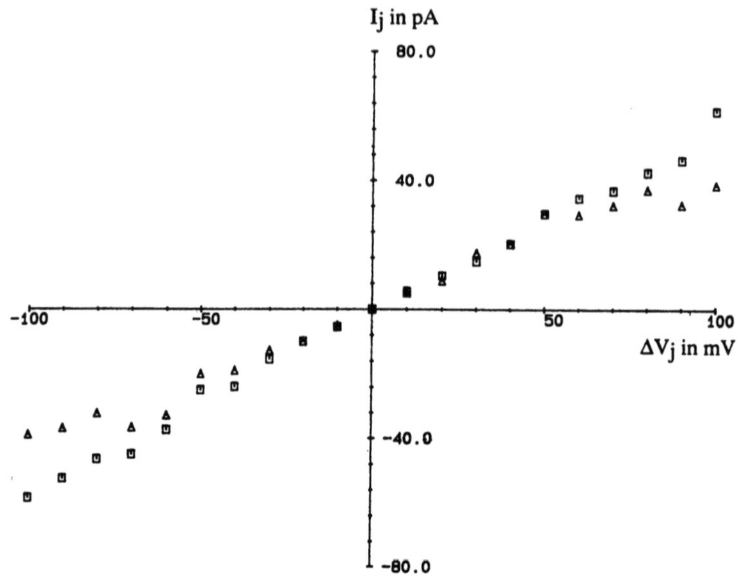

Fig. 3.Junctional current-transjunctional voltage relation of a weakly coupled myocyte pair. (♦) measured at beginning of pulse.; (Δ) measured at end of 2 s pulse. The slope of the relation between -50 and +50 mV is 1.6 GΩ

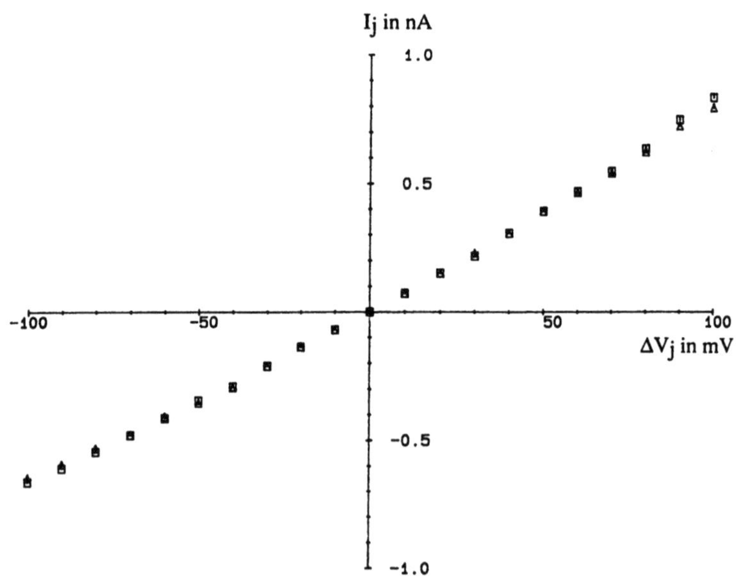

Fig. 4. Junctional current- transjunctional voltage relation of a well coupled pair of myocytes. (♦) measured at beginning of pulse; (Δ) measured at end of 2 s pulse. The slope of both relations is 135 MΩ

Figure 5 shows current clamp records from a pair of fibrocytes in a 24 h culture. A stimulus (50 pA, 10ms) was applied to cell A causing it to depolarise rather steeply. After the stimulus the membrane potential relaxes exponentially to its resting level of -20 mV indicating passive membrane behaviour. Cell B also depolarises but much less steep. There is no delay between the depolarisation in cell A and B as is to be expected for purely passive electrotonic coupling. Figure 6 shows a comparable experiment on a mixed pair consisting of a myocyte (M) and a fibrocyte (F). In the left panel the myocyte was stimulated (50 pA, 5 ms) eliciting an action potential in this cell. At the upstroke of the action potential a slow depolarisation started in the

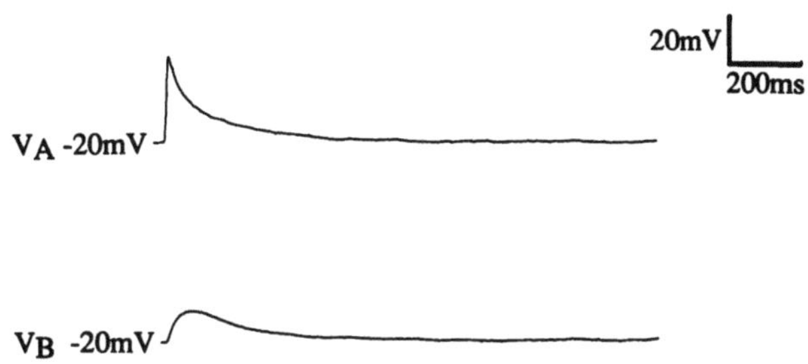

Fig. 5. Current clamp records from a weakly coupled pair of fibrocyte in a 24 h culture. A 50 pA 5 ms rectangular pulse was applied to cell A

fibrocyte which decayed again on repolarisation of the myoctye. In the right panel the stimulus (100 pA, 5ms) was applied to the fibrocyte. A depolarisation comparable to that in figure 5 ensued. The depolarisation caused the myocyte to depolarise also and after a considerable delay in which the threshold was reached to generate an action potential. This action potential in its turn causes the already repolarising fibrocyte to depolarise again as in the left panel. Clearly we observe coupling between a cell with a regenerative current system in its membrane to one without it.

Fig. 7 shows examples of voltage clamp records of a FF pair and a MF pair together with a record of a weakly coupled MM pair for comparison. The transjunctional potential difference was 100 mV in all cases and only the current trace of the cell clamped at its holding potential is shown (i.e. the junctional current trace). Single channel events can be distinguished clearly in each of the three recordings. As in Figure 2 the voltage dependence in the MM trace is evident. In the MF pair there is also voltage dependence but considerably less than in the MM pair, while in the FF pair voltage dependence is virtually absent. It should be noted that the amplitude of the

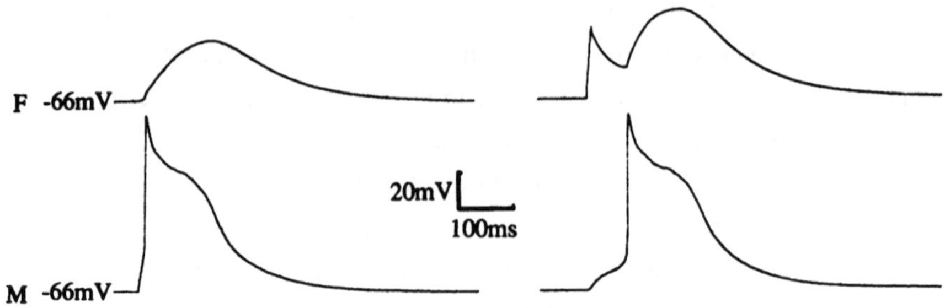

Fig. 6. Current clamp records of a myocyte-fibrocyte pair. In the left panel a rectangular pulse of 50 pA, 5 ms was applied to the myocyte. In the right panel a pulse of 100 pA, 5 ms was applied to the fibrocyte

single channel events in the three records is different, indicating a difference in single channel conductance in the three cases. To provide a more complete picture we collected a large number of records at different transjunctional voltage differences and constructed frequency histograms of the junctional current steps. The results are shown in Figure 8. The upper row depicts the histograms obtained in the different preparations by combining current measurements at all transjunctional voltages used (40 - 100 mV). In fibrocyte pairs the apparent single channel conductance (γ_j) is 21.2 pS, in MF pairs it is 29.4 pS and in MM pairs two peaks are present, one at 43.4 pS and the other at 18.4 pS. As we have discussed before (Rook *et al.*.1988) there are several reasons to believe that γ_j in myocytes is 43.4 pS and that the lower conductance is either a separate channel or a subconductance state of the 43 pS channel. From the rows of histograms at separate transjunctional voltages it can be seen that in MM pairs γ_j is not voltage dependent. Both the major and the minor peaks are identical at 100 and 75 mV transjunctional voltage difference. In FF pairs γ_j tends to increase slightly with decreasing transjuctional voltage but this effect is statistically not significant. The same holds true for the MF pairs. It may be concluded that voltage dependence of junctional conductance is not caused by voltage dependent changes in single channel conductance.

Discussion

Interest in the properties of gap junctions has substantially increased in recent years, due to the fact that with the new techniques of molecular biology, cell biology and electrophysiology it has

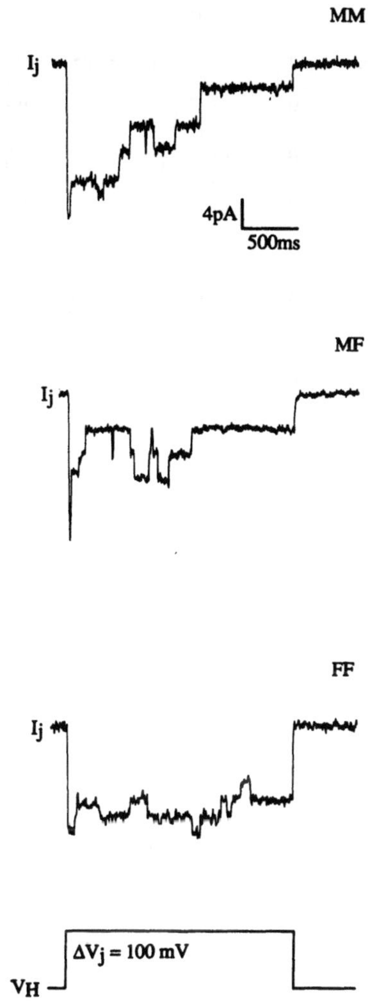

Fig.7. Junctional current tracings of a myocyte (MM) pair, of a myocyte-fibrocyte (MF) pair and of a fibrocyte (FF) pair. The holding potential (V_H) was -58 mV in the MM pair, -65 mV in the MF pair and -22 mV in the FF pair. Scalebars in the upper trace apply to the lower traces also

become possible to study the properties of these structures and especially the intercellular channels of which they are composed, at a molecular level.

In this paper we present electrophysiological data we have gathered on the conductance and gating properties of junctional channels between cells from the neonatal rat heart. We showed that these channels are not only present between myocytes (Fig. 1, 2) but also between fibrocytes (Fig. 5,7) and between fibrocytes and myocytes (Fig. 6,7). Functional interaction between myocytes mediated by a fibrocyte has been demonstrated before by Goshima (1970). This author showed synchronised beating of two myocytes intercalated by a fibrocyte and also

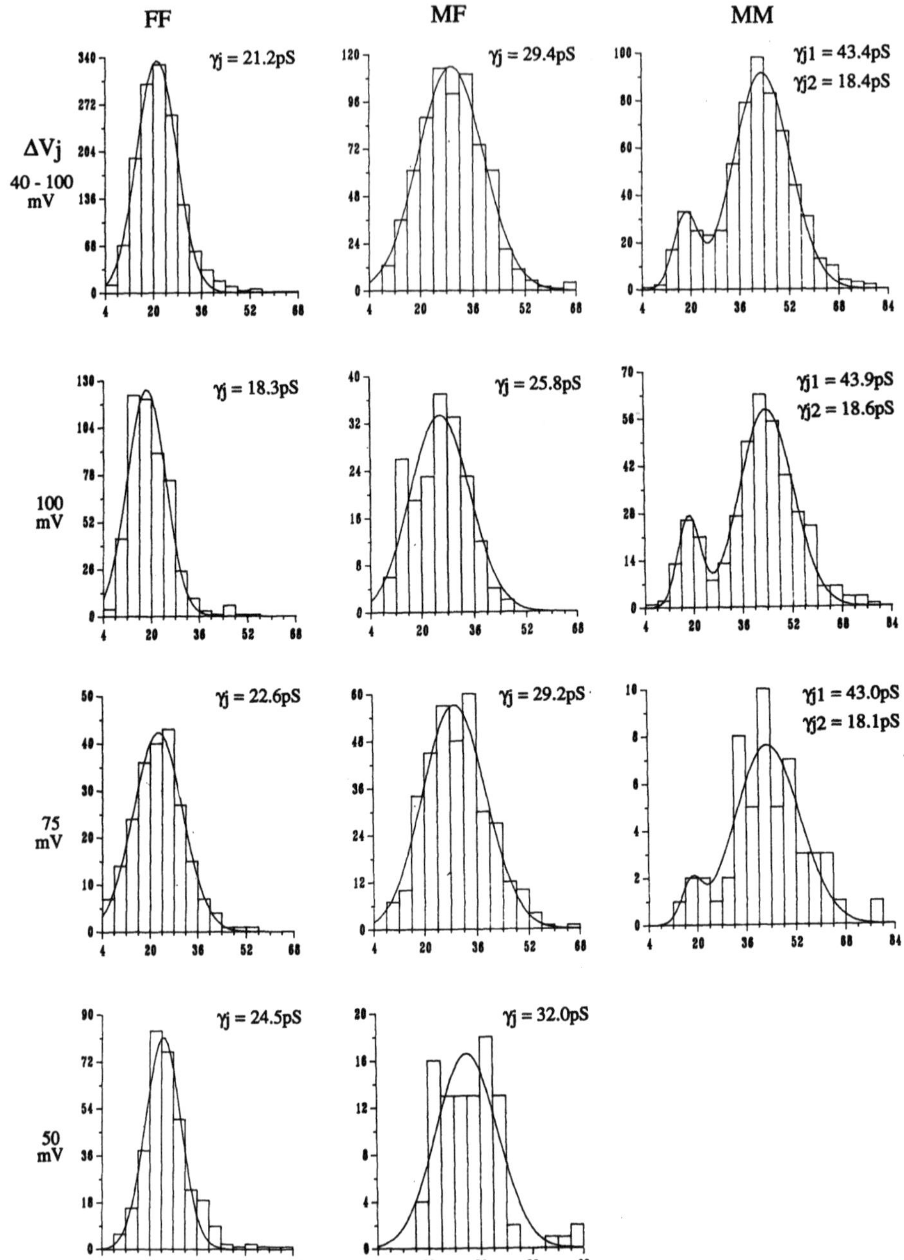

Fig. 8. Frequency histograms of steps in junctional current measured in myocyte (MM) pairs, in myocyte-fibrocyte (MF) pairs and in fibrocyte (FF) pairs. Vertically the number of events in each bin is plotted. Horizontally conductance is plotted with a binwidth of 4 pS. γ is the conductance at the peak of the best fitting gaussian curve. N_t is the total number of events in each histogram

depolarisation of the fibrocyte upon activation of a myocyte. From Figure 5 and 6 it can be seen that our results confirm and extend these observations. When a fibrocyte is coupled to another fibrocyte both show a resting potential of about -20 mV a value that we also find for isolated fibrocytes and a depolarisation induced in one cell of the pair is electrotonically conducted to the other cell. Fig. 7 (lower panel) shows that in such pairs single junctional channel events can be observed. In the case of coupling between a myocyte and a fibrocyte the resting potential of the fibrocyte is much more negative that that of an isolated fibrocyte and resembles closely that of the myocyte which in itself is an argument for electrical coupling. Stimulation of either the myocyte or the fibrocyte causes the generation of an action potential in the myocyte only and a passive depolarisation in the fibrocyte. In this case single channel events can be observed also (Fig. 7, middle panel). The single channel conductance which can be calculated from the records differs in the three cases shown. As indicated in the upper row of Figure 8, where all data for the three cases are combined, we find for MM pairs a γ_i of 43.4 pS with a minor peak of the best fitting gaussian at 18.4 pS. The major value agrees with the γ_i reported by Burt and Spray (1988) for the same preparation. These authors provided some theoretical arguments to explain the rather low junctional conductance in this case compared to values of at least 100 pS expected by others. A second lower conductance peak was not found by these authors. Veenstra and DeHaan (1986) however found two conductance peaks in gap junctions between enbryonic chick heart cells of 165 and 50-70 pS respectively. The smaller peak was not sensitive to octanol, an agent known to decrease junctional conductance (Délèze and Hervé, 1983; Niggli *et al.* 1989) possibly indicating that the conductance peaks represent different channels. Between FF pairs channels with one conductance only of 21.2 pS are found (Fig.8, left column). Although their γ_i is even lower than that between MM pairs they effectively couple such cells because the non-junctional membrane resistance of fibrocytes derived from the heart is very high (>20 GΩ at a potential ranging between -20 and -60 mV; Rook, unpublished observations).

MF pairs exhibit an intermediate junctional conductance of 29.4 pS (Fig. 8, middle column). When it is assumed that the junctional channels between myocytes and fibrocytes are composed of a connexon (hemichannel) of both the myocyte and the fibrocyte this is to be expected. A myocyte connexon would have a conductance of 86.8 pS and a fibrocyte connexon a conductance of 42.4 pS. The conductance of these two connexon in series would be (86.8 * 42.4)/ (86.8 + 42.4) = 28.5 pS which agrees very well with the value we actually measured. We conclude that the cytoplasmic oriented face of the connexons is cell specific while the outside face is not. The lower conductance found in myocytes is clearly not able to combine with fibrocyte connexons becauses a second peak expected at 19.7 pS in that case, is not observed. In our opinion this supports the notion that the smaller conductance found in MM pairs represents a separate channel.

The lower rows of Figure 8 show that the open channel conductance of myocyte and fibrocyte junctional channels is not voltage dependent. At all voltages tested the conductance peaks are at the same values for the respective channels. The voltage dependence of junctional current, which we demonstrated (fig. 3) and which has recently also been reported by Veenstra (1989) in embryonic chick heart gap junctions must therefore reside in voltage dependence of the mean channel open time. This conclusion is supported by evidence we showed elsewhere (Rook *et al.*1988). As shown in Figure 4 voltage dependence is absent in larger gap junctions. This disappearance of voltage dependence may be understood by considering that the conductance of a gap junction is proportional to its area (i.e. the number of separate channels it contains). This means that the voltage applied between the tips of the electrodes drops to a progressively larger extent over the cytosolic resistance with which the junctional resistance is in series. As can be seen from Figure 3 small gap junctions show voltage dependence when the transjunctional voltage is greater that 50 mV. It might be expected therefore that once a junction is so large that a 100 mV interelectrode voltage difference drops to a voltage difference of 50mV or less over the junction proper, voltage dependence will disappear. Theoretical considerations and experimental evidence (Rook *et al.* 1988) support this viewpoint.

Acknowledgements

This work was supported by grant no 86.030 of the Netherlands Heart Foundation (NHS).

References

Burt JM, Spray DC (1988) Single-channel events and gating behaviour of the cardiac gap junction channel. Proc Natl Acad Sci 85:3431-3434

De Bruijne J, Jongsma HJ (1980) Membrane properties of aggregates of collagenase dissociated rat heart cells. In: Tajuddin M, Das PK, Tariq M, Dhalla NS (eds) Advances in Myocardiology. University Park Press Baltimore Md, pp 231-243

Délèze J, Hervé JC (1983) Effect of several uncouplers of cell to cell communication on gap junction morphology in the mammalian heart. J. Membr. Biol. 74:203-215

Fozzard HA (1979) Conduction of the actionpotential. In: Berne RM, Sperelakis N Geiger SR (eds) Handbook of Physiology. Section 2 The cardiovascular system Vol 1 The heart. Am Physiol Soc Bethesda Md. pp 335-356

Goshima K (1970) Formation of nexuses and electrotonic transmission between myocardial and FL cells in monolayer culture. Exp. Cell Res. 63:124-130

Hamill OP, Marty A, Neher E, Sakmann B, Sigworth FJ (1981) improved patch-clamp techniques for high resolution current recording from cells and cell-free membrane patches. Pflüg. Arch. 391:85-100

Herzberg EL, Johnson RG (eds) (1988) Gap Junctions. Alan R. Liss New York.

Loewenstein WR (1981) Junctional intercellular communication: the cell-to-cell membrane channel. Physiol. Rev. 61:829-913

Neyton J, Trautmann A (1985) Single channel currents of an intercellular junction. Nature 317:331-335

Niggli E, Rüdisüli A, Maurer P, Weingart R (1989) Effects of general anesthetics on current flow across membranes in guinea pig myocytes. Am. J. Physiol. 256:C273-C281

Noma A, Tsuboi N (1987) Dependence of junctional conductance on proton, calcium and magnesium ions in cardiac paired cells of guinea-pig. J. Physiol. 382:193-211

Reverdin EC, Weingart R (1988) Electrical properties of the gap junctional membrane studied in rat liver cell pairs. Am. J. Physiol. 254:C226-C234

Rook MB, Jongsma HJ, Van Ginneken ACG (1988) Properties of single gap junctional channels between isolated neonatal rat heart cells. Am. J. Physiol. 255:H770-H782

Rook MB, Jongsma HJ, De Jonge B (1989) single channel currents of homo- and heterologous gap junctions between cardiac fibroblasts and myocytes. Pflüg. Arch 414:95-98

Simpson I, Rose B, Loewenstein WR (1977) size limit of molecules permeating the junctional membrane channels. Science 197:294-296

Sjöstrand FS, Anderson E (1954) Electron microscopy of the intercalated disks of cardiac muscle tissue. Experientia 10:369-374

Somogyi R, Kolb HA (1988) Cell-to-cell channel conductance during loss of gap junctional channel conductance in pairs of pancreatic acinar and Chinese hamster ovary cells. Pflüg.Arch. 412:54-65

Spray DC, Harris AL, Bennett MVL (1979) Voltage dependence on junctional conductance in early amphibian embryos. Science 204:432-434

Veenstra RD (1989) Voltage-dependent gating of embryonic cardiac gap junction channels. Biophys. J. 55:152a

Veenstra RD, DeHaan RL (1986) Measurements of single channel currents from cardiac gap junctions. Science 233:972-974

Weingart R (1986) Electrical properties of the nexal membrane studied in rat ventricular cell pairs. J. Physiol. 370:267-284

White RL, Spray DC, Campos de Carvalho AC, Wittenberg BA, Bennett MVL (1985) Some electrical and pharmacological properties of gap junctions between adult rat ventricular myocytes. Am. J. Physiol. 249:C447-C455

PATTERNS OF GAP JUNCTIONAL PERMEABILITY IN DEVELOPING INSECT TISSUES

S Caveney
Department of Zoology
University of Western Ontario
London Ontario
Canada N6A 5B7

INTRODUCTION

The gap junction is an essential membrane component of animal tissues. Arising early in animal evolution, gap junctions are found in the most primitive multicellular organisms and in most Metazoa (reviewed in Revel, 1987).

The cell-to-cell membrane channels comprising the gap junction permit a diffusion-driven, non-specific and direct movement of ions and small cytoplasmic metabolites among tissue cells. The primary, and presumably the ancestral, role of gap junctions is to allow developing and differentiated tissue cells to coordinate their ionic and metabolic activity and to expel metabolic byproducts; this function in developing cells is apparently very important, as shown by the extensive coupling between a developing oocyte and the follicular epithelium that surrounds it. The non-selective permeability of gap junctional channels presumably preadapted them for serving a secondary function, namely as a communication route for the cell-to-cell spread of regulatory molecules such as second messengers; and possibly even a third and higher order function, namely for the signalling of as yet unidentified cytoplasmic morphogens thought to be involved in spatial patterning in embryonic and postembryonic tissues (Caveney, 1985; Green, 1988). This third proposed function is much debated, since the evidence available to support it is limited (Fraser *et al.* 1987).

Are Gap Junctions Needed To Transmit Morphogens During Early Embryonic Development?

Gap junctions appear early in embryonic development and junctional permeability frequently changes at critical moments in tissue differentiation. Gap junctions are detectable in the insect embryo the moment the cellular blastoderm forms (Eichenberger-Glinz, 1979) and persist in most tissues until the end of adult life. Spatial restrictions in junctional communication in the embryo and in postembryonic tissues delineate communication compartments with boundaries that

NATO ASI Series, Vol. H 46
Parallels in Cell to Cell Junctions in Plants and Animals
Edited by A. W. Robards et al.

frequently coincide in space with those of embryonic germ layers or developmental compartments (Caveney, 1985; Serras and van den Biggelaar, 1987; Kalimi and Lo, 1988). Nevertheless, the role of gap junctions in developing cells remains unclear. In mature tissues it is generally accepted that gap junctions assist in keeping the metabolic activity of the interconnected cells the same, i.e., stabilizing their differentiated state. In the embryo, on the other hand, it is proposed that gap junctions allow the movement of morphogenetic signals that induce cells to become different. Yet metabolic cooperation is no doubt important to rapidly developing cells, and these conflicting roles make it difficult to determine precisely why gap junctional permeability changes in time and space in developing cells. Recent successful experiments that disrupted development in vertebrate embryos after blocking gap junctional permeability with the injection of anti-connexin antibodies into selected blastomeres (Warner *et al.* , 1984; Lee *et al.*, 1987), may have done so by stopping the passage of developmental signals. They would also have disrupted normal metabolic interactions between embryonic cells (such as setting up ionic gradients and trophic pathways) required for further development. Developmental signals are presumably minor components of the molecular traffic diffusing through gap junctions and would be swamped by the normal flux of cytoplasmic ions and metabolites. In other words, only a few gap junctional channels would be sufficient for developmental signalling were it the only function for gap junctions in the embryo.

The developing insect epidermis illustrates this point. The level of junctional coupling in this secretory tissue fluctuates in a cyclical fashion as the growth of the larva progresses through successive moult cycles. Peak levels of junctional coupling coincide with times of maximal secretory activity in the tissue (at the moult), and not with times when patterning occurs in the tissue (in the early premoult). Furthermore, it is ionic coupling between the cells that rises during secretion; tracer coupling remains relatively constant. If cytoplasmic morphogens exist in this tissue to set up morphogenetic gradients, they simply 'go with the flow' rather than being the basis for changes in junctional communication.

Several reports have demonstrated that gap junctions may not even be required for some inductive processes that initiate cell differentiation. It is claimed that cell polarization in the mouse blastula occurs before gap junctions form (Goodall and Johnson, 1982), and functional gap junctions are not needed to induce muscle gene activation in frog mesoderm cells (Warner and Gurdon, 1987). The progress of myotube differentiation is indifferent to the presence of gap junctions, since the time at which they are lost varies considerably in the developing muscles of different vertebrates (reviewed in Caveney, 1985). Here cell surface molecules involved in cell adhesion and recognition must operate in the absence of membrane junctions.

The non-selective permeability of the gap junction is a severe challenge to the investigator wishing to demonstrate that the channels are a transmission route for developmental signals, particularly since the signals remain to be discovered.

A Reduction in Gap Junctional Communication may be a Prerequisite for Differentiation

The global pattern of junctional coupling present in the early embryo soon transforms into smaller domains of coupled cells. After gastrulation, the cells of embryonic germ layers display a progressive loss of coupling both within and between themselves (Serras and van den Biggelaar, 1987; Kalimi and Lo, 1988). In the gastrulating mouse embryo, dye coupling, but not ionic coupling, is apparently lost between the germ layers (Kalimi and Lo, 1988). This may be due to a selective reduction in the number of gap junctions connecting cells of different germ layers, or a change in their permeability properties. Either way, the germ layers become partially restricted communication compartments, and this may be essential for subsequent diversification. The cells of the dermis of the mouse skin, for example, although strongly coupled with each other, are not coupled to those of the overlying epidermis (Kam *et al.*, 1986; Pitts *et al.*, 1986). The neurectoderm uncouples from the surrounding ectodermal layer early in amphibian development (Warner, 1973). In this case, the excitable nature of the plasma membrane of mature neurons (as well as many receptor cells and skeletal muscle) precludes the presence of gap junctions, since these cell types perform normally in electrical isolation. A limitation in junctional coupling appears to be important for further differentiation in these cells.

The segmented insect epidermis is a good model to study the relationship between developmental compartments (determined by genetic means) and communication compartments (detected by tracer micro-injection). The formation of a segmented body plan in the embryo of *Drosophila* involves the sequential expression of different sets of patterning genes (reviewed in Ingham, 1988). The embryonic ectoderm (which gives rise to the epidermis, among other tissues) is first divided into reiterated units called parasegments through the overlapping activity of two sets of pair-rule genes (Martinez-Arias and Lawrence, 1985). Anterior and posterior compartments in each parasegment are in turn determined by segment polarity genes, which are regionally active in each parasegment. Each segment border occurs at an interface between a zone expressing the engrailed (en) gene and an area posterior to it that does not. Recently we have mapped the posterior compartment of the abdominal segment in the larva of the milkweed bug *Oncopeltus fasciatus* through the compartment-specific expression of en. Using an antibody that recognizes the nuclear-located product of the en gene, the posterior compartment can be shown to consist of a strip of cells about 10-20% the length of the segment (Campbell and Caveney, 1989). Whereas the margin between the posterior compartment of one segment and the anterior compartment of the segment behind is sharply defined by a linear and abrupt transition between cells positive or negative for the en gene product, the anterior margin of the posterior compartment is more irregular in shape where it blends with cells of the anterior compartment of its own segment (Fig.1). At the former site a selective reduction in the cell-to-cell spread of organic tracers (Warner and Lawrence, 1982) has been shown to be due to a special population of border cells with low junctional permeability to organic tracers (Blennerhassett and Caveney, 1984). At the latter site we have been unable to demonstrate a discrete diffusion barrier to tracer movement, suggesting that while compartment borders may coincide in space with communication barriers, this is not essential for normal spatial patterning to occur. Ionic coupling is not attenuated across either the anterior or the posterior limit of the posterior compartment in each segment.

The interface between the most posterior cells of one anterior segment and the most anterior cells of the next appears to be rather special in that it is only at this site that border cells are

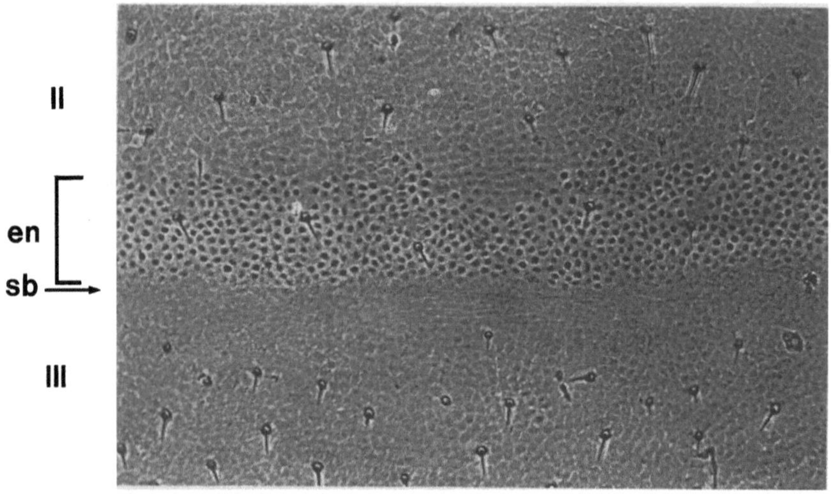

Fig. 1. Expression of engrailed is limited to a narrow band of cells just anterior to the segment border. Shown here is a whole mount of the integument of *Oncopeltus fasciatus*, centred on the border between segments II and III, labelled with antibody to the en gene product and stained with peroxidase-conjugated secondary antibody. The interface between en-expressing and non-expressing cells at the segment border (sb) forms a reasonably straight line, whereas the interface with cells in the same segment is less regular. Only a the segment border is a barrier limiting cell-to-cell tracer diffusion detectable at certain times in development (from Campbell and Caveney, 1989)

generated.Even here, however, the barrier to the movement or organic tracers is neither absolute nor permanent. This row of cells probably only restricts the diffusion of cytoplasmic signals to within their own segment at specific times during epidermal development (Blennerhassett and Caveney, 1984).

The diffusible morphogens controlling aspects of spatial development in the epidermis, when identified, will probably turn out to be small organic metabolites (rather than inorganic ions which pass across the segment border with ease at all times). Consequently, the challenge now is to predict when such pattern mechanisms are active and then to search at these times for the likely signals involved. Finding out when during a moult cycle the segment border acts most strongly as a permeability barrier is a logical step in the right direction. It is unlikely that the primary morphogenetic processes are active during periods of intense synthetic activity, such as during cuticle deposition, because at these times functional integration in the epidermis must extend beyond the confines of individual segmental compartments. That is, pattern specification and intense metabolic activity are most likely temporally in anti-phase. Preliminary studies (Campbell and Caveney, unpublished) suggest that the border cells form the most effective barrier at the start of the moult cycle, as the epidermis begins to respond to rising blood titres of 20-hydroxyecdysone. It is at this time that cell patterning mechanisms can be shown to operate in the insect epidermis (Locke, 1985).

Many Cells become Different while Retaining Junctional Coupling with Other Cells

Many tissues contain cells of more than one functional type. Little or no reduction in junctional coupling is seen during the differentiation of many of these complex tissues. The level to which coupling is lost between different cell types as a tissue matures appears to be related to the extent to which the regulation and nature of their specialized activities differ. Clearly the loss of gap junctions is not needed for these tissues to differentiate; on the contrary, coupling is very likely essential for them to function normally.

This principle can be demonstrated in the epidermis of the developing insect. Most epidermal cells are strongly coupled and function to secrete an overlying cuticle. These cells respond to developmental hormones in a tightly coordinated manner. Specialized epidermal cells have a more limited synthetic and secretory activity, often not in synchrony with that of the general epidermis. These cells have reduced coupling with one another and the general epidermis in which they are embedded (Caveney and Berdan, 1982).

On the other hand, coupling is frequently present between very different cells. The coupling between the follicular epithelium and developing oocyte in both the vertebrate and invertebrate ovary (Caveney, 1985) and in the epithelium and fibre cells of the vertebrate crystalline lens during differentiation (Miller and Goodenough, 1986) may also serve as a pathway for cytoplasmic regulators (Caveney, 1985). Clearly a causal relationship between loss of coupling *per se* and differentiation is difficult to discern.

Does Gap Junctional Coupling Allow the Embryo to Survive Ionic and Metabolic Stress?

Extensive ionic coupling between developing cells may allow them to overcome the metabolic stresses associated with rapid growth and development. This may be the reason why mouse blastomeres become strongly coupled in the 8-cell embryo (McLachlin *et al.* , 1983). Tight ionic coupling appears to be needed to maintain the compacted stage and to form the embryonic blastocoel (Buehr *et al.*, 1987). It fails to develop in poorly coupled mouse embryos. During blastocoel expansion, extracellular Na^+ is transported across the trophectoderm by a Na^+ pump located on the basal surface of the trophoectodermal cells (Watson and Kidder, 1988). Since the trophectoderm is a conventional fluid transporting epithelium (Manejwala *et al.*, 1989), it is likely that the strong cell-to-cell coupling present in the mouse trophectoderm contributes towards setting up the strong electro-chemical gradients within and across the mouse blastocyst needed for blastocoel expansion, or 'cavitation'. Na^+-dependent amino acid uptake also starts at the blastocyst stage (van Winkle *et al.*, 1988), and ionic coupling is likely important in holding the membrane potential stable during the Na^+ dependent uptake of cell metabolites (Petersen, 1985). Junctional coupling may have little direct role in morphogenesis at this time. Embryos that do not undergo active cavitation become extensively coupled much later in their development - at the 24-cell stage in *Limnaea*, (Serras and van den Biggelaar, 1987) and 32-cell stage in *Ciona* (Serras *et al.*, 1988).

Ionic coupling in the epidermis rises under a variety of stresses *in vitro*, including an "equilibration" response immediately after the cells are excised into culture medium, on exposure to the developmental hormone, 20-hydroxyecdysone (Caveney and Blennerhassett, 1980) or inhibitors of protein synthesis (Caveney *et al.*, 1980). Coupled cells in general have a better chance of surviving in tissue culture than uncoupled or isolated cells, but why this is so is not known. For example, cells on excised squares of metamorphosing beetle integument survive for many days in culture, yet oenocytes (lipid synthesizing cells) loosely scattered in the epidermis do not (Caveney, unpublished). The oenocytes are not junctionally coupled to each other or the general epidermis.

Recently we have found that the amino acid L-glutamate, the main excitatory neuromuscular transmitter in insects, raises junctional coupling in the epidermis. It appears to do so by placing the cells under ionic stress. Epidermal cells have a powerful low-affinity Na^+-dependent uptake system for L-glutamate (McLean and Caveney, in preparation) that holds blood glutamate levels in the micromolar range. Na^+-dependent uptake saturates at about 100 uM glutamate. On exposure to higher concentrations of glutamate, the epidermis swells and ionic coupling, but not tracer coupling, doubles (Caveney, 1988). Exposure to 10mM glutamate in the absence of external Na^+ has no effect on the epidermis, although the amount of glutamate taken up is similar to that seen at 100uM glutamate in the presence of Na^+. Our interpretation is that, at high glutamate levels in the presence of Na^+, influx overload the efflux mechanisms, causing the tissue to swell as water follows Na^+ into the cells. The increased junctional conductance evoked by Na^+-dependent glutamate uptake in the epidermis may reflect a general cellular response to a prolonged rise in cytoplasmic Na^+. In addition to being expelled from a cell by membrane Na^+/K^+ pumps, excess Na^+ in a cell may be lost through diffusion-driven ionic coupling with its neighbours. Consequently, junctional conductance would be predicted to be high at times of greatest Na^+ flux across a tissue.

There can be little doubt that most changes in junctional coupling seen in developing cells are in response to metabolic demands.

Are Gap Junctions Needed for Normal Tissue Growth?

The regulation of tissue size and shape during growth requires mechanisms that measure the number and/or density of the constituent cells. Cells at the free edges of tissue explant proliferate more rapidly than those at the centre, suggesting that contact- and density-dependent mechanisms of growth control (employing cell surface molecules and concentration gradients in external growth factors) operate in vitro. Akin to tissue wound repair mechanisms in vivo, these mechanisms probably operate across the extracellular space between the cells.

Growth of a compact tissue in vivo may be more complex, since there are no free edges, the tissue is bounded by a basal lamina, and growth frequently occurs uniformly throughout the tissues. Since functional gap junctions connect both dividing and interphase cells in growing tissue, it is not unreasonable to assume that growth control involves diffusion of cytoplasmic regulatory signals (either growth inhibitors or activators) that use gap junctional channels as the signal transit route. A tissue connected by gap junctions has a "finite volume with respect to diffusible intracellular molecules ... and this volume increases as the number of connected cells increases during growth" (Loewenstein, 1979). The control of cell growth (or, in morphogenesis, the relative spatial position of cells with different developmental fates) may be expressed in terms of the concentration gradient of a growth regulator (or in morphogenesis, a diffusible morphogen) produced at specific sites in the tissue. Simple dilution models, however, do not explain adequately size-invariance in body patterns (Cooke, 1981), since it requires that the cells communicate among themselves over large distances, and, importantly, in a tissue size-dependent manner (Othmer and Pate, 1980). Several reaction-diffusion models for size-invariance have a requirement that the rate of diffusion of regulatory molecules increases in proportion to the square of the tissue area (in two-dimensional tissues). An obvious way this might be achieved is for more channels to be added to the junctions connecting the individual cells as the tissue grows. This actually occurs during postembryonic growth in the larval insect.

Segment growth during larval development in the insect is near-isometric and pattern 'scale-invariant'. Segment proportions change only slightly during larval development because epidermal growth occurs uniformly across the segment and is not concentrated at its margins (segment borders and pleural regions). Dividing epidermal cells do not uncouple from surrounding cells. Epidermal core (tissue) resistance drops 4-5 times during segment growth, whether measured immediately at ecdysis, or at the midinstar phases of successive moult cycles (Fig.2) (Caveney and Safranyos, 1989). Although the 5-fold increase in ionic coupling seen is relatively small compared to the 64-fold increase in segment areas, the time available to set up the diffusion-dependent growth sensor may partially compensate for this. Successive stadia of larval development take longer to complete.

The rate at which larger organic molecules move in the epidermis also rises during epidermal growth. This is determined by injecting fluorescent tracers into individual cells. In well-coupled epidermal preparations, carboxyfluorescein is detectable within seconds in cells adjacent to that injected with tracer, and up to eight cells away from the source in 80 seconds (Caveney and Safranyos, 1989). An effective diffusion coefficient may be calculated from videotapes of the spreading fluorescence caused by tracer diffusion (Safranyos and Caveney, 1985; Caveney *et al.*, 1986). Organic tracers diffuse more slowly in the epidermis of young larvae than in that of mature larvae. Presumably natural growth regulators and morphogens would behave the same way, and this may be important to normal tissue growth.

In addition, it appears that the properties of epidermal junctions, as determined from these two independent measures of junctional permeability, change in a complex way as the epidermis grows through successive moult cycles. By measuring the core ionic resistance and diffusion coefficient for organic tracers in the same area of epidermis in many preparations, junctional resistance and permeability may be estimated for different stages of epidermal growth (after accounting for the cytoplasmic contribution to total pathway resistance to ion and organic tracer movement (Caveney and Blennerhassett, 1980; Safranyos *et al.*, 1987). Whereas the rate at which organic tracers pass among the cells increases smoothly during epidermal growth, the rate at which inorganic ions move peaks at the moult, and drops during the intermoult phase of successive larval stadia, as mentioned earlier.

Junctional Permeability may be Regulated Selectively During Development

The diffusion of microinjected cell-to-cell tracers among the blastomeres of the amphibian embryo depends on the position of the injected cell and the stage of embryonic development (Guthrie *et al.*, 1988). Gap junctions in cells destined to form ventral structures differ from future dorsal cells in their relative permeability to a series of tracers. This suggests that gap junctional permselectivity in the frog embryo is under cellular control. Whether this is due to structural (i.e., different connexins) or physiological differences in the cells is unclear.

A temporal pattern of junctional permselectivity also occurs during the development of the insect epidermis. Not only does junctional permeability rise as the segment grows, as mentioned

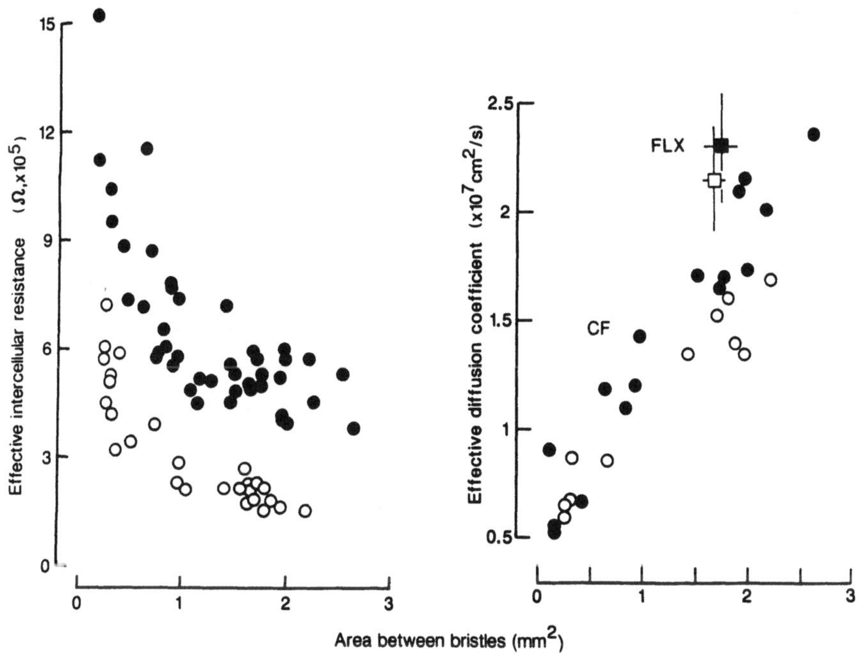

Fig.2. Developmental changes in ionic coupling and tracer coupling in the epidermis during a greater than 50-fold increase in larval segment area in *Tenebrio molitor*. Values for effective core resistance to inorganic ions and effective diffusion coefficients for carboxyfluorescein (CF) and fluorexon (FLX) for both newly moulted (o) and midinstar (o) tissue are shown. Core resistance drops during the first half of the larval stadium (shown by the vertical displacement of the two sets of data for segments of comparable size) whereas diffusion of organic tracers remains constant (the non-overlap in the two sets of values for CF at large segment sizes is due to CF binding non-specifically to cytoplasmic glycogen granules, a property not shared by FLX, which has a similar diffusion coefficient in both newly-moulted and midinstar epidermis from segments of similar size). (From Caveney and Safranyos, 1989)

earlier, but ionic conductance fluctuates in each larval instar so as to change the relationship between junctional conductance and tracer permeability. The mechanism behind this is not known. If only one class of junctional channel exists in the epidermis, permeable to selected ions and organic tracers, then one has to propose developmentally-regulated mechanisms that (i) selectively affect the access to, or passage through, the channels of certain classes of charged permeant molecules, or (ii) affect the dynamic nature of the channels (time spent open, frequency of opening, time spent opening and closing), or (iii) affect the stability of different substates of channel aperture. Alternatively, separate classes of channels could exist with different permeability characteristics and different mechanisms of regulation. Changes in the fraction of each of these channels open at any one moment would alter the permselectivity of the junctional membrane.

The changes in junctional permselectivity detected during normal development can also be stimulated in the epidermis *in vitro* . Firstly, the ionic conductance of epidermal gap junctions approximately doubles in the tissue exposure to 20-hydroxyecdysone *in vitro* (Caveney and Blenerhassett, 1980). This response does not appear to involve the synthesis of new junctional channels, since it is not blocked by inhibitors of protein synthesis (Caveney *et al.*, 1981). No equivalent increase in junctional permeability was seen. Secondly, epidermal exposure to the excitatory neurotransmitter L-glutamate results in instantaneous membrane depolarization followed within two hours by a doubling in junctional conductance without any increase in junctional permeability to organic tracer (Caveney, 1988).

These findings in the frog embryo and insect epidermis clearly demonstrate an unsuspected complexity in the regulation of junctional coupling in developing tissues. Do all gap junctional channels have low selectivity to inorganic ions and cytoplasmic metabolites, or are there ion-specific channels too? And to which ions? To study this problem, the dynamics of single gap junctional channels need to be examined. This has been accomplished in a few vertebrate cells using double whole-cell voltage clamp methods (Neyton and Trautmann, 1985; Veenstra and DeHaan, 1986) and similar methods (Chow and Young, 1987). Analysis of single channels in these cells has revealed a single conductance state (of about 100 pS) or multiple conductance states (substates with conductance less than 100 pS) as well as considerable variation in the transition time from the open to closed state (less than 1ms to greater than 10ms), depending on the tissue of origin of the cells. Only one class of junctional channel was detected in each of the cell types. The situation in the insect epidermis is probably not as straightforward. We plan to examine changes in the kinetic behaviour of single gap junctional channels, looking in particular for shifts in the temporal stability of different conductance states in one class or more of junctional channel in the epidermal cells, both before and after exposure to 20-hydroxyecdysone, L-glutamate and second messengers. It may now be possible to obtain unitary values - at least for channel conductance - for the one or more classes of junctional channel in the epidermis.

Gap junctional permeability is extensively modulated during growth and differentiation in postembryonic insect tissues. However, no general statement can be made as to the relationship between these changes and normal development. Even if there were ways of specifically disrupting junctional communication without affecting other activities in insect cells, the metabolic consequences may mask any direct effects of inhibiting developmental signalling between the cells.

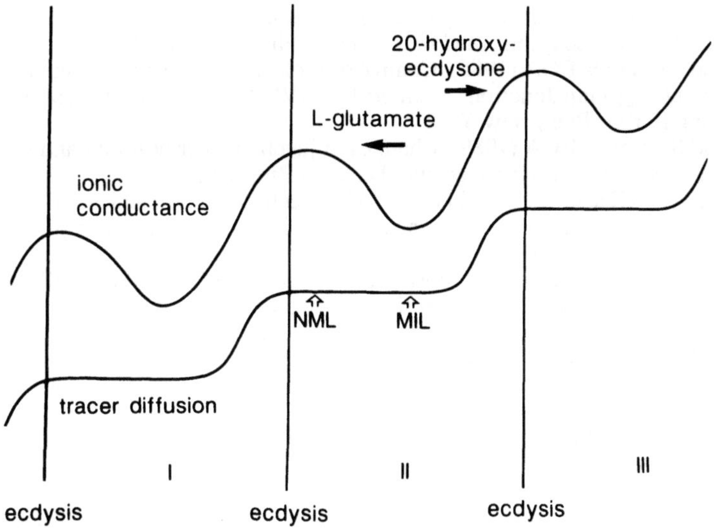

Fig.3. Scheme outlining the possible control of ionic conductance and tracer permeabililty of epidermal gap junctions in *Tenebrio* through successive larval stages of development. The rate of tracer diffusion in the epidermis rises in a steplike fashion at each moult. Ionic conductance rises in an oscillatory fashion with peak conductance in phase with moult-related activities. 20-Hydroxyecdysone *in vivo* is thought to raise both junctional conductance and permeability at the moult by stimulating the formation of new gap junctional channels. L-glutamate, on the other hand, is thought to raise junctional conductance selectively *in vivo* by a mechanism related to that controlling junctional conductance during the transition from the newly-moulted (NML) to mid-instar (MIL) phase of each larval stage. This change appears to be independent of junction growth, and therefore presumably depends on changes in the kinetic properties of a pre-existing population of junctional channels

References

Blennerhassett MG and Caveney S (1984) Separation of developmental compartments by a cell type with reduced junctional permeability. Nature 308: 361-364

Buehr M, Lee S. McLaren A and Warner A (1987) Reduced gap junctional communication is associated with the lethal condition characteristic of DDK mouse eggs fertilized by foreign sperm. Development 101: 449-459

Campbell GL and Caveney S (1989) Engrailed gene expression in the abdominal segment of *Oncopeltus* : gradients and cell states in the insect segment. Development, in press

Caveney S (1985a) The role of gap junctions in development. Ann Rev Physiol 47: 319-335

Caveney S (1988) Developmental physiology of insect gap junctions. In Gap Junctions. Edited by EL Hertzberg and RG Johnson. pp 495-504. Alan Liss New York

Caveney S and Berdan RC (1982) Selectivity in junctional coupling between cells of insect tissues. In Insect Ultrastructure Vol 1. Edited by RC King and H Akai. pp 434-465

Caveney S and Blennerhassett MG (1980) Elevation of ionic conductance between insect epidermal cells by -ecdysone *in vitro* . J. Insect Physiol 26: 13-25

Caveney S and Safranyos RGA (1985) Control of molecular movement within a developmental compartment. In Gap Junctions. Edited by MVL Bennett and DC Spray. pp 265-273. Cold Spring Harbor Press New York

Caveney S and Safranyos RGA (1989) Changes in junctional permeability and selectivity in the epidermal cells of the growing segment. Develop Biol, submitted

Caveney S, Berdan RC and McLean S (1980) Cell-to-cell ionic communication stimulated by 20-hydroxyecdysone occurs in the absence of protein synthesis and gap junction growth. J Insect Physiol 26: 557-567

Caveney S, Berdan RC, Blennerhassett MG and Safranyos RGA (1986) Cell-to-cell coupling via membrane junctions: methods that show its regulation by a developmental hormone in an insect epidermis. In Techniques in the Life Science, C2: In Vitro Invertebrate Hormones and Genes. Edited by E Kurstak. pp 1-23. Elsevier Scientific Publishers Ireland

Chow I and Young SH (1987) Opening of single gap junction channels during formation of electrical coupling between embryonic muscle cells. Develop Biol 122: 332-337

Cooke J (1981) Scale of body pattern adjusts to available cell number in amphibian embryos. Nature 290: 775-778

Eichenberger-Glintz S (1979) Intercellular junctions in development and tissue cultures of *Drosophila melanogaster* : an electron-microscope study. Wilhelm Roux Arch Entwicklungsmech Org 186: 333-349

Fraser SE, Green CR, Bode HR and Gilula NB (1987) Selective disruption of gap junctional communication interferes with a patterning process in *Hydra* . Science 237: 49-55

Goodall H and Johnson MH (1984) The nature of intercellular coupling within the preimplantation mouse embryo. J. Embryo Exp Morph 79: 53-76

Green CR (1988) Evidence mounts for the role of gap junctions during development. Bioessays 8: 7-10

Guthrie S, Turin L and Warner A (1988) Patterns of junctional communication during development of the early amphibian embryo. Development 103: 769-783

Ingham PW (1988) The molecular genetics of embryonic pattern formation in *Drosophila* . Nature 335: 25-34

Kalimi GH and Lo CW (1988) Communication compartments in the gastrulating mouse embryo. J Cell Biol 107: 241-255

Kam E, Melville L and Pitts JD (1986) Patterns of junctional communication in the skin. J Invest Dermatol 87: 748-753

Lee S, Gilula NB and Warner AE (1987) Gap junctional communication and compaction during preimplantation stages of mouse development. Cell 51: 851-860

Locke M (1985) The structure of epidermal feet during their development. Tissue and Cell 17: 901-921

Loewenstein WE (1979) Junctional intercellular communication and the control of growth. Biochem Biophys Acta 560: 1-65

Manejwala FM, Cragoe EJ and Schultze RM (1989) Blastocoel expansion in the mouse embryo: Role of extracellular sodium and chloride and possible apical routes of their entry. Develop Biol 133: 210-220

Martinez-Arias A and Lawrence PA (1985) Parasegments and compartments in the *Drosophila* embryo. Nature 313: 639-642

McLachlin JR, Caveney S and Kidder GM (1983) Control of gap junction formation in early mouse embryos. Develop Biol 98: 155-164

Miller TM and Goodenough DA (1986) Evidence for two physiologically distinct gap junctions expressed by the chick lens epithelial cell. J Cell Biol 102: 194-199

Neyton J and Trautmann A (1985) Single channel currents of an intercellular junction. Nature 274: 133-136

Othmer HG and Pate E (1980) Scale-invariance in reaction-diffusion models of spatial pattern formation. Proc Natl Acad Sci USA 77: 4180-4184

Petersen OH (1985) Importance of electrical cell-cell communication in secretory epithelia. In Gap Junctions (MVL Bennett and DC Spray eds). pp 315-324. Cold Spring Harbor New York

Pitts JD, Kam E, Melville E and Watt FM (1986) Patterns in junctional communication in animal tissues. In Junctional Complexes of Epithelial Cells. Ciba Foundation Symposium 125. pp 140-153. Wiley Chichester

Revel J-P (1987) The oldest multicellular animal and its junctions. In Gap Junctions (EL Hertzberg and R Johnson eds) Modern Cell Biology 7. pp 135-149. Alan Liss New York

Safranyos RGA and Caveney S (1985) Rates of diffusion of fluorescent molecules via cell-to-cell membrane channels in a developing tissue. J Cell Biol 100: 736-747

Safranyos RGA, Caveney S, Miller JG, and Petersen NO (1987) Relative roles of gap junction channels and cytoplasm in cell-to-cell diffusion of fluorescent tracers. Proc Natl Acad Sci USA 84: 2272-2276

Serras F and van den Biggelaar JAM (1987) Is a mosaic embryo also a mosaic of communication compartments? Developmental Biology 120: 132-138

Serras F, Baud C, Moreau M, Cuerrier P and van den Biggelaar JAM (1988) Intercellular communication in the early embryo of the ascidian *Ciona intestinalis*. Development 102: 55-63

Van Winkle LJ, Haghighat N, Campione AL and Gorman JM (1988) Glycine transport in mouse eggs and preimplantation conceptuses. Bioch et Biophys Acta 941: 241-256

Veenstra RD and De Haan RL (1986) Measurement of single channel currents from cardiac gap junctions. Science 233: 972-974

Warner, AE (1973) The electrical properties of the ectoderm during induction and early development of the nervous system. J. Physiol 235: 267-286

Warner A and Gurdon JB (1987) Functional gap junctions are not required for muscle gene activation by induction in *Xenopus* embryos. J Cell Biol 104: 557-564

Warner AE and Lawrence PA (1982) Permeability of gap junctions at the segmental border in insect epidermis. Cell 28: 243-252

Warner AE, Guthrie SE and Gilula NB (1984) Antibodies to gap junctional protein selectively disrupt junctional communication in the early amphibian embryo. Nature 311: 127-131

Watson AJ and Kidder GM (1988) Immunofluorescence assessment of the timing and appearance and cellular distribution of Na/K-ATPase during mouse embryogenesis. Developmental Biology 126: 80-90

GAP JUNCTIONAL INTERCELLULAR COMMUNICATION AND CARCINOGENESIS

Hiroshi Yamasaki
Unit of Mechanisms of Carcinogenesis
International Agency for Research on Cancer
150, cours Albert Thomas
69372 Lyon cedex 08
France

INTRODUCTION

It is widely accepted that the process by which a normal cell becomes a malignant tumour is multistage (Doll, 1978; Yamasaki and Weinstein, 1985). It is possible that cell growth is such a tightly controlled process that multiple events are needed in order for a normal cell to deviate so markedly as to be potentially cancerous. Since one of the most important control mechanisms of cell growth and differentiation is intercellular communication, it is reasonable to suppose that aberrant intercellular communication is involved at certain stages of carcinogenesis. Gap junction-mediated intercellular communication is considered to be the sole means by which low molecular weight factors inside a cell can pass directly into the interior of neighbouring cells (Pitts and Finbow, 1986; Loewenstein, 1979). By equalizing the intracellular levels of growth-related factors, gap junctions are considered to play an essential role in the maintenance of homeostasis (Pitts and Finbow, 1986; Loewenstein, 1979).

By contrast to normal cells, tumour cells do not maintain homeostasis with neighbouring cells in a given tissue, reinforcing the idea that gap junctional intercellular communication (GJIC) is disrupted during carcinogenesis. Evidence is indeed accumulating that reduced GJIC is involved in the process of carcinogenesis and also in the maintenance of transformed phenotypes (Trosko <u>et al.</u>, 1983; Yamasaki, 1988). Since disrupted GJIC is considered to be involved in carcinogenesis, it is also

NATO ASI Series, Vol. H 46
Parallels in Cell to Cell Junctions in Plants and Animals
Edited by A. W. Robards et al.

possible that intact GJIC plays an important role in tumour suppression; several lines of evidence also exist to support this view(Yamasaki,1989b).

ROLE OF REDUCED GAP JUNCTIONAL INTERCELLULAR COMMUNICATION IN THE PROCESS OF CARCINOGENESIS

The first evidence suggesting a possible role of reduced GJIC during the process of carcinogenesis came from the discovery that the skin tumour promoting agents, phorbol esters, inhibited GJIC in cultured cells (Yotti et al., 1979; Murray and Fitzgerald, 1979). This finding has since been confirmed and extended, and, as summarized in Table 1, there are indeed many tumour promoting stimuli that inhibit GJIC. Most of these studies have been performed in cultured cells but we have recently shown that the liver-specific tumour promoter phenobarbital, upon administration to rats, can drastically decrease the mRNA level of connexin 32, a major gap junction protein; this did not occur in kidney or stomach (Mesnil et al., 1988). Altered connexin gene expression presumably reflects disturbed GJIC, as indicated by recent in vitro experiments of effects of phenobarbital exposure on rat hepatocytes (unpublished observations). Therefore, as has been suggested, tumour promoting agents may release "initiated" cells from the growth restraint exerted from surrounding normal cells by reducing intercellular communication and allowing initiated cells to clonally expand to form a tumour.

Further evidence has come from the use of agents which inhibit tumour promotion. Glucocorticoids, retinoic acid and cyclic AMP have been shown to inhibit mouse skin promotion by 12-O-tetradecanoylphorbol 13-acetate (TPA) (Slaga, 1984). If the inhibition of GJIC is involved in tumour promotion, it is reasonable to suggest that these tumour promoter antagonists abrogate the effect of promoters on GJIC. When BALB/c 3T3 cells were treated with these anti-promoters and TPA, there were indeed antagonistic effects on the TPA inhibition of GJIC (Yamasaki and Enomoto, 1985). Furthermore, when we studied the effect of these compounds on DMBA/TPA two-stage transformation of BALB/c 3T3 cells, we found that they inhibited transformation of these cells (Yamasaki and Katoh, 1988). These results therefore further support the hypothesis that gap junction inhibition is involved in the promotion or enhancement of cell transformation.

Table 1. Inhibition of gap-junctional intercellular communication by tumour promoting stimuli (modified from Yamasaki, 1989a)

Method of measuring communication	Promoting stimulus
Metabolic cooperation	
3H-uridine metabolite transfer	Phorbol esters, chlordane
$HGPRT^+/HGPRT^-$(a)	Phorbol esters, and many other tumour promoting agents
ASS^-/ASL^-(b)	Phorbol esters, DDT
AK^+/AK^- (c)	Phorbol esters
Electrical coupling	
	Phorbol esters
	Skin wounding
Dye transfer	
Microinjection	Phorbol esters, cigarette smoke condensate, PCB, diacylglycerol, and certain other tumour promoting agents
	Partial hepatectomy
Photobleaching	TPA, dieldrin, PBB
Scrape loading	TPA, dieldrin and other tumour promoting agents
Gap junction structure analysis	
Electron microscope	Phorbol esters, mezerein
	Phenobarbital, DDT
Gel electrophoresis analysis	Phorbol esters
Analysis with gap junction antibody	Partial hepatectomy
Gap junction gene expression	
Connexin 32mRNA level	Phenobarbital
	Partial hepatectomy

(a) HGPRT, hypoxanthine guanine phosphoribosyltransferase
(b) ASS^-, argininosuccinate synthetase-deficient;
ASL^-, argininosuccinate lyase-deficient
(c) AK, adenosine kinase

Kakunaga has isolated various variants of BALB/c 3T3 cells which have different susceptibilities to induction of cell transformation (Kakunaga and Crow, 1980). A31-1-13 is a variant which is highly transformable in the presence of polycyclic hydrocarbons or after irradiation with UV light. By contrast, the variant A31-1-8 was very resistant to transformation induction by these agents. These two cell variants have shown similar properties in terms of their initiating events, i.e., metabolism of carcinogens, DNA adduct formation, rate of DNA repair, and induction of mutation (Lo and Kakunaga, 1982). Therefore, we have postulated that their transformation inducibility difference may arise from a later stage, namely tumour promotion. When we compared the capacity of GJIC of these two cell lines, the following was observed: cells in growing phase did not show any difference in GJIC; however, at confluence, the transformation-sensitive cell line (A31-1-13) displayed a drastic decrease in GJIC capacity, whereas there was no such change in the transformation-resistant cells (A31-1-8). It thus appeared that A31-1-13 cells exhibit a TPA-like effect at growth confluence, and we postulated that this may be the mechanism by which these cells are proficient in tumour promotion expression (Yamasaki et al., 1985).

Although these results suggest a relationship between blocked intercellular communication and promotion of cell transformation, there are also lines of evidence which do not support this hypothesis. For example, TGF-β and TCDD can enhance transformation of BALB/c 3T3 cells and C3H10T1/2 cells, respectively (Hamel et al., 1988; Abernethy et al., 1985). However, these two chemicals did not inhibit GJIC (Hamel et al., 1988; Boreiko et al., 1986). Similarly, Kam and Pitts (Kam and Pitts, 1988) have recently shown that TPA painting of mouse skin does not inhibit GJIC of epidermal cells. These results are in contrast to our in-vitro results; TPA inhibited communication of cultured mouse epidermal cells (Fitzgerald and Yamasaki, 1989).

ANALYSIS OF GAP JUNCTIONAL INTERCELLULAR COMMUNICATION IN TRANSFORMED OR TUMOURIGENIC CELLS - IMPORTANCE OF SELECTIVE COMMUNICATION

Since the characteristic behaviour of tumour cells is to proliferate beyond the confines of neighbouring normal cells, it is reasonable to speculate that intercellular communication capacity of already-transformed or tumourigenic cells should have been altered. This notion had first been proposed by Loewenstein and Kanno who showed that certain tumour cells do not communicate among themselves and who postulated that this lack of communication may be the mechanism for disturbed growth of cancer cells (Loewenstein and Kanno, 1966; Loewenstein, 1979). However, subsequent studies have shown that not all tumours have an altered communication capacity (Weinstein and Pauli, 1987).

We have examined various different transformed and tumour cells in their communication function and/or gap junction protein mRNA level. From cumulative results of such an analysis, we propose a new hypothesis, namely, that what is of primary importance is not the intrinsic GJIC capacity of tumour cells, but instead the lack of communication with surrounding normal cells. This can be attained by one of two different ways: 1) homologous communication of tumour cells is blocked or decreased so that they naturally do not communicate with surrounding normal cells; or 2) transformed cells maintain their communication capacity but they do not communicate with surrounding normal cells, i.e., they express a selective communication capacity (Fig. 1). Many tumour cells including mouse skin tumours, human mesothelioma cells, and rat liver tumours belong to the first category (Klann <u>et al.</u>, 1989; Fitzgerald <u>et al.</u>, 1989; unpublished results). Transformed BALB/c 3T3 rat liver epithelial cells belong to the second category (Enomoto and Yamasaki, 1984; Mesnil and Yamasaki, 1988).

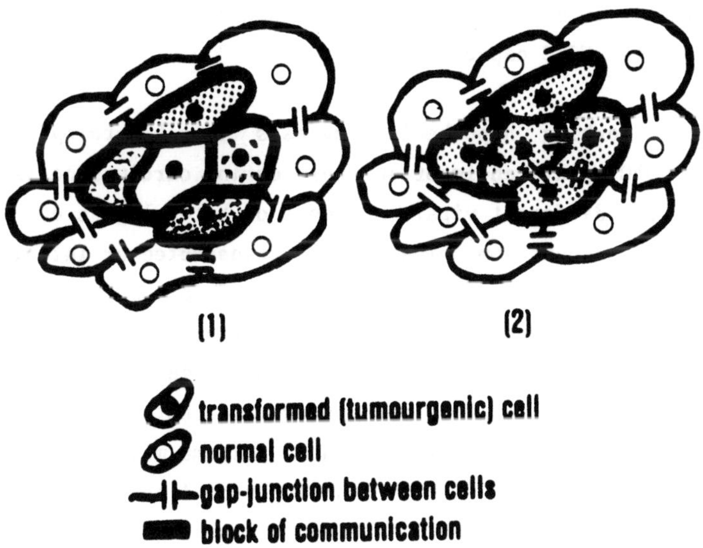

Fig. 1 Selective intercellular communication and maintenance of transformed phenotypes - schematic view. Scheme (1) shows a tumour in which individual cells cannot intercommunicate and do not communicate with the surrounding normal cells (e.g., mouse skin tumours, rat liver tumours, human mesothelioma). In scheme (2), the cells in the tumour communicate among themselves but not with surrounding normal cells (e.g., BALB/c 3T3 cells, rat liver epithelial cell lines). In neither case is there communication between the tumour and the surrounding normal cells

This hypothesis is therefore consistent with the idea that tumour or transformed cells do not communicate with surrounding normal cells since, if they did so, the normal cells may transfer growth controlling factors through gap junctions and thus result in suppression of transformed phenotypes. This also implies that if we reestablish the communication between transformed and surrounding normal cells, then transformed phenotypes may disappear (see below).

ONCOGENE EXPRESSION AND GAP JUNCTIONAL INTERCELLULAR COMMUNICATION

There is little doubt that activated cellular oncogenes and inactivated cellular anti-oncogenes are critically involved during carcinogenesis (Bishop, 1987; Bos, 1988; Harris, 1988). The relationship between intercellular communication and oncogenes has been studied quite extensively; Table 2 summarizes the reported effects of different oncogenes on intercellular communication. Clearly, some but not all oncogene products are able to alter GJIC capacity.

Since our working hypothesis suggests that heterologous rather than homologous GJIC plays an important role during carcinogenesis, we studied the effect of different viral oncogenes on heterologous communication as well as on homologous communication. We transfected or infected different viral oncogenes into NIH 3T3 cells and measured homologous communication and heterologous communication with non-transfected control cells. Our results show that v-*myc*-, v-*fos*- or polyoma large T gene-containing cells did not show any alteration in homologous or heterologous communication capacity. However, v-*src*-, v-*ras*- and polyoma middle T gene-containing cells did not communicate with surrounding normal cells, although they did communicate among themselves. It is interesting to note that only the oncogenes with cytoplasmic/membrane products, i.e., *src*, *ras*, and polyoma MT, could alter heterologous communication; the oncogenes with nuclear products had no such effect. It is also important to emphasize that the cells with membrane/cytoplasmic oncogenes, but not those with nuclear oncogenes, can form foci in the presence of an excess number of normal cells, suggesting a strong correlation between the lack of heterologous communication and the appearance of transformed foci (Bignami *et al.*, 1988; unpublished results).

Table 2. Effect of oncogenes on homologous and heterologous gap junctional intercellular communication (taken from Yamasaki, 1989b)

Oncogene	Cells	Homologous GJIC[1]	Heterologous GJIC[2]
v-src	NRK	↓	NT
	NIH 3T3	↓ or →	-
	Quail and chick embryo fibroblasts	↓	NT
	NIH3T3	→	-
c-src	NIH 3T3	↓	NT
v-ras	NIH 3T3	→	-
EJ-ras^H	BALB/c 3T3	→	-
	Rat liver epith. cell line IAR20	→	+
	Rat liver epith. cell line	↓	NT
v-myc	NIH 3T3	→	+
v-fos	NIH 3T3	→	+
v-mos	C3H10T1/2	→ or ↑	NT
PyMT	Rat F cells	↓	NT
	NIH 3T3	→	-
PyLT	NIH 3T3	→	+
SV40T	Human hepatocytes	↓	NT
	Human keratinocytes	↓	NT
	Human fibroblasts	↓	NT

[1]Homologous communication is the communication among oncogene-containing cells; their communication capacity was compared with that of normal counterparts. ↓, decreased; →, no change; ↑, enhanced;

[2]Heterologous communication is the presence (+) or absence (-) of communication between oncogene-containing cells and normal cells measured in coculture. NT, not tested.

ROLE OF INTERCELLULAR COMMUNICATION IN TUMOUR SUPPRESSION

The phenomenon of tumour suppression has been studied at the genetic level, and such studies have suggested that phenotypes of normal cells are usually dominant over transformed/tumourigenic cells (Harris, 1988). This notion of normal cell dominance is common also for the role of intercellular communication during carcinogenesis. In other words, it is proposed that tumourigenic cells or potential tumourigenic cells may be suppressed if they are communicating with surrounding normal cells (Yamasaki, 1989b). Therefore it is not surprising to see several lines of evidence which suggest a role for GJIC in tumour suppression.

Table 3 summarizes the evidence that normal cells can suppress the growth of cancer cells when these two types of cells are in direct contact. In most of these studies, GJIC was not measured so that it is unclear whether cell-to-cell contact alone was enough to suppress tumour cell growth or whether the suppression involved GJIC. The first suggestion that GJIC may be involved in such a suppression came from the work of Stoker (Stoker, 1967). More detailed analysis has come from the work of Mehta and coworkers in which various chemically- and virally-transformed C3H10T1/2 cells were cocultured with normal counterparts (Mehta et al., 1986). In cell combinations where heterologous communication was weak or absent, there was no detectable growth inhibition, while inhibition of transformed cells was apparent in coculture combinations where heterologous communication was strong or where heterologous communication was induced by treatment with cAMP (Mehta et al., 1986).

We have recently demonstrated a similar correlation between heterologous communication and growth inhibition of transformed cells using an in situ transformation system. Transformed foci of BALB/c 3T3 cells in monolayer culture were generated by 3-methylcholanthrene. These transformed cells did not communicate with surrounding normal cells although they communicated among themselves. When, however, cultures containing transformed foci were treated with upregulators of GJIC, i.e., retinoic acid, cyclic AMP or glucocorticoids, there was gradual resumption of GJIC between transformed and normal cells. In the continuous presence of these chemicals, there occurred a marked decrease in the number of transformed foci (Yamasaki and Katoh, 1988).

Table 3. Suppression of transformed phenotypes by contact with normal counterparts (see Yamasaki, 1989b for details and individual references)

Cell type	Evidence of GJ communication[1] involved in the suppression
Polyoma virus-BHK21 cells	-
SV40-Swiss 3T3	Rescue of transformed foci by croton oil
Chemically-trasformed C3H10T1/2	-
UV+TPA transformed C3H10T1/2	Rescue of transformed foci by TPA
Tumorigenic rat tracheal epithelial cells (in vivo transplant)	-
Harvey sarcoma virus-transformed mouse epidermal cells (+ dermal fibroblasts in vivo grafting)	- -
Chemically-transformed mouse epidermal cells	-
Chemically- and virally-transformed C3H10T1/2 cells	Dye transfer
Chemically-transformed BALB/c 3T3 cells (with dbcAMP, retinoic acid, glucocorticoids)	Dye transfer
C-myc- or N-myc-transformed 3T3 cells	-
v-myc-transformed NIH 3T3 cells	-
v-myc-, v-fos-, PL-LT-transformed BALB/c 3T3 cells	Dye transfer

[1]This was obtained either by adding GJIC blocking agents (croton oil or TPA) to rescue transformed foci, or by direct measurement of GJIC between transformed and normal cells.

The above results suggest that normal cells are dominant over phenotypes of transformed cells, and that there is a possibility that some gap junction-permeable factors may be involved in tumour suppression. Evidence for the existence of tumour suppressing factors in cytoplasm has been recently provided by Shay and his colleagues (Shay and Werbin, 1988; Shay, 1983) using "recons" of transformed cell nuclei and normal cell cytoplasm. Karyoplasts (nuclei surrounded by a thin layer of cytoplasm within a plasma membrane envelope) from tumourigenic NIH 3T3 cells were fused to nontumourigenic NIH 3T3 cells to produce such recons. Ten clones were isolated and all of them were nontumourigenic (Shay and Werbin, 1988). These results confirm the earlier finding with hybrids between tumour cell and normal cell cytoplasms suggesting the presence of cytoplasmic factors in nontumourigenic cells that can suppress tumourigenicity (Shay, 1983). Whether such factors are the same as those putative regulatory factors that pass through gap junctions is presently unknown.

CONCLUSION AND DISCUSSION

There is strong theoretical premise to believe that altered GJIC is involved in the process of carcinogenesis and there are now numerous lines of evidence to support this hypothesis. However, as discussed above, there are also several important experimental results which are not supportive. What we need at present is a few crucial experiments to truly establish whether or not there is a causal relationship between disturbed GJIC and carcinogenesis. One such experiment would involve inducing expression of the gene which codes for the gap junction channel. Such expression after transfection of cells undergoing carcinogenic transformation may alter the process of transformation; molecular probes are now available to test this.

Although it is widely believed that connexins isolated from different cell types are the channels of gap junctions (Paul, 1986; Kumar and Gilula, 1986; Milks et al., 1988; Goodenough et al., 1988), a recent study suggests that a 16Kd protein which is common to the gap junction from various cell types forms the gap junction channel and that connexins are cell type specific regulators of the function of this channel (M. Finbow, this volume). The recent work on expression of gap junction protein genes is based on the use of probes from connexins, but it would be interesting to see whether this 16Kd protein gene expression is also modulated during carcinogenesis.

Further studies of the involvement of GJIC in carcinogenesis will need: (i) a better understanding of GJIC control mechanisms, and (ii) identification of gap junction-permeable growth controlling factors.

It is likely that a complex chain of cell-cell recognition processes precedes the formation of gap junctions between adjacent cells. Such cellular recognition could include connexin molecules themselves, cadherins (cell adhesion molecules) (Mege et al., 1988), extracellular matrix

(Spray et al., 1987), and other factors (Spray and Saez, 1987). In spite of the intrinsic difficulty in identifying gap junction-diffusible messengers, it has been shown that certain members of two major signal transducing pathways (cAMP kinase and protein kinase C), such as calcium (Saez et al., 1989), cAMP (Lawrence et al., 1978) and inositol triphosphate (Saez et al., 1989), can diffuse through gap junctions. Considering the availability of molecular biology techniques and probes, further understanding of gap junction structure and function will emerge and such knowledge, in turn, should advance our understanding of the role of GJIC in carcinogenesis.

ACKNOWLEDGEMENTS

I would like to thank Drs D.J. Fitzgerald and D. Spray for their critical reading of the manuscript, and Ms C. Fuchez for her skillful secretarial help. Part of the work of my own laboratory was supported by a grant of the USA-NCI No. R01 CA40534.

REFERENCES

Abernethy DJ, Greenlee WF, Huband JC, Boreiko CJ (1985) 2,3,7,8-tetrachlorodibenzo-p-dioxin (TCDD) promotes the transformation of C3H/10T1/2 cells. Carcinogenesis 6:651-653

Bishop JM (1987) The molecular genetics of cancer. Science 235:305-311

Boreiko CJ, Abernethy DJ, Sanchez JH, Dorman BH (1986) Effect of mouse skin tumor promoters upon (3H)uridine exchange and focus formation in cultures of C3H/10T1/2 mouse fibroblasts. Carcinogenesis 7:1095-1099

Bos JL (1988) The ras gene family and human carcinogenesis. Mutat Res 195: 255-271

Doll R (1978) An epidemiological perspective of the biology of cancer. Cancer Res 38:3573-3583

Enomoto T, Yamasaki H (1984) Lack of intercellular communication between chemically transformed and surrounding nontransformed BALB/c 3T3 cells. Cancer Res 44:5200-5203

Fitzgerald DJ, Mesnil M, Oyamada M, Tsuda H, Ito N, Yamasaki H (1989) Changes in gap junction protein (connexin 32) gene expression during rat liver carcinogenesis. J Cell Biochem, in press

Fitzgerald DJ, Yamasaki H (1989) Tumor promotion: models and assay systems. Teratogen Carcinogen Mutagen, in press

Goodenough DA, Paul DL, Jesaitis L (1988) Topological distribution of two connexin 32 antigenic sites in intact and split rodent hepatocyte gap junctions. J Cell Biol 107:1817-1824

Hamel E, Katoh F, Mueller G, Birchmeier W, Yamasaki H (1988) Transforming growth factor β as a potent promoter in two-stage BALB/c 3T3 cell transformation. Cancer Res 48:2832-2836

Harris H (1988) The analysis of malignancy by cell fusion: the position in 1988. Cancer Res 48:3302-3306

Kakunaga T, Crow JD (1980) Cell variants showing differential susceptibility to ultraviolet light-induced transformation. Science 209:505-507

Kam E, Pitts JD (1988) Effects of the tumor promoter 12-O-tetradecanoyl phorbol-13-acetate on junctional communication in intact mouse skin: persistance of homologous communication and increase of epidermal -dermal coupling. Carcinogenesis 9:1389-1394

Klann RC, Fitzgerald DJ, Piccoli C, Slaga TJ, Yamasaki H (1989) Gap-junctional intercellular communication in epidermal cell lines from selected stages of SENCAR mouse skin carcinogenesis. Cancer Res 49:699-705

Kumar NM, Gilula NB (1986) Cloning and characterization of human and rat liver cDNAs coding for a gap junction protein. J Cell Biol 103:767-776

Lawrence TS, Beers WH, Gilula NB (1978) Hormonal stimulation and cell communication in cocultures. Nature 272:501-506

Lo KY, Kakunaga T (1982) Similarities in the formation and removal of covalent DNA adducts in benzo(a)pyrene-treated BALB/c 3T3 variant cells with different induced transformation frequencies. Cancer Res 42: 2644-2650

Loewenstein WR (1979) Junctional intercellular communication and the control of growth. Biochim Biophys Acta 560: 1-65

Loewenstein WR, Kanno Y (1966) Intercellular communication and tissue growth. Lack of communication between cancer cells. Nature 209:1248-1249

Mege RM, Matsuzaki F, Gallin WJ, Goldberg JI, Cunningham BA, Edelman GM (1988) Construction of epithelioid sheets by transfection of mouse sarcoma cells with cDNAs for chicken cell adhesion molecules. Proc Natl Acad Sci USA 85: 7274-7278

Mehta PP, Bertram JS, Loewenstein WR (1986) Growth inhibition of transformed cells correlates with their junctional communication normal cells. Cell 44:187-196

Mesnil M, Fitzgerald DJ, Yamasaki H (1988) Phenobarbital specifically reduces gap junction protein mRNA level in rat liver. Molecular Carcinogenesis 1: 79-81

Mesnil M, Yamasaki H (1988) Selective gap-junctional communication capacity of transformed and nontransformed rat liver epithelial cell lines. Carcinogenesis 9:1499-1502

Milks LC, Kumar NM, Houghten R, Unwin N, Gilula NB (1988) Topology of the 32-kd liver gap junction protein determined by site-directed antibody localizations. EMBO J 7:2967-2975

Murray AW, Fitzgerald DJ (1979) Tumor promoters inhibit metabolic cooperation in cocultures of epidermal and 3T3 cells. Biochem Biophys Res Commun 91:395-401

Paul DL (1986) Molecular cloning of cDNA for rat liver gap junction protein. J Cell Biol 103:123-134

Pitts JD, Finbow ME (1986) The gap junction. J Cell Sci (Suppl) 4:239-266

Saez JC, Connor JA, Spray DC, Bennett MVL (1989) Hepatocyte gap junctions are permeable to the second messenger, inositol 1,4,5-trisphosphate, and to calcium ions. Proc Natl Acad Sci USA 86:2708-2712

Shay JW (1983) Cytoplasmic modification of nuclear gene expression. Mol Cell Biochem 57: 17-26

Shay JW, Werbin H (1988) Cytoplasmic suppression of tumorigenicity in reconstructed mouse cells. Cancer Res 48:830-833

Slaga TJ (1984) Can tumour promotion be effectively inhibited? In: Borzsonyi M, Day NE, Lapis K, Yamasaki H (eds) Models, Mechanisms and Etiology of Tumour Promotion. IARC Scientific Publications No. 56, IARC, Lyon, pp. 497-506

Spray DC, Fujita M, Saez JC, Choi H, Watanabe T, Hertzberg E, Rosenberg LC, Reid LM (1987) Proteoglycans and glycosaminoglycans induce gap junction synthesis and function in primary liver cultures. J Cell Biol 105:541-551

Spray DC, Saez JC (1987) Agents that regulate gap junctional conductance: sites of action and specification. In: Milman HA, Elmore E (eds) Biochemical Mechanisms and Regulations of Intercellular Communication, Princeton Scientific Publishing Co., Inc., Princeton, New Jersey, p. 1-26

Stoker MGP (1967) Transfer of growth inhibition between normal and virus-transformed cells: autoradiographic studies using marked cells. J Cell Sci 2:293-304

Trosko JE, Chang CC, Metcalf A (1983) Mechanisms of tumor promotion: potential role of intercellular communication. Cancer Invest 6:511-526

Weinstein RS, Pauli BU (1987) Cell junctions and the biological behaviour of cancer. Ciba Found Symp 125:240-260

Yamasaki H (1988) Tumor promotion: from the viewpoint of cell society. In: Iversen OH (ed) Theories of Carcinogenesis, Hemisphere Publishing Corp., Washington, pp. 143-157

Yamasaki H (1989a) Short-term assays to detect tumor-promoting activity of environmental chemicals. In: Slaga TJ, Klein-Szanto AJP, Boutwell RK, Stevenson DE, Spitzer HL, D'Motto B, eds, Skin Carcinogenesis: Mechanisms and Human Relevance, pp. 265-279

Yamasaki H (1989b) Role of cell-cell communication in tumour suppresson. In: Klein G (ed) Tumor Suppressor Genes, Marcel Dekker, New York, in press

Yamasaki H, Enomoto T (1985) Role of intercellular communication in BALB/c 3T3 cell transformation. In: Barrett JC, Tennant RW (eds) A Comprehensive Survey, Mammalian Cell Transformation - Mechanisms of Carcinogenesis and Assays for Carcinogens, Carcinogenesis, Vol. 9, Raven Press, New York, pp. 179-194

Yamasaki H, Enomoto T, Shiba Y, Kanno Y, Kakunaga T (1985) Intercellular communication capacity as a possible determinant of transformation sensitivity of BALB/c 3T3 clonal cells. Cancer Res 45:637-641

Yamasaki H, Katoh F (1988) Further evidence for the involvement of gap-junctional intercellular communication and maintenance of transformed foci in BALB/c 3T3 cells. Cancer Res 48:3490-3495

Yamasaki H, Weinstein IB (1985) Cellular and molecular mechanisms of tumor promotion and their implications for risk assessment. In: Vouk VB, Butler GC, Hoel DG, Peakall DB (eds) Methods for Estimating Risk of Chemical Injury: Human and Non-Human Biota and Ecosystems. John Wiley, New York, pp. 155-180

Yotti LP, Chang CC, Trosko JE (1979) Elimination of metabolic cooperation in Chinese hamster cells by a tumor promoter. Science 206:1089-1091

PROGRESSIVE RESTRICTIONS IN GAP JUNCTIONAL COMMUNICATION DURING DEVELOPMENT

Florenci Serras and Jo A.M. van den Biggelaar
Department of Experimental Zoology
University of Utrecht
3584 CH Utrecht
The Netherlands

One of the most fascinating challenges in developmental biology is to elucidate the mechanisms that control the organization of the body plan during embryonic development. In this regard intercellular transduction of signals is of particular importance for a coordinated pattern of cell differentiation. Because of their ability to allow the exchange of ions and small molecules up to a molecular weight of about 1200 D (Simpson et al., 1977), gap junctions have been proposed as a putative pathway for the intercellular transfer of developmentally important signals (Furshpan and Potter, 1968). In most of the embryos so far studied gap junctions appear in the early stages of development, when important developmental decisions take place (see Caveney, 1985; Guthrie, 1987 for reviews). More striking results have shown that differences in junctional communication may be found between embryonic tissues with different developmental programmes. The changes of functional gap junctional communication between cells with divergent developmental fates suggest an involvement of gap junctional-mediated signal transduction in the specialization and organization of the different domains of the embryo (Van den Biggelaar, 1988).

We have approached the changes of patterns of cell communication during the successive stages in the early development of molluscs and ascidians by microinjection of either dyes (dye-coupling) or currents (electrical-coupling) and monitored their subsequent spread to adjacent cells. For these studies both the molluscs and the ascidians are suitable systems, first because the cell lineage of the early cells has been extensively studied and their fates mapped (Nishida and Satoh, 1985; Van den Biggelaar, 1977; Verdonk and Van den Biggelaar, 1983; Zalokar and Sardet, 1984). Secondly, due to the regular pattern and short duration of the early cell cycles the cells can be easily identified and thirdly, the state of commitment of individual cells is known. The aim of this paper is to review our most recent results on the developmental pattern of intercellular communication in both systems.

NATO ASI Series, Vol. H 46
Parallels in Cell to Cell Junctions in Plants and Animals
Edited by A. W. Robards et al.

EARLY STAGES OF DEVELOPMENT: THE FORMATION OF A SINGLE COMMUNICATION COMPARTMENT

The Molluscan Embryo

In the mollusc *Lymnaea stagnalis* gap junctions have been found from the 4-cell stage without any indication of differences in their distribution up to the 24-cell stage (Dorresteijn et al., 1981). Iontophoretic injections of the hydrophilic fluorescent dye Lucifer Yellow CH (LY; MW 479 D) into one cell of a 4-cell stage embryo, results in transfer of the dye to the other three cells, as observed in an epifluorescence microscope. In contrast, injection of high molecular weight dyes, as FITC-Dextrans (MW 41 kD), shows no transfer at all or to only the sister cell of the injected one (Van den Biggelaar and Serras, 1988). This suggests that in the case of LY dye-coupling with other cells occurs via gap junctions, whereas after injection of dextrans dye-coupling occurs via cytoplasmic bridges remaining after incomplete cytokinesis. From the 4-cell stage gap junctional coupling, using LY as a tracer, has also been found in the molluscs *Bithynia tentaculata* (Fig. 1 A) and from the 8-cell stage of *Dentalium vulgare*.

In the embryos of *Patella vulgata* the appearance of gap junctions, at the 2-cell stage (Dorresteijn et al., 1982) is not correlated with the transfer of low molecular weight dyes. Dye-coupling between non-sister cells does not occur before the 32-cell stage (De Laat et al., 1980; Dorresteijn et al., 1983). At the 32- and 64-cell stage extensive junctional coupling between all cells of the embryo has been monitored with different low molecular weight

Figure 1. Pattern of gap junctional communication in early molluscan and ascidian embryos. (A) Fluorescence image of a four-cell stage embryo of *Bithynia* in which a cell has been injected with Lucifer Yellow CH. (B,C,D) Fluorescence images of embryos of *Patella vulgata* in which a cell has been injected with Fluorescein Complexon. (B) Two-cell stage. The dye does not spread to the adjacent cell. (C) Four-cell stage injected after cleavage. The sister cell is also labelled. (D) 32-cell stage. Dye spreads to all adjacent cells. (E) Electrical-coupling between two non-sister cells of the 32-cell stage after injection of a 2 nA 300 msec depolarizing current pulse; Upper trace: voltage deflection in the cell in which the current has been injected. Lower trace: voltage deflection in the other cell. Both traces were obtained and photographed from a pen recorder. Bar: 10 mV. (F) Dye-coupling after injection of Lucifer Yellow CH into an animal cell of the 64-cell *Patella coerulea* embryo. (G,H) Oscilloscope images showing electricalcoupling between non-sister cells of the 16-cell *Patella vulgata* embryo after injection of (G) 2 nA 200 msec hyperpolarizing and (H) 1 nA 100 msec depolarizing current pulses. Upper traces: voltage deflection in the injected cell; lower traces: voltage deflection in the other cell. (I) Oscilloscope image showing electrical-coupling between non-sister cells of the 8-cell *Phallusia mammillata* embryo after injection of a 2 nA 100 msec hyperpolarizing current pulse. Lower trace: voltage deflection in the injected cell; upper trace: voltage deflection in the other cell. (J-K) Injection of Lucifer Yellow CH into single cells of *Phallusia*. (J) Fluorescence and fluorescence-bright-field image of an 8-cell embryo; dye-coupling occurs only with the sister cell. (K) Fluorescence image of a 16-cell embryo; dye spreads to all cells of the embryo. Scale bars in fluorescence micrographs: 50 µm. Arrows indicate the injected cell fluorescent dyes (i.e. below 1000 D: LY dilithium salt, LY dipotassium salt, LY ammonium salt, Fluorescein, Fluorescein complexon (FC) and Rhodamine B). In contrast, injections of LY-Dextrans (MW 10kD) and FITC-Dextrans (MW 9 kD and 41 kD) in any cell of the 32- or 64-cell stage *Patella* embryo results in no spread or exclusively to the respective sister cell. These results indicate that all cells of 32- and 64-cell embryos are well coupled via functional gap junctions (Fig. 1 B-F). This has been confirmed by a reduction of dye-coupling after injection with antibodies against gap junctions (Serras et al., 1988 a; Finbow et al., 1988). The absence of dye-coupling before the 32-cell stage does not exclude the possibility that between the 2- and 32-cell stage gap junctional communication, below the resolution of dye-coupling, exists. Monitoring electrical-coupling between cells can give a more accurate indication of the

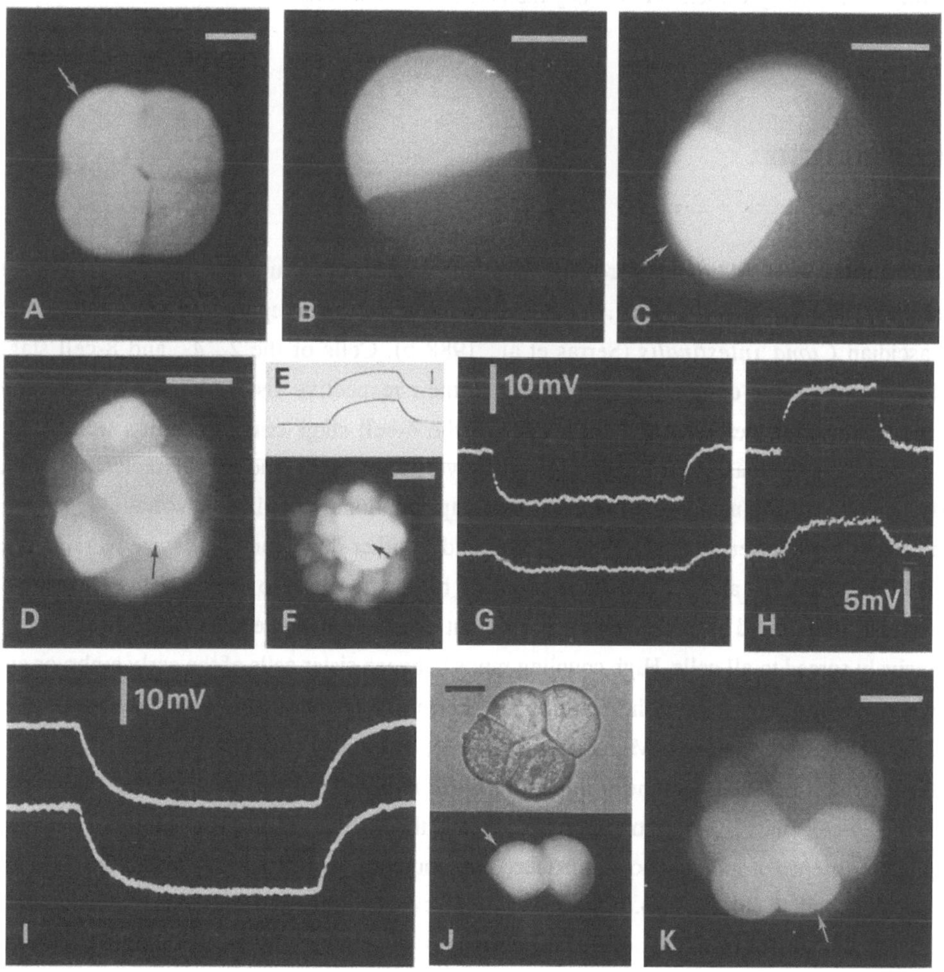

level of gap junctional-mediated coupling. In order to measure electrical-coupling two microelectrodes were impaled into different cells. Injection of electrical current into one cell generates a voltage deflection in this cell proportional to the membrane resistance. When cells are electrically coupled, a voltage deflection is detected in the second cell. We found that between non-sister cells of the 4-, 8-, and 16-cell stage, coupling ratios (the ratio of the voltage deflection in the second cell to the voltage deflection in the current-injected cell) range from almost zero to about 0.6; the highest ratios are found at the 16-cell stage (Fig. 1 G,H). At the 32-cell stage coupling ratios are close to one (Fig. 1 E). This apparent increase in junctional communication is in accordance with the observations that at the 32-cell stage gap junctions are more abondant than in earlier stages (Dorresteijn et al., 1982). In *Patella, Lymnaea, Bithynia* and *Dentalium*, the species studied so far, it appears that as soon as dye-coupling can be detected for the first time, all cells of the embryo are mutually coupled, forming one single communication compartment without any indication of restrictions.

The Ascidian Embryo

Gap junctions in ascidians appear to be identical with the gap junctions of the vertebrates (Georges, 1979; Lane et al., 1986). Functional coupling has been reported in early embryos of the ascidian *Ciona intestinalis* (Serras et al., 1988 b). Cells of the 2-, 4-, and 8-cell stage embryos are not dye-coupled, or transiently to their respective sister cell (via cytoplasmic bridges). However, between non-sister cells of the 4-cell stage coupling ratios were found lower than 0.6, whereas at the 8-cell stage ratios were above 0.6. Another characteristic of the early coupling of the ascidian embryos is that gap junctional coupling is voltage dependent (Knier et al., 1986; Serras et al., 1988 b), as found in other developmental systems (Harris et al., 1983). In the 16- and 32-cell stage coupling ratios are close to 1.0. In a few cases dye-transfer has been found at the 16-cell stage. Without exception dye injected at the 32-cell stage is extensively spread to all cells. High coupling ratios between sister cells of the early embryos and between non-adjacent cells of the blastula and gastrula stages of *Ciona intestinalis* and other ascidians have been reported (Miyazaki et al., 1974; Takahashi and Yoshii, 1981; Dale et al., 1982; Merritt et al., 1986). In the ascidian *Phallusia mammillata* we have recently shown that from the 16-cell stage up to gastrulation LY spreads through all cells of the embryo. However, despite the absence of gap junctional-mediated dye-coupling, all cells of the preceding stages are electrically coupled (Fig. 1 I-K). In summary, in all cases studied so far it appears that in early ascidian development from the onset of the cleavages all cells are electrically coupled, whereas from the 16- or 32-cell stage dye-coupling occurs between all cells of the embryo. This indicates that early in ascidian development blastomeres become highly coupled and form one single junctional communication compartment.

THE MOLLUSCAN BLASTULA: SUBDIVISION OF THE SINGLE COMMUNICATION COMPARTMENT INTO TWO COMPARTMENTS BY LOW PERMEABILITY CELLS

The first changes in gap junctional communication within the uniformly coupled molluscan embryo occur in the animal hemisphere. In the 48-cell *Lymnaea* embryo the four clones of primary trochoblasts (two cells per clone) are situated interradially. These cells will arrest their cell cycle and will contribute to the prototroch as well as to other larval organs, whereas cell proliferation in the rest of the surrounding cells continues (Verdonk and Cather, 1983). After the 48-cell stage injection of LY into a primary trochoblast never results in dye-transfer, whereas dye-injection into any other cell of the embryo results in transfer to all cells except to the four clones of primary trochoblasts. The same result was found using FC as a tracer (Fig 2a). The dye-uncoupling of the four interradial clones of primary trochoblasts is followed by uncoupling of the later formed radial trochoblasts. Together, both groups of trochoblasts form a ring of uncoupled cells, which separates two domains of undifferentiated cells with different developmental fates: the most anterior (pretrochal) will develop the head, and the posterior (posttrochal) will develop the rest of the body. Injection of LY into any cell of the pre- or post-trochal domain results in extensive dye-spread through all cells of the injected domain but not to the surrounding trochoblasts or to the other domain. Both domains can be considered as different gap junctional communication compartments, separated from each other by a boundary of dye-uncoupled trochoblasts (Van den Biggelaar and Serras, 1988).

The pattern of junctional communication of the trochoblasts has been extensively studied in *Patella* (Serras et al., 1989a). At the 72-cell stage, the four interradial clones of primary trochoblasts are formed by four cells each. From the 72-cell stage these primary trochoblasts do not divide further and, like in the 48-cell *Lymnaea* embryo, become uncoupled, as shown by injection of LY or FC. In contrast, injections into any other cell show dye-spread to all neighbouring cells except to the primary trochoblasts (Fig. 2 C,D). With the scanning electron microscope we found that dye-uncoupling of the trochoblasts coincides with the cell cycle arrest and differentiation of the trochoblasts into ciliated cells (Fig. 2 B). Additional cells become cell cycle arrested and dye-uncoupled at the 88-cell stage, when the radial clones of accessory trochoblasts become uncoupled from the rest of the embryo (about 5 hr 30 min after the first cleavage). The set of clones of accessory trochoblasts becomes ciliated after the 88-cell stage.

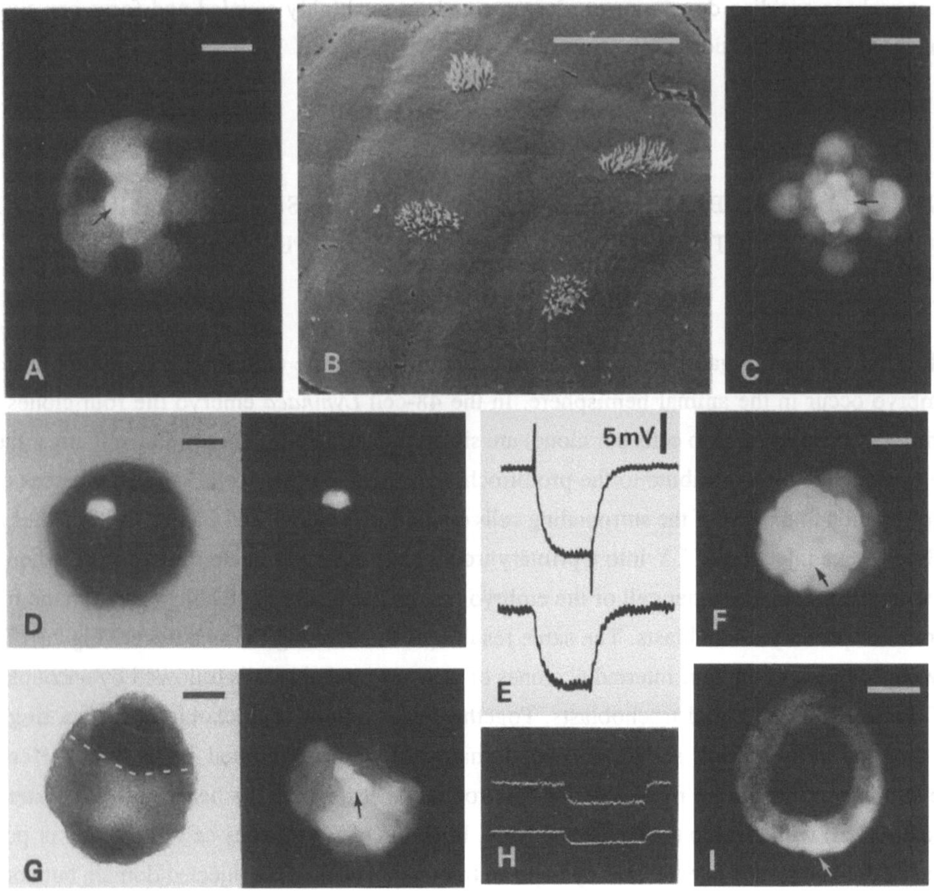

Figure 2. (A) Fluorescence image of a *Lymnaea* embryo injected with FC into a non-trochoblast cell of the 48-cell stage. (B) Scanning electronmicrograph of a clone of trochoblasts of *Patella vulgata* at the end of the 72-cell stage. All four trochoblasts begin to develop cilia. Scale bar: 25 µm. (C) Fluorescence image of an embryo of *Patella vulgata* injected with LY into a non-trochoblast cell at the 72-cell stage. (D) Fluorescence-bright field images of a *Patella vulgata* embryo injected with LY into a trochoblast at the 72-cell stage. (E) Electrical-coupling traces recorded between two primary trochoblasts of different clones at the 72-cell *Patella vulgata* embryo after injection of a 1 nA 300 msec hyperpolarizing current pulse. Upper trace: voltage deflection in the injected cell; lower trace: voltage deflection in the second cell. (F) Fluorescence image of a *Dentalium* embryo injected with LY into a cell of the pretrochal domain showing dye-compartmentalization into this domain (more intensively fluorescent cells), and reduction of dye-transfer to the surrounding trochoblasts. (G) Fluorescence-bright field images of an embryo of *Patella vulgata* injected with Lucifer Yellow CH into a single cell of the posttrochal domain after the 88-cell stage. Dashed line: border between trochoblasts and posttrochal cells. (H) Oscilloscope image of electrical-coupling between a pre- and a posttrochal cell of a young *Patella* trochophore after injection of 3 nA 200 msec hyperpolarizing current pulse. Upper trace: voltage deflection in the injected cell; lower trace: voltage deflection in the second cell. (I) Fluorescence image of a *Patella vulgata* embryo after injection of LY into a prototroch cell (i.e. the former trochoblasts). The photograph was taken from the animal pole to show the ring of prototroch cells. Scale bars: 50 µm. Arrows: injected cells

Despite of the dye-uncoupling, electrical coupling between trochoblasts of different clones or between trochoblasts and any other cell of the embryo is maintained (Fig. 2 E), although in many cases in lower ratios (when measured after 5 hr 30 min after the first cleavage) than in the dye-coupled stage. Injections into any non-trochoblast cell results in transfer of dyes, as well as in high electrical-coupling ratios. Like in embryos of *Lymnaea stagnalis*, after the 88-cell stage of *Patella vulgata* embryos the clones of trochoblasts form a closed ring of low permeability cells which subdivides the embryo into two different communication compartments: the pre- and the posttrochal domains. The same pattern of communication has been found in embryos of *Patella coerulea* and *Dentalium vulgare* (Fig. 2 F,G). Electrical coupling between cells of the pre- and the posttrochal compartment is maintained (Fig. 2 H).

In *Patella* the closed ring of dye-uncoupled cells consists of the sets of clones of trochoblasts. These cells will not divide further, and later in development, they will differentiate into ciliated cells which together form the prototroch, the larval locomotory organ of the trochophore larva. In contrast to the preceding stages, in embryos with a fully differentiated prototroch injections of LY or FC into a prototroch cell results in dye spread to all other prototroch cells (Serras et al., 1985) as shown in figure 2I. These results suggest that concomitantly with the cell cycle arrest and onset of differentiation, each set of clones of trochoblasts first reduces their gap junctional permeability, but once they together form one functional organ, the prototroch, they become mutually re-coupled. The re-coupling of the cells of the prototroch might have a physiological rather than a developmental importance for the coordination of the ciliary movement or for the transfer of metabolites throughout this differentiated organ.

FURTHER COMPARTMENTALIZATION OF THE EMBRYO

The Molluscan Trochophore Stage

Subsequent analysis of the pattern of dye-coupling in the following stages in the development of *Lymnaea* has shown that in the central part of the pretrochal ectoderm, from dorsal to ventral, a band of other larval cells (apical plate cells) becomes dye-uncoupled. Consequently, the pretrochal compartment is divided into a right and a left communication compartment (the two cephalic plates). These cephalic plates are formed by frequently dividing ectodermal cells and represent the units that will form the right and left sides of the head. Injection of LY into a single cell of any of the cephalic plates results in spread within that domain but never to the surrounding uncoupled larval cells (Serras and Van den Biggelaar, 1987; Van den Biggelaar and Serras, 1988).

In the posttrochal compartment dye-coupling becomes further restricted as well (Serras et al., in prep.). In the about 3 days old embryo different communication compartments can be observed within the posttrochal ectoderm: the domain which surrounds the stomodaeum, the ectoderm of the foot anlage and the domain from which the shell will develop (shell field).

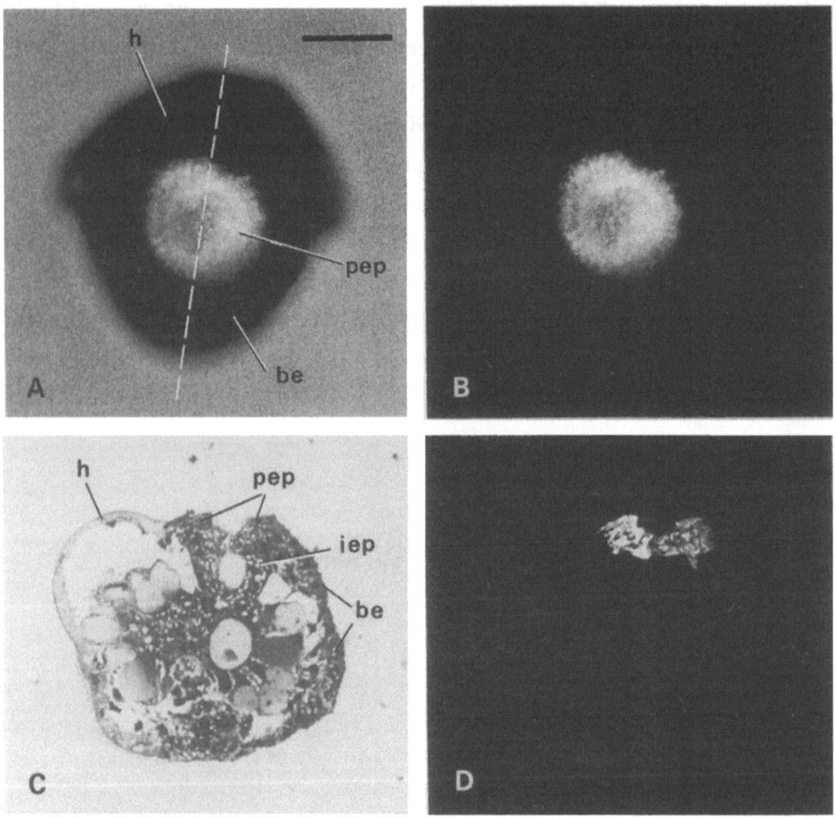

Figure 3. Peripheral epithelium communication compartment of a *Lymnaea* embryo. Fluorescence and bright field images of an embryo (age: 77 hr after the first cleavage at 25°C) injected with Lucifer Yellow CH into a cell of the peripheral epithelium of the shell field. (A,B) *In toto* viewed from the dorsal side. (C) Bright field image of a semithin section of the same embryo after fixed and embedded in Spurr resin and stained with Methylene Blue. *h* head, *pep* peripheral epithelium, *be* body ectoderm, *iep* invaginated epithelium. (D) Fluorescence image of that semithin section before Methylene blue staining. Dashed line: plane of the section shown in C and D. Scale bar: 50 μm

Further restrictions can be found between differently specialized groups of cells within these compartments: e.g. within the epithelium of the shell field the peripheral cells form a separate compartment which is uncoupled from the cells of the invaginated epithelium as well as from the rest of the body ectoderm (Fig. 3 A-D). In summary our results show that in *Lymnaea* embryos concomitantly with the specification of the different developmental programmes communication between these developmental domains becomes progressively restricted. Differences in gap junctional communication between the germ layers are found earlier in development. For instance, in the two days old *Lymnaea* embryo injections of LY into any cell of the ectoderm does not show dye-spread to the underlying germ layers as demonstrated by semithin sections of dye-injected embryos. Also in the early trochophore of *Patella* the pattern of cell coupling

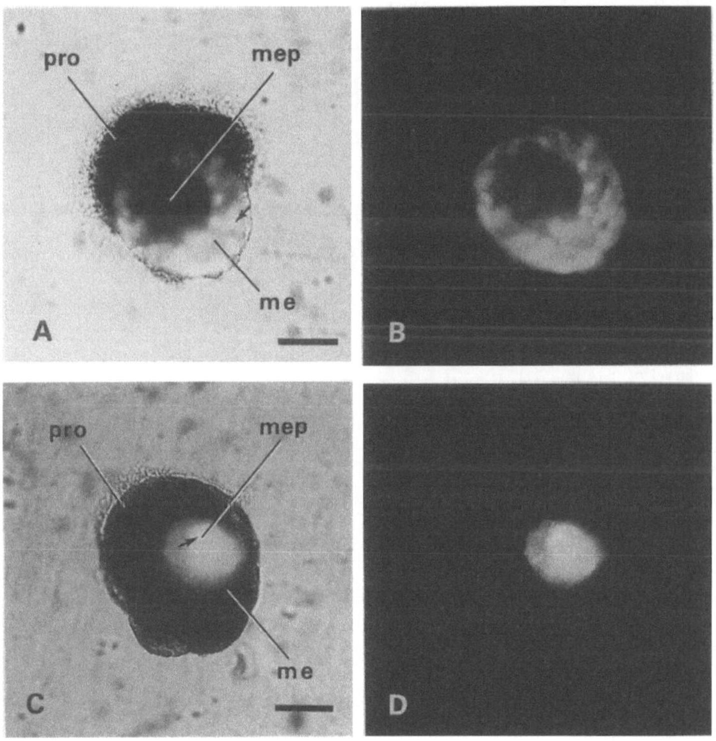

Figure 4. Communication compartments within the dorsal domain of the posttrochal ectoderm of the 24 hr old *Patella vulgata.* Fluorescence and bright field images of embryos after injection of Lucifer Yellow CH into single cells of the posttrochal ectoderm. (A,B) The injected cell belongs to the mantle edge. (C,D) The injected cell belongs to the mantle epithelium. Arrows indicate the injected cells. *pro* prototroch, *mep* mantle epithelium, *me* mantle edge. Scale bar: 50 μm

within both the pre- and posttrochal domain has been studied (Serras et al., 1989b). We found that the posttrochal region of the ectoderm of the 15 hr old embryo is subdivided first into a ventral and a dorsal communication compartment. Double injections into both compartments have shown that there are no uncoupled cells in between the dorsal and ventral domain. In the 24 hr old embryo both compartments are subdivided into smaller compartments. An example is shown in figure 4 where two different compartments are found within the former dorsal compartment: (1) the mantle edge surrounding (2) the mantle epithelium. In all cases so far studied electrical coupling has been detected across the compartment-restriction boundaries.

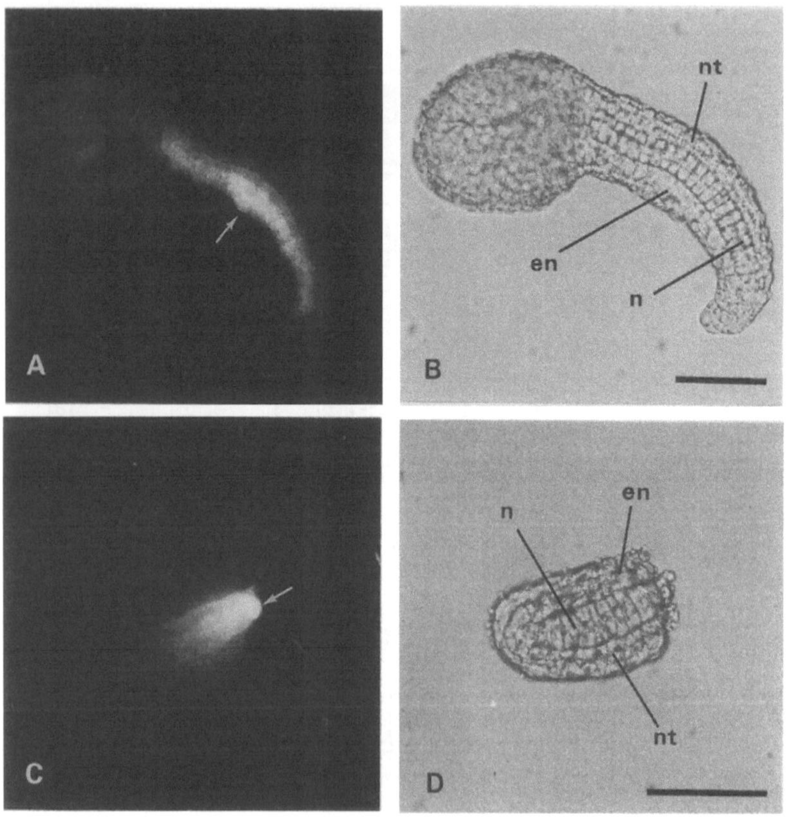

Figure 5. Communication compartments in embryos of the ascidian *Phallusia mammillata*. (A,B) Fluorescence and bright field images of tailbud embryos after injection into a cell of (A,B) the tail endoderm, (C,D) the notochord. In the last example the embryo was dissected in order to impale the cell. *n* notochord, *nt* neural tube, *en* endoderm. Arrows: injected cells. Scale bars: 50 μm

The Ascidian Tailbud Stage

The tailbud stage of the ascidian *Phallusia mammillata* is covered by transparent epidermal cells and the different primordia can be recognized *in vivo* under a conventional optic microscope. Under the epidermal layer, from dorsal to ventral, the neural tube, the notochord and the endoderm can be observed. In the tail three rows of muscle cells are located at either side of the notochord. Each of the primordia has been injected with LY to analyse the possible restrictions between the different organ anlagen (Serras et al., in prep.). The results revealed extensive coupling between cells of the epidermis. Injections into a single cell of the endoderm or of the neural tube resulted in preferential spread to the cells of the injected primordium. Both primordia appeared to form a separate communication compartment. When an embryonic muscle cell was injected with LY, dye spread to other muscle cells. Inside of the notochord the spread of LY from an injected cell to other occurs a few minutes after injection, suggesting a lower diffusion rate to the adjoining notochord cells (Fig.5).

DISCUSSION

From the results reviewed here we conclude that initially in the early molluscan and ascidian development, the embryo consists of one single dye- and electrical-coupling compartment and that during development junctional communication becomes progressively restricted to communication compartments which correspond to domains with different developmental programmes. Gap junctional communication compartments allow the spread of dyes to cells within the borders of a domain with a specific developmental programme but not across the boundaries that separate domains with different developmental programmes. Electrical coupling, however, may persist across such boundaries. These results suggest that gap junctional-boundaries between compartments may selectively restrict the passage of small molecules, metabolites and ions (i.e. of a size limit comparable to LY). These junctional communication-restriction boundaries can be understood in two ways: (1) either some low permeability cells separate the compartments or (2) direct membrane-contacts between the adjacent borders that delineate different compartments, restrict junctional communication. The separation of the pre- and posttrochal communication compartments by interposed trochoblast cells with low coupling abilities in the molluscs so far studied, and the further separation of the pretrochal compartment in *Lymnaea* into the right and left cephalic plate by uncoupling of interposed cells, are examples of the first possibility. The progressive separation of the pre- and posttrochal gap junctional communication compartments into more specified compartments are examples of the second possibility since we could not find interposed communication-restriction boundaries of low permeability cells. Similarly, we could not find uncoupled cells separating the different compartments of the ascidian embryo. It is likely that the membranes of the adjacent border cells of different compartments have changed their coupling abilities. No interposed dye-uncoupled cells have been found in mouse gastrulating embryos in which the different germ layers form separate communication compartments, but also within the germ layers one can distinguish further compartments (Kalimi and Lo, 1988; Lo, 1988).

Of all the possible functions that have been attributed to gap junctional communication its role in the regulation of cell proliferation is one of the best documented (Loewenstein, 1981; Pitts and Finbow, 1986; Sheridan, 1987; Yamasaki, 1988). The early development of molluscs is a suitable model for studying the correlation between changes in cell cycle and in cell-to-cell communication. In *Patella* the first five cell cycles are short (duration about 30 minutes each) and all cells cleave synchronous. During these early cycles gap junctions are present and they are functional as detected by ionic-coupling but not by transfer of dyes. In the stages both duration and synchrony of cell cycles are independent of cell contacts as shown by dissociation experiments (Van den Biggelaar and Guerrier, 1979). In intact *Patella* embryos, after the fifth cleavage (i.e. from the 32-cell stage onwards) cell cycles become longer and asynchronous (Van den Biggelaar, 1977). However, in dissociated embryos or in embryos in which cell contacts have been interfered, the cell cycle asynchrony and long duration characteristic of the stages after the fifth cleavage, are lost (Van den Biggelaar and Guerrier, 1979; Kühtreiber and Van Dongen, 1989). These results indicate that after the fifth cleavage the differences in the periodicity of the cell cycles between the different cells of the embryo, depend upon cell-to-cell interactions. At the same stage we found that ionic-coupling increases and dye-coupling is extended to all cells of the embryo.

Although the determination of the primary trochoblasts (presumably by segregation of cytoplasmic determinants) occurs before their junctional uncoupling, it is intriguing that they reduce coupling when they arrest their cell cycle. Cell cycle arrest might be important for the differentiation of the trochoblast into ciliated cells. The uncoupling of the trochoblasts can be interpreted as an "isolation" which is a prerequisite for the realization of a restricted programme of development and differentiation, different from the rest of the embryo.

In addition, the uncoupling of the trochoblasts can be significant for the onset of the compartmentalization of the embryo. The uncoupling of the sets of trochoblasts, as well as of other larval cells, may restrict the transfer of developmentally important molecules within a single undifferentiated compartment. These results resemble the insect communication compartments, which are separated by border cells with low junctional permeability (Warner and Lawrence, 1982; Blennerhasset and Caveney, 1984; Weir and Lo, 1984). Uncoupling of cells during development has also been described in other embryos (Kimmel et al., 1984; Lo and Gilula, 1979).

The above examples demonstrate that after a certain level of developmental specification changes in gap junctional communication occur. Different populations of cells become separated into communication compartments by communication-restriction boundaries. It has been suggested that within a group of coupled cells a gradient of small molecules might be generated, which according to the position would dictate the specification of the developmental programmes of the cells (Michalke, 1977; Wolpert, 1978). Recently evidence has been given for the involvement of gap junctions in the development of a gradient of positional information in *Hydra* (Fraser et al., 1987; 1988). As long as a group of cells functionally communicates via gap junctions, a gradient of morphogenetic molecules might be generated and specify the developmental fate of the cells. Once this has become irreversible, the differently specified cells might mutually reduce intercellular communication resulting in communication-restriction boundaries that delineate newly formed communication compartments. Subsequently, during further development, a similar phenomenon might be repeated within each compartment and progressively communication compartments with more specified developmental programmes will be formed.

ACKNOWLEDGEMENTS

This work was supported by a grant to F.S. from BION/NWO.

REFERENCES

Blennerhasset MG, Caveney S (1984) Developmental compartments are separated by a cell type with reduced junctional permeability. Nature 309: 361-364

Caveney S (1985) The role of gap junctions in development. Annu Rev Physiol 47: 319-335

Dale B, De Santis A, Ortolani G, Rassotto M, Santella L (1982) Electrical coupling of blastomeres in early embryos of ascidians and sea urchins. Exptl Cell Res 140:457-461

De Laat SW, Tertoolen LGJ, Dorresteijn AWC, Van den Biggelaar JAM (1980) Intercellular communication patterns are involved in cell determination in early molluscan development. Nature 287:546-548

Dorresteijn AWC, Van den Biggelaar JAM, Bluemink JG, Hage WJ (1981) Electron microscopical investigations of the intercellular contacts during the early cleavage stages of *Lymnaea stagnalis*. Roux's Arch Dev Biol 190:215-220

Dorresteijn AWC, Bilinski SM, Van den Biggelaar JAM, Bluemink JG (1982) The presence of gap junctions during early *Patella* embryogenesis: An electron microscopical study. Dev Biol 91: 397-401

Dorresteijn AWC, Wagemaker HA, De Laat SW, Van den Biggelaar JAM (1983) Dye-coupling between blastomeres in early embryos of_*Patella vulgata* (Mollusca, Gastropoda): Its relevance for cell determination. Roux's Arch Dev Biol 192: 262-269

Finbow ME, Buultjens TEJ, Serras F, Kam E, John S, Meagher L (1988) Immunological and biochemical analysis of the low molecular weight gap junctional proteins. In: Hertzberg EL, Johnson RG (eds) Modern Cell Biology Vol 7, pp 53-67. Alan R Liss, Inc, New York

Fraser SE, Green CR, Bode HR, Gilula NB (1987) Selective disruption of gap junctional communication interferes with a patterning process in hydra. Science 237:49-55

Fraser SE, Green CR, Bode HR, Bode PM, Gilula NB (1988) A perturbation analysis of the role of gap junctional communication in developmental patterning. In: Hertzberg EL, Johnson RG (eds) Modern Cell Biology Vol 7, pp 515-526. Alan R Liss, Inc, New York

Furshpan EJ, Potter DD (1968) Low-resistance junctions between cells in embryos and tissue culture. In: Moscona AA, Monroy A (eds) Current topics in developmental biology, Vol 3 pp 95-127 Academic Press New York

Georges D (1979) Gap and tight junctions in tunicates. Study in conventional and freeze-fracture techniques. Tiss Cell 11:781-792

Guthrie SC (1987) Intercellular communication in embryos. In: De Mello WC (ed) Cell to cell communication. pp 223-244 Plenum Press, New York

Harris AL, Spray DC, Bennett MVL (1983) Control of intercellular communication by voltage dependence of gap junctional conductance. J Neurosci 3:79-100

Kalimi GH, Lo CW (1988) Communication compartments in the gastrulating mouse embryo. J Cell Biol 107:241-255

Kimmel CB, Spray DC, Bennett MVL (1984) Developmental uncoupling between blastoderm and yolk cell in the embryo of the teleost *Fundulus*. Dev Biol 102:483-487

Knier JA, Merritt MW, White RL, Bennett MVL (1986) Voltage dependence of junctional conductance in ascidian embryos. Biol Bull 171:495

Kühtreiber WM, Van Dongen CAM (1989) Microinjection of lectines, hyaluronidase and hyaluronate fragments interferes with cleavage deley and mesoderm induction in embryos of *Patella vulgata*. Dev Biol 132:436-441

Lane N, Dallai R, Burighel P, Martinucci GB (1986) Tight and Gap junctions in the intestinal tract of tunicates (Urochordata): a freeze-fracture study. J Cell Sci 84:1-17

Lo CW, Gilula NB (1979) Gap junctional communication in the post- implantation mouse embryo. Cell 18:411-422

Lo CW (1988) Communication compartments in insect and mammalian development. In: Hertzberg EL, Johnson RG (eds) Modern Cell Biology Vol 7, pp 505-514. Alan R Liss, Inc, New York

Loewenstein WR (1981) Junctional intercellular communication: The cell to cell membrane channel. Phyciol Res 61(4): 829-913

MW, Knier JA, Bennett MVL (1986) Communication compartments in ascidian embryos at the blastopore stage. Biol Bull 171:474

Michalke W (1977) A gradient of diffusible substances in a monolayer of cultured cells. J Memb Biol 33:1-20

Miyazaki S, Takahashi K, Tsuda K, Yoshii M (1974) Analysis of non linearity observed in the current-voltage relation of the tunicate embryo. J Physiol 238: 55-77

Nishida H, Satoh N (1985) Cell lineage analysis in ascidian embryos by intracellular injection of a tracer enzyme.II. The 16- and 32-cell stages. Dev Biol 110:440-454

Pitts JD, Finbow ME (1986) The gap junction. J Cell Sci Suppl 4:239-266

Serras F, Van den Biggelaar JAM (1987) Is a mosaic embryo also a mosaic of communication compartments?. Dev Biol 120:132-138

Serras F, Kühtreiber WM, Krul MRL, Van den Biggelaar JAM (1985) Cell communication compartments in molluscan embryos. Cell Biol Int Rep 9:731-736

Serras F, Buultjens TEJ, Finbow ME (1988 a) Inhibition of dye-coupling in *Patella* (Mollusca) embryos by microinjection of antiserum against *Nephrops* (Arthropoda) gap junctions. Exptl Cell Res 179:282-288

Serras F, Baud C, Moreau M, Guerrier P, Van den Biggelaar JAM (1988 b) Intercellular communication in the early embryo of the ascidian *Ciona intestinalis*. Development 102:55-63

Serras F, Dictus WJAG, Van den Biggelaar JAM (1989a) Changes in junctional communication in relation with cell cycle arrest and differentiation of trochoblasts in embryos of *Patella vulgata*. Dev Biol (in press)

Serras F, Damen P, Dictus WJAG, Notenboom RGE, Van den Biggelaar JAM (1989b) Communication compartments in the ectoderm of embryos of *Patella vulgata*. Roux's Archiv Dev Biol (in press)

Sheridan JD (1987) Cell Communication and growth. In: De Mello WC (ed) Cell to Cell communication, pp 187-222. Plenum Press, New York

Simpson I, Rose B, Loewenstein WR (1977) Size limit of molecules permeating the junctional membrane channel. Science 195:294-296

Takahashi K, Yoshii M (1981) Development of the sodium, calcium and potassium channels in the cleavage arrested embryo of an ascidian. J Phisiol 315:515-529

Van den Biggelaar JAM (1977) Development of dorsoventral polarity and mesentoblast determination in *Patella vulgata*. J Morphol 154:157-186

Van den Biggelaar JAM, Guerrier P (1979) Dorsoventral polarity and mesentoblast determination as concomitant results of cellular interactions in the mollusk *Patella vulgata*. Dev Biol 68:462-471

Van den Biggelaar JAM (1988) Intercellular communication and development. In: Hertzberg EL, Johnson RG (eds) Modern Cell Biology Vol 7, pp 469-471. Alan R Liss, Inc, New York

Van den Biggelaar JAM, Serras F (1988) Determinative decisions and dye-coupling changes in the molluscan embryo In: Hertzberg EL, Johnson RG (eds) Modern Cell Biology Vol 7 pp 483-493. Alan R Liss, Inc, New York

Verdonk NH, Cather JN (1983) Morphogenetic determination and differentiation. In: Verdonk NH, Van den Biggelaar JAM, Tompa AS (eds) The Mollusca Vol 3, pp 215-252 Academic Press, New York

Verdonk NH, Van den Biggelaar JAM (1983) Early development and the formation of the germ layers. In: Verdonk NH, Van den Biggelaar JAM, Tompa AS (eds) The Mollusca Vol 3, pp 91-122 Academic Press, New York

Warner AE, Lawrence PA (1982) Permeability of gap junctions at the segmental border in insect epidermis. Cell 23:247-252

Weir MP, Lo CW (1984) Gap-junctional communication compartments in the *Drosophila* wing imaginal disk. Dev Biol 102:130-146

Wolpert L (1978) Gap junctions: channels for communication in development. In: Feldman J, Gilula NB, Pitts JD (eds) Intercellular junctions and synapses. pp 81-96 Wiley, New York

Yamasaki H (1988) Role of gap junctional communication in malignant cell transformation. In: Herzberg El, Johnson RG (eds) Modern Cell Biology Vol 7, pp 449-465. Alan R. Liss, Inc, New York

Zalokar M, Sardet C (1984) Tracing of cell lineage in embryonic development of *Phallusia mammillata* (Ascidia) by vital staining of mitochondria. Dev Biol 102:195:205

THE NECK REGION OF PLASMODESMATA:

GENERAL ARCHITECTURE AND SOME FUNCTIONAL ASPECTS

P Olesen and A W Robards*
Biotechnology Research Division
Danisco A/S
1 Langebrogade
DK-1001 Copenhagen K
Denmark

*Institute for Applied Biology
The University of York
York YO1 5DD
England

Introduction

For more than 100 years plasmodesmata have been known as fine channels, of a cytoplasmic nature, that connect neighbouring plant cells through the prominent and rigid carbohydrate walls that separate the cells (Tangl 1879). However, our understanding of the structure and function of plasmodesmata is surprisingly poor compared with their anticipated major roles in intercellular communication between plant cells, i.e. the symplastic transport of water and solutes as well as the channelling of biophysical and biochemical signals from cell to cell (Gunning and Robards 1976a).

Thus, although plasmodesmata certainly play a key role in plant cell biology, they are most commonly included among the so-called unique plant cell structures that are hardly mentioned in textbooks on general cell biology. There are several reasons for the detailed structure of plasmodesmata to be so comparatively poorly resolved even with the highly advanced electron microscopical technology of today. Firstly, plasmodesmata are tightly embedded in complex cell wall macromolecular structures; this inevitably causes considerable superimposition of non-relevant information in electron micrographs from ultra-thin sections. Furthermore, ultra-thin sections are usually as thick as, or thicker than, the diameter of plasmodesmata and these circumstances form the most serious obstacles to obtaining high resolution images of plasmodesmatal components (Robards 1976). Secondly, compared with the planar, repetitive structure of animal junctions, plasmodesmata are cylindrical in nature and vary dramatically in

NATO ASI Series, Vol. H 46
Parallels in Cell to Cell Junctions in Plants and Animals
Edited by A. W. Robards et al.

structure along their length, so that different structures are included in one and the same ultra-thin section (Fig. 1). Thirdly, internal subunit structures of plasmodesmata cannot be observed by freeze-fracture microscopy because the fracture plane typically "jumps off" at the plasmodesmatal entrance (Robards and Clarkson 1984, Willison 1976, Thomson and Platt-Aloia 1985). Finally, no procedures have yet been developed to isolate plasmodesmatal components from their cell wall encasings which, again, precludes detailed analysis by negative staining and high resolution shadowing.

The intention of the present review is to integrate structural data published since the comprehensive book of Gunning and Robards (1976c) into a model describing the general architecture of plasmodesmata with special reference to their neck constrictions which may be considered functional plant analogues to animal gap junctions (Gunning and Overall 1983). We have by no means, however, attempted to provide a comprehensive survey of all publications on plasmodesmatal structure since the Gunning and Robards book. In the context of the present workshop, some lines will be drawn on possible analogies between structural components of plasmodesmata and those of some other junctional complexes.

Structure and Cytochemistry

In the following discussion we will concentrate on the neck regions of simple, “basic-type” plasmodesmata (Robards 1976, Olesen 1979) which are diagrammatically shown in Figure 2 with relevant dimensions in Table 1. This structural assembly can be found in almost all known plasmodesmata of higher plants ranging from very young ones formed in the cell plate by the constriction of irregular pores and strands of endoplasmic reticulum (ER) (Jones 1976) to complicated, branched plasmodesmata of more differentiated cells and even older, partly occluded plasmodesmata. Thus, the neck region is considered a basic standardised structure, perhaps even an assembly of defined macromolecular constructions (Gunning and Overall 1983). The only known exceptions are the "open" straight, linear plasmodesmata of certain

---

Fig.1. High magnification transmission electron micrographs of plasmodesmata between mesophyll and bundle sheath cells in leaves of *Salsola kali* fixed with tannic acid/glutaraldehyde (Olesen 1979). Cross sections of the neck region (A,B) clearly show the subunit structure of the cytoplasmic sleeve between the desmotubule and the plasmamembrane. Longitudinal sections (C,D) show the continuity between the desmotubule and elements of the endoplasmic reticulum (D). Large arrows indicate the outline of the external sphincter below the raised collar while the small arrows show the particulate nature of the sphincter structure (see also Fig.3). Scale markers = 25 nm

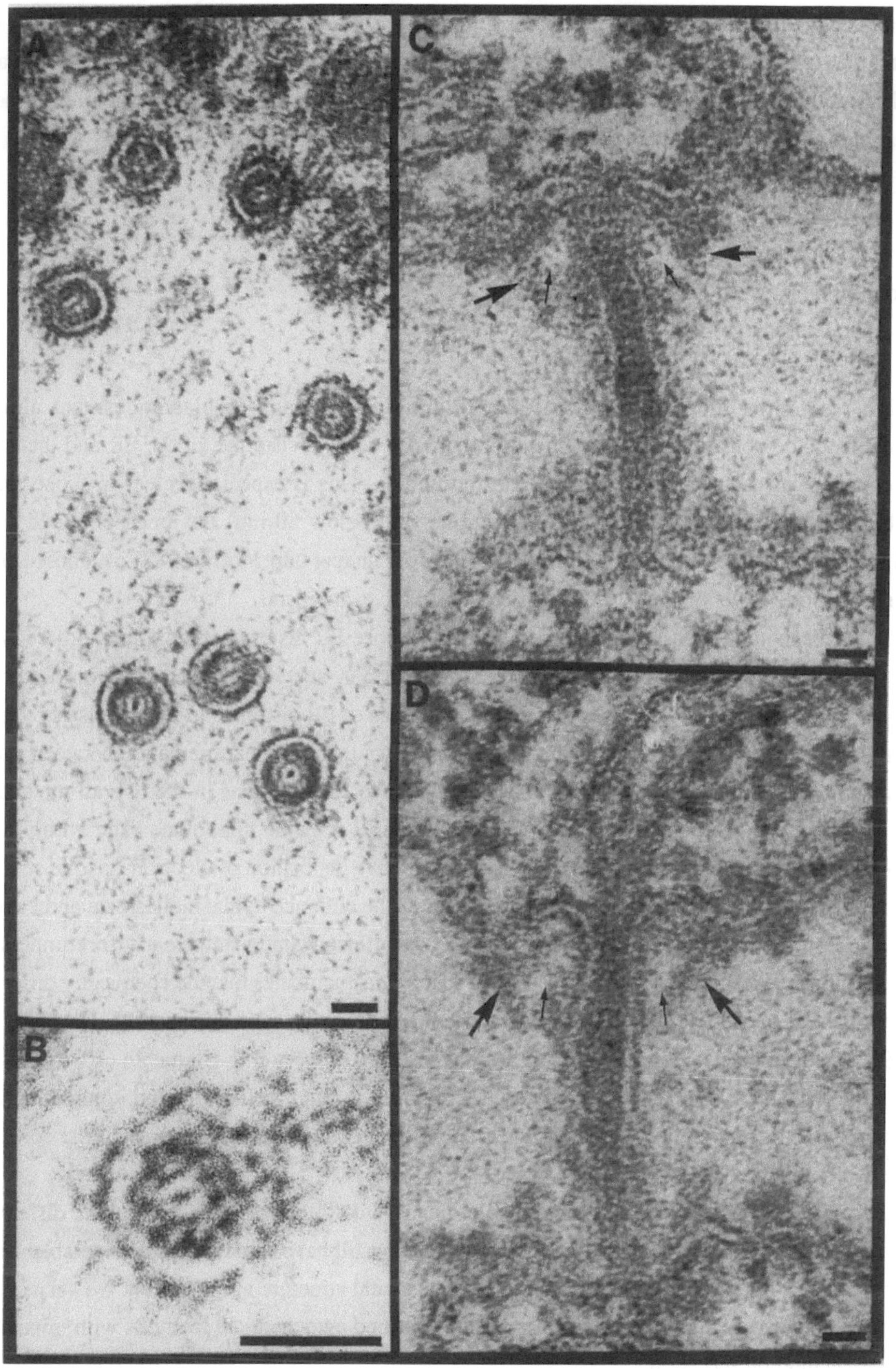
A
B
C
D

gland cells (Gunning and Hughes 1976) apparently lacking a defined neck region. Here, the situation may resemble that found in newly formed plasmodesmata of very young cell walls where nothing is yet known about how the differentiation from simple linear structures to more complex plasmodesmata with defined neck regions at each end takes place.

Connection between ER and Plasmodesmata

High resolution micrographs from thin longitudinal sections of plasmodesmata almost universally demonstrate a direct association between cortical cisternae of the ER and the axial component of each plasmodesma, the desmotubule. This is especially clear in those cases where improved contrast in ultra-thin sections (i.e. below 50 nm) has been obtained (e.g. Fig.1) by the use of tannic acid (Olesen 1979, Mollenhauer and Morré 1987) or tannic acid - ferric chloride (Overall *et al.* 1982) as improved fixatives/mordants.

It has been suggested (e.g. Robards 1971) that those ER cisternae that connect directly to desmotubules through funnel-like extensions form a specific symplastic compartment continuous from cell to cell. Whether different classes of ER cisternae in plant cells may be so different in structural composition, and overall function, that a compartmentation between water transport, solute transport and other more well-defined ER functions in the overall metabolic and synthetic machinery of the cell would exist, still remains unknown. The addition of potassium ferrocyanide to fixatives results in a selective deposition of electron-opaque stain in the lumen of ER cisternae which, however, stops at the entrance to plasmodesmata and cannot be found inside desmotubules (Hepler 1982). Since ferrocyanide stains early forerunners of desmotubules in cell plates, this indicates that mature desmotubules may be chemically different from the connected ER cisternae. Using another ER-selective stain, zinc iodide, Hawes *et al.* (1981) and Stephenson and Hawes (1986) have described very thin filaments (10-30 nm) of positively stained membrane-like material stretching through cell walls and connecting ER cisternae on both sides. Presumably the zinc iodide positive structures represent the cores of the desmotubules.

Taken together, these cytochemical data indicate that, although the desmotubule differs in some respects from its connected ER cisternae, it may still have some chemical resemblance to ER. Clearly, in terms of understanding plasmodesmatal structure and function, we very much need quantitatively integrated studies combining refined cytochemical methods with improved ultrastructural preservation.

Structure of the Axial Component: the Desmotubule

Of all the component parts of a plasmodesma, the axial core, the desmotubule (Robards 1968) is the most difficult to interpret. As a result, the structure of the desmotubule and the nature and significance of its relation to the ER have been the objects of repeated reinterpretations. López-Sáez (1966) first described the axial component as a cylindrical, tightly curved extension of ER without any lumen. Later, Robards (1971, 1976) developed models with the prevalent view that the desmotubule is an open pathway delineated by a proteinaceous, microtubule-like wall formed from an extended and much-modified ER membrane. This was at a time when current theories of cell membrane structure (e.g. Robertson 1964) seemed to rule out the possibility that a lipid bilayer could be curved about a radius as small as that of the desmotubule.

Based upon electron micrographs from ultra-thin cross sections as the one and only source of information, any model of desmotubule structure must accommodate the following four structural components (Fig. 3B, seen from the centre): a central dark dot (about 1.0 nm) presumably representing a central rod along the axis is surrounded by a pale ring about 12 nm in diameter. This pale zone is covered on its outer surface by an electron-opaque layer which is a few nm in thickness. The area between these inner structures and the inner face of the plasma membrane has the form of a sleeve or tube which, depending upon conditions of fixation, sectioning and staining, appears as an annulus being pale and electron-translucent (López-Sáez 1966), filled with electron-opaque stain or containing particulate material (many observers, see below). This fourth component is generally called the cytoplasmic annulus which, however, is an obvious misnomer and should rather be named a cytoplasmic *sleeve* (Esau and Thorsch 1985): a term that we recommend is adopted.

The different interpretations of desmotubule structure have been amply reviewed by Gunning and Overall (1983) who also developed a new model for plasmodesmatal ultrastructure which is now generally accepted among plant cell biologists. This model is based on data obtained by Olesen (1979) in *Salsola* leaves and Overall *et al.* (1982) in *Azolla* roots using tannic acid or tannic acid-ferric chloride for contrast enhancement. From such images Gunning and co-workers concluded that the inner translucent ring of the desmotubule is equivalent to the central layer of a membrane bilayer derived from membranes of ER tubules connected to cortical ER cisternae. Here, the ER membranes merge together in the desmotubule to form a more or less solid cylinder where the hydrophilic (and electron-opaque) parts of the inner leaflet are close-packed to form the dark central rod. This rod was difficult to explain in the Robards model which depicted the innermost translucent ring as the desmotubule lumen.

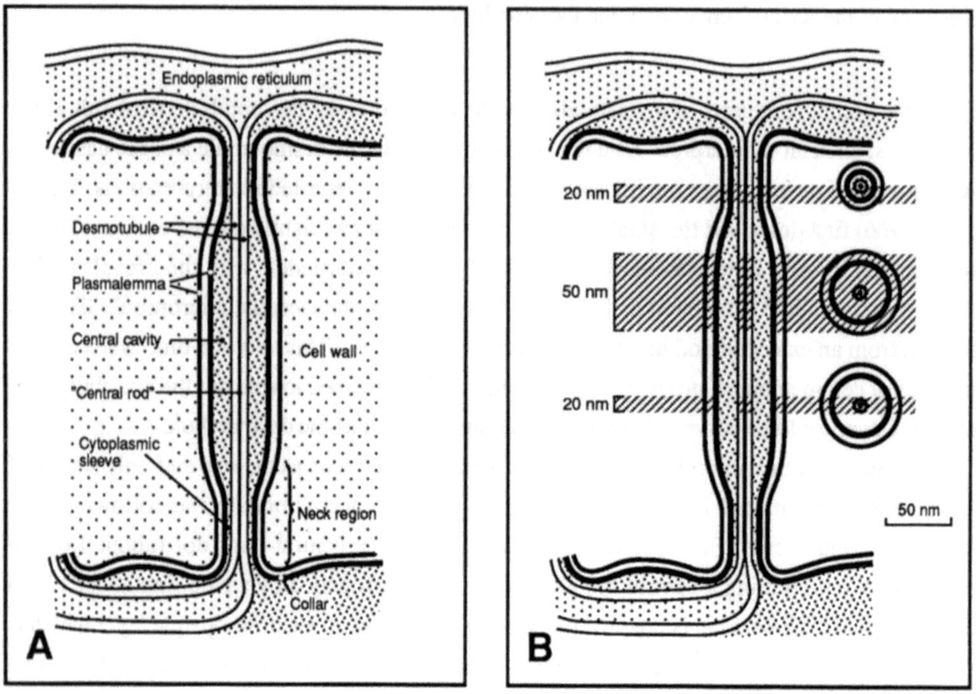

Fig. 2. Diagrams illustrating the components of a single plasmodesma (A) and the relationship between plasmodesmatal dimensions and section thickness (B): both diagrams are redrawn and updated from Robards (1976) and show general features of "basic-type" plasmodesmata as determined from transmission electron micrographs; for the detailed structure and dimensions see Fig. 3 and Table 1. The different section thicknesses (20 or 50 nm) indicated in B clearly demonstrate the limitations in resolution that arise from the considerable superimposition of excessive non-uniform information, even in very thin sections

Table 1. Dimensions (all values in nm) of the various components of plasmodesmatal neck regions (cf. Figs. 2 & 3)

	Range*	*Salsola kali*
Central rod:	2-4	3
Desmotubule:	10-13	12
Ring of subunits in cytoplasmic sleeve:	17-22	21
Average subunit size:	4-5	4.5
Space between subunits:	1-2	1.5
Plasma membrane, inner diameter:	21-44	22-31[a]
Plasma membrane, outer diameter	35-60	38-47[a]
External sphincter, outer diameter:	97-117[b]	110
Particulate subunits of external sphincter:	-	27

*Variation according to Robards (1976), Olesen (1979) and Overall *et al.* (1982).
[a]Large variation caused by superposition of variable diameters; lower values represent mainly the constricted area whereas the higher values mainly represent the more open tube below the neck constriction.
[b]Variation in *Salsola kali* only (Olesen 1979 and unpublished).

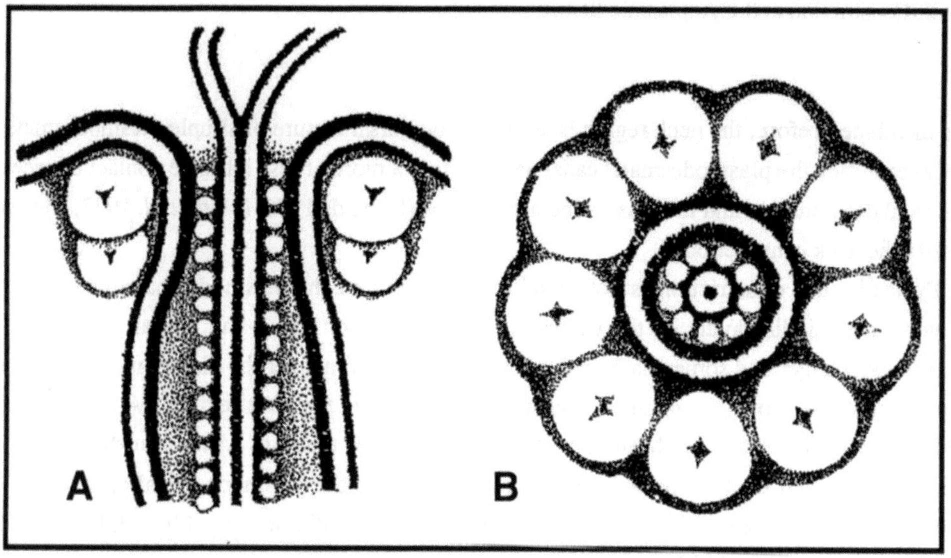

Fig. 3. Diagrammatic representation of a plasmodesma drawn approximately to scale, based on electron micrographs of plasmodesmata from the mesophyll-bundle sheath interface in *Salsola kali* leaves (such as Figs. 1, 5). Components are shown in longitudinal (A) and transverse sections (B); dimensions are given in Fig.2 and Table 1. The relationships between ER, desmotubule and the 5 nm subunits of the cytoplasmic sleeve are fully compatible with the view of Overall *et al.* (1982, Figs 19-20). Due to superposition effects and ill-defined boundaries between positively and negatively stained areas, it cannot be resolved with certainty whether the 5 nm subunits are in direct contact with the inner plasmamembrane surface. Below the neck constriction the cytoplasmic sleeve is significantly distended, which is often to some degree, also the case for the desmotubule (not shown here). The large, 27 nm, subunits of the external sphincter are depicted with a central dark spot (depression) which may indicate a subunit structure (see also Figs. 1D and 5B,D). Longitudinal views sometimes indicate the presence of more than one ring of the large particles (A, see also Fig 1C,D)

As stated above, a key question to this desmotubule model is whether the ER membranes can sustain the high degree of curvature that was considered not possible in earlier models (Robards 1971, 1976). On the basis of more recent data concerning relevant biophysical parameters for major lipid head groups and tails found in ER membranes of plant cells, Wolfe (Overall *et al.* 1982) made a theoretical calculation of the possibilities for the necessary close packing of membraneous components in the proposed closed desmotubule model. The conclusion was that the desmotubule can indeed be constructed of lipids known to exist in the ER, though it does not necessarily mean that both structures must have the same composition. A rather interesting aspect of these calculations is the probability that, if the electron-opaque central rod truly reflects the space available for head groups of the inner membrane leaflet, there would still be room for a small number of water molecules per head group; in this sense the desmotubule would still be a tubule - if only just.

Neck Constriction and Cytoplasmic Sleeve

As mentioned before, the neck region is an almost universal feature of simple plasmodesmata, i.e. at each side the plasmodesmatal canal narrows into a more or less intimate contact between the axial desmotubule and the plasma membrane tube (Robards 1976, Evert *et al.* 1977, Olesen 1979), the neck constriction.

Typically, this constriction is not found in those parts of the plasmodesmatal canals that traverse the middle layers of the cell walls they penetrate. Here, the overall diameter is significantly larger and somewhat variable (Olesen 1979) and neither the plasma membrane nor the desmotubule shows the parallel-sided appearance seen in the neck regions. A special case seems to exist in the bundle sheath-mesophyll interface in C_4 grasses where a median suberin lamella in the cell walls imposes a similar constriction on the plasmodesmatal canal, i.e. an intimate contact between desmotubule and plasma membrane (Olesen unpublished, Evert *et al.* 1977, Botha and Evert 1988).

Basically, two pathways for transport through plasmodesmata are possible (Robards 1971 and 1976, Olesen 1975 and 1979, Gunning 1976), one through the desmotubule between the ER cisternae, the other through the cytoplasmic sleeve between the two cytosolic compartments - or even a combination of these (Robards 1976). With the introduction of the model of Overall *et al.* (1982) and the discussions by Gunning and Overall (1983) on the efficacy of either pathway, the cytoplasmic sleeve pathway and its possible analogy to gap junctions in animal cells has received much more attention as the major pathway for intercellular communication and symplastic transport between plant cells.

Prior to the work of Overall *et al.* (1982) several authors had discussed the occurrence of subunits or spokes across the cytoplasmic sleeve area (e.g. Burgess 1971, Robards 1976, Olesen 1979) but generally believed these to be elements of the desmotubule wall in the Robards model. In the present models (Fig. 3 and Overall *et al.* 1982) these subunits are components of a partially occluded cytoplasmic sleeve and somehow bridge the gap between desmotubule and plasma membrane. Originally Robards (1968) suggested 11 subunits for the desmotubule in *Salix* while Zee (1969) showed some evidence for 14 subunits in plasmodesmata of *Vicia* phloem. Later, tannic acid mediated contrast enhancement facilitated the more precise determination of nine lucent subunits in the cytoplasmic sleeve of plasmodesmata in *Salsola* leaves and *Epilobium* roots (Olesen 1979) as well as in the grass *Sporobolus* (Olesen, unpublished). Subsequent reports have typically cited nine subunits to be the general feature of the cytoplasmic sleeve (Overall *et al.* 1982, Gunning and Overall 1983, Terry and Robards 1987). The only exception is found in a freeze-fracture study of chemically fixed and cryoprotected *Tamarix* gland cells where Thomson and Platt-Aloia (1985) obtained some evidence for a cluster of six subunits being present in the desmotubule at the point where the E-fracture face breaks off at the ER-desmotubule connection. However, this observation cannot necessarily be considered representative for the internal structure of the cytoplasmic sleeve inside the neck constriction and, in any case, must be regarded with caution in view of the fixed and cryoprotected nature of the material.

Although it cannot be excluded, at least on theoretical grounds, that radial spokes in the cytoplasmic sleeve are being positively stained so that the lucent spots are not particles but spaces, all available data on staining patterns do support the notion that the nine lucent subunits represent negatively-stained particles (Overall *et al.* 1982). The estimation of accurate dimensions (including particle size) in the cytoplasmic sleeve is made difficult by the relatively ill-defined surfaces which are the positively-stained layers belonging to the outer half of the desmotubule bilayer and to the equivalent plasma membrane bilayer. Particle size measurements are further hampered by uncertainties in delineating the precise point of their surface relative to the dense material in between the particles which, again, depends on the actual relative contrast in the micrographs (i.e. section thickness versus staining intensity etc.). However, best estimates (Overall *et al.* 1982, Terry and Robards 1987) are 5-6 nm for the width of the cytoplasmic sleeve, 4.5 - 5 nm for the diameter of each of the nine lucent particles and, accordingly, 1.5 - 3 nm radial gaps between the particles.

Very little is known about the anchorage of the subunit particles which, theoretically at least, might be bound either to the desmotubule or to the plasma membrane in the neck constriction. In *Salsola* plasmodesmata displaying a very conspicuous neck constriction (Olesen 1979), cross sections located more deeply in the cell wall (i.e. below the neck constriction) show an increasingly greater distance between the particle ring and the plasma membrane, but a constant association between the desmotubule and the particles is maintained. Here, although the desmotubule expands significantly in diameter, lucent subunits can still be seen at its outer surface - a situation identical to Figs. 11-18 of Overall *et al.* (1982). Therefore, the subunit particles are most likely bound to the desmotubule surface, but nothing is known about possible changes in their number outside the neck constriction - i.e. whether or not the extended desmotubule would be covered by more than nine particles. Similarly, it is not known whether the particle ring extends from the neck constriction into the funnel-like connection to the ER cisternae. In each of these cases high resolution images are impossible to obtain because of the non-parallel course of the structures in question and the inclusion of excess non-relevant information from other structures - especially compared to the more ideal situation in the parallel-sided neck constriction.

Nothing is known about how the 5 nm subunits are arranged along the desmotubule surface, but normally their visualisation depends strongly on the section plane being exactly transverse and, even then, all nine subunits do not appear in focus simultaneously. This can be illustrated very nicely using a tilting stage in the electron microscope (Fig. 4) where clear particle outlines can be seen to appear and disappear as the result of rather small changes in tilt angle (5°-10°). This tilting series also demonstrates the superposition of different structures in one and the same section. In this case the section includes part of the narrow neck constriction as well as a deeper portion of the plasmodesma where the plasma membrane tube extends significiantly in diameter. Furthermore, these micrographs have been selected because of the appearance of an extra particulate component in the cytoplasmic sleeve, i.e.between the plasma membrane and the desmotubule with its associated nine-subunit ring. As shown in Fig. 4, tilting experiments clearly resolve this material to be an extra ring-like structure which may possibly be caused by the superposition of two different plasma membrane diameters within the section thickness. An alternative explanation of this second ring structure is discussed later in relation to the possible existence of an internal sphincter in plasmodesmata in leaves of C_4 -grasses.

The tilting experiments demonstrate the possibility that all nine subunits in the cytoplasmic sleeve do not occur in exactly the same transverse plane, but sometimes overlap. Therefore, although the average number of subunits is nine, these may not appear in absolute register but rather in slightly oblique rows or in a helical arrangement (such as the protofilaments of microtubules).

The Plasma Membrane Tube

In most transverse sections of plasmodesmata the plasma membrane stands out in high contrast and clarity, mainly because of its parallel-sided nature within the section thickness. In neck regions, the traditional asymmetry of outer and inner leaflet opacity in the plasma membrane, with the thicker or more opaque leaflet seen on the extracellular side, often appears reversed (Olesen 1979). However, this is not always the case (Overall *et al.* 1982) and it is most probably related to difficulties in defining the outer surface of the inner leaflet which more or less merges into the opaque material separating the subunits in the cytoplasmic sleeve. Whether the appearance of closely-packed, more or less globular, subunits in the plasma membrane tube

---

Fig. 4. Longitudinal sections of plasmodesmata between mesophyll and bundle sheath cells in leaves from the C_4 grasses *Spartina townsendii* (A) and *Sporobolus rigens* (B) after tannic acid-glutaraldehyde fixation. In both cases a similar constriction of the plasmodesmatal canal is seen in the neck regions as well as in the suberin lamella in the median part of the cell wall. (C) shows transverse sections of *Sporobolus* plasmodesmata between the neck constriction and the suberin lamella; note the clear association between the nine subunits of the cytoplasmic sleeve and the desmotubule in this distended area. (D) shows the association between ER and plasmodesmatal neck regions in *Spartina* plasmodesmata. (E) shows a series of tilted micrographs of the same cross-sectioned *Salsola* plasmodesma (cf. Fig. 1). From left to right the tilt angles were: -10°, -5°, 0°, +5°, +10°. Note the appearance and disappearance of an extra ring of particulate material between the plasmamembrane and the ring of nine cytoplasmic sleeve subunits. All scale markers = 50 nm in the neck region (Olesen 1979) indicates a different structure and composition relative to other parts of the plasma membrane inside and outside plasmodesmata, or just reflects a pronounced vertical superposition in the parallel-sided neck region, is unknown. Interestingly, however, the monoclonal antibody MAC 207 which specifically recognises an arabinogalactan-rich epitope in all higher plant plasma membranes (Pennell *et al.* 1989), was shown by immunogold

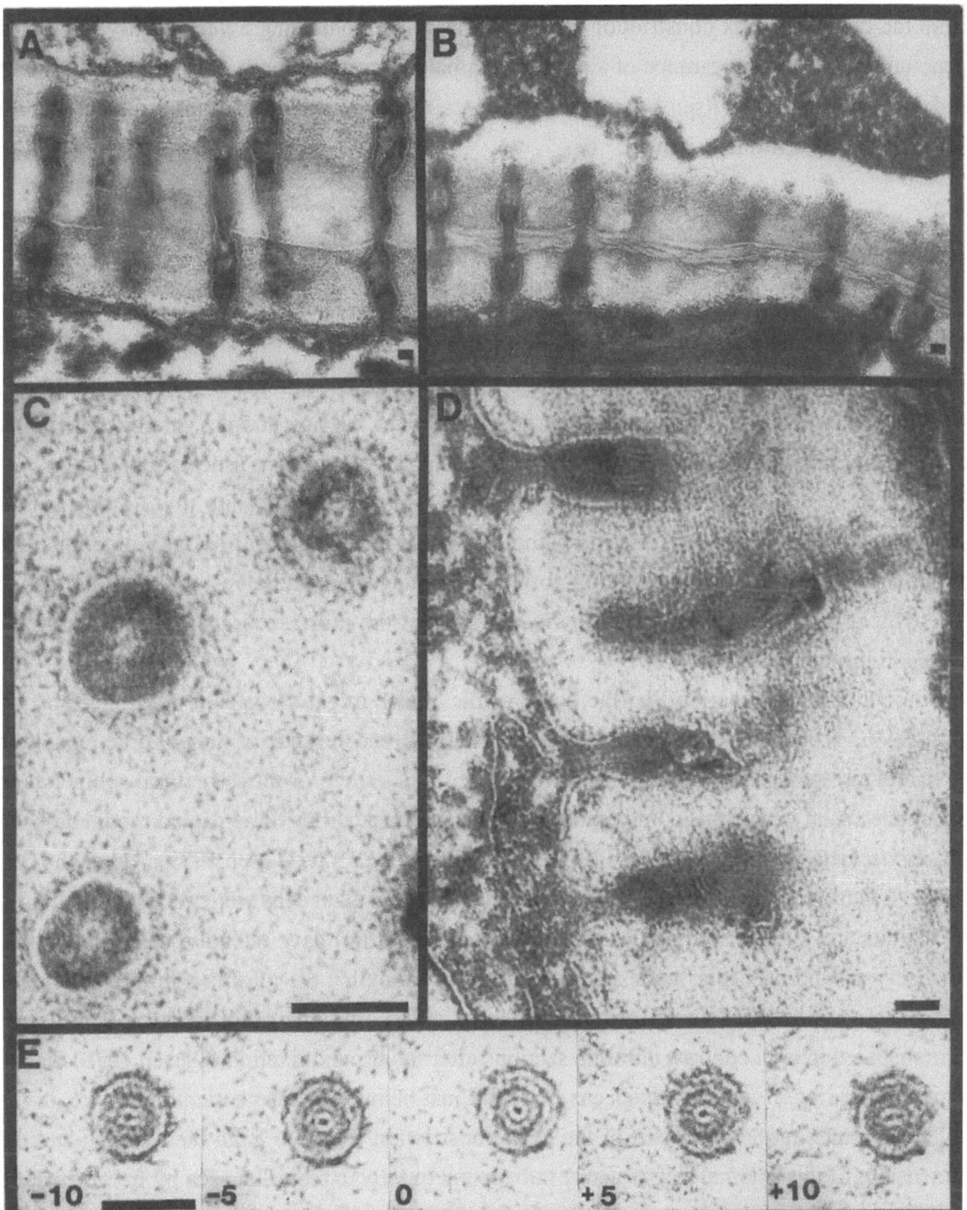

cytochemistry to be excluded from plasmodesmata in frozen ultrathin sections. If, as suggested by Gunning and Overall (1983), the neck constriction could be seen as an assembly of defined macromolecular construction, a different structure and composition of the plasma membrane in this area would, by no means, be a surprising feature. In fact, such a concept is supported by data from newly formed plant protoplasts where remnants of plasmodesmata can be observed at the surface, apparently locked into the plasma membrane (Olesen, unpublished). Here the complete neck constriction structure was intact, indicating a very tightly integrated structure. Also, the appearance of a nine-fold radial symmetry in both the cytoplasmic sleeve and the outer sphincter (see below) substantiates the view that the whole structure is of a defined, almost invariable, nature.

The Sphincter Concept

The functional implication of the Gunning model (Overall *et al.* 1982, Gunning and Overall 1983) is that intercellular transport through plasmodesmata takes place via the cytoplasmic sleeve - and more specifically through the spaces between the ring of 5 nm subunits that partly occlude this sleeve. The cross-sectional area of each of these spaces was calculated to be 2.6 - 3.6 nm^2 in *Azolla* (Overall *et al.* 1982) which is remarkably similar to the 3.1 nm^2 cross-sectional area of the pores in the macromolecular components of animal gap junctions, the connexons. This raises the interesting possibility that even small dimensional modulations of the cytoplasmic sleeve might significantly influence the cross-sectional area available for transport through this area.

Since the first realisation that the cytoplasmic sleeve might serve as a possible transport pathway (Olesen 1975, Gunning and Robards 1976a,b), the question of the possible regulation of solute transport through plasmodesmata has led to a search for ultrastructural sphincters or plasmodesmatal valves that could regulate symplastic transport by dimensional modulations of plasmodesmata (Gunning 1976, Willison 1976, Evert *et al.* 1977, Olesen 1979). Since the first freeze-fracture images of non-fixed and non-cryoprotected plant cells were published (Willison 1976) this and similar studies (Robards and Clarkson 1984) have unequivocally shown that precisely around each plasmodesmatal entrance or "mouth", the plasma membrane appears raised above its normal level forming a distinct collar. At this exact location it has been demonstrated in high contrast ultrathin sections after tannic acid fixation (Olesen 1979) that an extracellular ring of large particles can be found just below the plasma membrane - i.e. in the collar surrounding the outer part of the neck constriction (Fig. 5). Both the location and the dimensions of this external ring structure correspond exactly to the collar seen by freeze-fracture

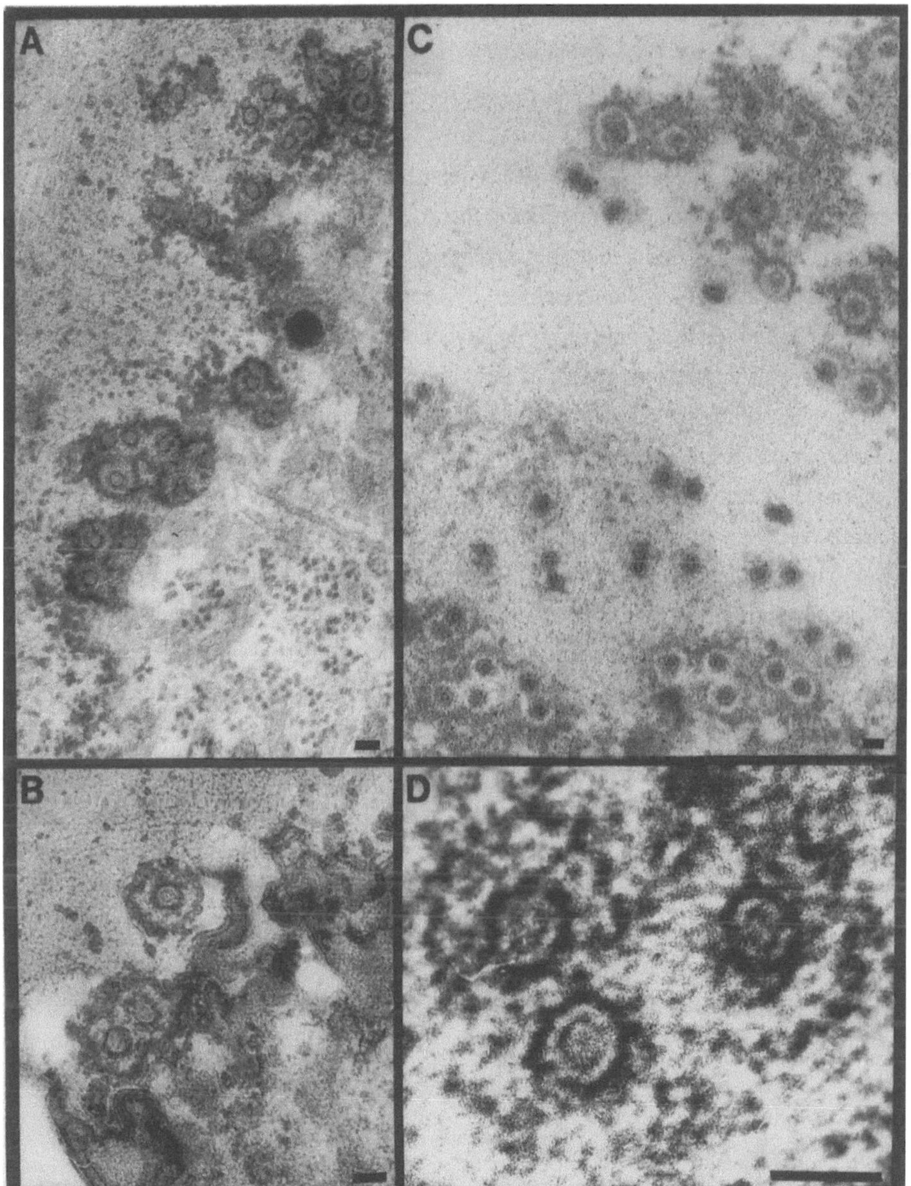

Fig. 5. Cross sections of plasmodesmata in the mesophyll-bundle sheath interface of *Salsola kali* leaves illustrating the appearance of the external sphincter after fixation with tannic acid added together with glutaraldehyde (A,B,D) or added after aldehyde-osmium fixation (C). The latter causes a significantly lower contrast of the sphincter material. A particulate substructure of the sphincter is seen in B,D correlating with the nine large (27 nm) particles seen after image enhancement (Olesen 1979). In D an opaque spot is seen in the centre of each 27 nm particle indicating a subunit structure of these components. All scale markers = 50 nm

and, furthermore, previous studies on enzyme cytochemistry and ion localisation (Olesen 1979 and references therein) have shown this are to be relatively more active than other parts of plasmodesmata and plasma membrane (i.e. enzyme activity and/or the presence of strongly reducing substances).

Bearing in mind that the cytoplasmic sleeve seems to contain nine subunit particles in radial symmetry (Olesen 1979), it is most interesting that rotational image enhancement experiments in the same study also indicated a nine-fold symmetry of the external ring structure (Fig. 3 - cf. Figs. 9-11 in Olesen 1979). It was concluded by Olesen (1979) that these ring structures might well be ultrastructural equivalents of hypothetical sphincters involved in the control of rates and direction of symplastic transport of solutes through plasmodesmata.

Although this outer sphincter structure can be seen in many published electron micrographs without tannic acid fixation (references in Olesen 1979), in most such cases visualisation seems to be facilitated by endogeneous contents of tanniferous substances. Certainly, the addition of tannic acid to fixatives strongly emphasises the sphincter structure (Robards 1976, Olesen 1979 and 1986, Mollenhauer and Morré 1987). Interestingly, the sphincter structure seems to be absent from the *Azolla* root plasmodesmata which were the subject of the meticulous study by Overall *et al.* (1982), whereas Robards (1976) published a micrograph from roots of another *Azolla* species showing distinct sphincter structures. A possible explanation for this apparent discrepancy is that conspicuous sphincters first develop at a certain stage of cellular differentiation since, as judged from the published micrographs, the tissue used by Overall *et al.* (1982) appears younger than that of Robards (1976). Another interesting possibility is that sphincters may not be present in all types of tissue, but may reflect a specialisation in physiological conditions where a very high capacity for symplastic transport is needed (Evert *et al.* 1977). In fact, the mesophyll-bundle sheath interface of *Salsola* leaves (Olesen 1979) represents a site of remarkably rapid cell-to-cell transport occurring during C_4 photosynthesis, where sphincters controlling flow through the cytoplasmic sleeve could be considered appropriate.

Internal Sphincters of Plasmodesmata in Grasses

A completely different internal sphincter structure has been proposed in grasses - or, at least, in leaves of the C_4 grass *Zea mays* (Evert *et al.* 1977). Here, very electron-opaque material was demonstrated in a specific area just below the neck constriction, but no fine structural details were resolved. Similar structures have been repeatedly seen (Olesen, unpublished) in leaves of other C_4 grasses *(Sporobolus rigens, Spartina townsendii)* and in *Zea mays* (Fig. 5). Here, a narrow invagination of the plasma membrane appears to form a diaphragm-like structure across the cytoplasmic sleeve. The location of this diaphragm corresponds exactly to the lower part of the electron-opaque sphincter structure suggested by Evert *et al.* (1977) and may well be the structure causing the sharp boundary that exists between the dense sphincter structure and the lucent area below it (Evert *et al.* 1977). Certainly, the cytoplasmic sleeve in neck regions from grass plasmodesmata contains the desmotubule-associated nine-particle ring (Fig. 5) but the extracellular outer ring structure, so typically seen in dicotyledonous species, is very difficult to resolve, even after tannic acid fixation, and has only been observed in a few cases (Olesen, unpublished). Whether the internal diaphragm-like structure represents some kind of sphincter or valve as suggested by Evert *et al.* (1977) is an open question but it is tempting to speculate that the extra internal ring structure seen at a certain level in *Salsola* plasmodesmata (Fig. 4) might be a structure analogous to the grass diaphragm. In any event, their rather precise localisation in the plasmodesmatal canal in both cases argues against the structures being merely fixation artifacts.

Sphincter Dynamics

Although the use of additional electron cytochemical stains or contrast enhancers has not thrown further light on detailed sphincter structure, some further data have been obtained about the plasmodesmata-cell wall interface (Olesen 1986). Here, the use of ruthenium red and lanthanum nitrate as carbohydrate stains revealed the sphincters as unstained structures against a granular wall material surrounding the plasmodesmata. This granular material of the pit field cell wall contrasted with the fibrillar nature of non-pore wall areas (Figs. 1-2 in Olesen, 1986). That the morphological distinction between pit fields containing plasmodesmata and non-pore wall areas is matched also by chemical differences is demonstrated by staining with Calcofluor White which causes fluorescence in only non-pore wall areas (Fig. 6). This indicates that fibrillar crystalline ß-glucans such as cellulose seem to be absent from the pits containing plasmodesmata.

Sphincter Fluorescence

Aniline Blue induced yellow fluorescence in pit fields (especially in the phloem) is a well-recognised phenomenon generally interpreted as a deposition of wound callose mediated by a labile, reversible enzymatic system (Currier and Strugger 1956) and with the possible effect of a constricting plasmodesmata (Currier 1957). In the pit fields of the mesophyll-bundle sheath interface of *Salsola* leaves a strong Aniline Blue induced fluorescence associated with plasmodesmata (Fig. 6) could be traced down to their sphincter regions by careful comparison of data from fluorescence microscopy and transmission as well as scanning electron microscopy (Olesen 1975, 1986). Physiological experiments with the Aniline Blue fluorescence have shown that this reaction is totally dependent upon light (i.e. photosynthetic activity) because it is completely, but fully reversibly, inhibited in darkness (Olesen 1986 and Table 2).

This study also showed that the fluorescent reaction is sensitive to metabolic inhibitors and displays diurnal and temperature dependent variation (Table 2). Most interestingly, it was later shown that, if different substrates are added to preparations inhibited by darkness (i.e. no Aniline Blue fluorescence detectable), this inhibition can easily be overcome by UDP-glucose (Table 2). Here, the addition of UDP-glucose results in fluorescence intensities even higher than during normal light conditions .

Table 2. Physiology of sphincter function in plasmodesmata of the mesophyll-bundle sheath interface in leaves of *Salsola kali*

Aniline Blue (AB 0.01% w/v) directly	+++	Light → Darkness (5 min) → AB	(+)
Potassium Cyanide (KCN, 5 mM) → AB	++	Light → Darkness (20 min) → AB	-
Sodium Azide (NaN_3, 5 mM) → AB	++	Darkness (20 min) → UDP Glucose → AB	++++
Dinitrophenol (2,4-DNP, 5 mM) → AB	+	Darkness → Light (5 min) → AB	-
Diethylstilbestrol (DES, 1 mM) → AB	+	Darkness → Light (20 min) → AB	+++
Sodium Vanadate (Na_3VO_4, 1 mM) → AB	+	Temperature Dependency 20°C	(+)
Boiling Water → AB	-	10 - 14°C	+++
UDP-Glucose → AB	++++	3 - 5°C	++

The dynamics of Aniline Blue (AB) induced point fluorescence in pit fields was analysed on thin, fresh hand sections incubated directly in 0.067M phosphate buffer (pH 8.0) with added chemicals and/or Aniline Blue. Number of + indicates relative strength of AB fluorescence.

Aniline Blue induced fluorescence is normally considered indicative for the visualisation of ß-1, 3-glucans such as callose (Eschrich and Currier 1964), but may also indicate local changes in the physical arrangement of polysaccharide molecules (Smith and McCully 1977) which might be the case also for the sphincter fluorescence. Taken together, however, the cytochemical data strongly indicate the presence of amorphous (i.e. non-fibrillar) glycoproteins apparently indentical to, or in a complex with, a callose-like polysaccharide containing ß-(1,3)-glucosidic linkages.

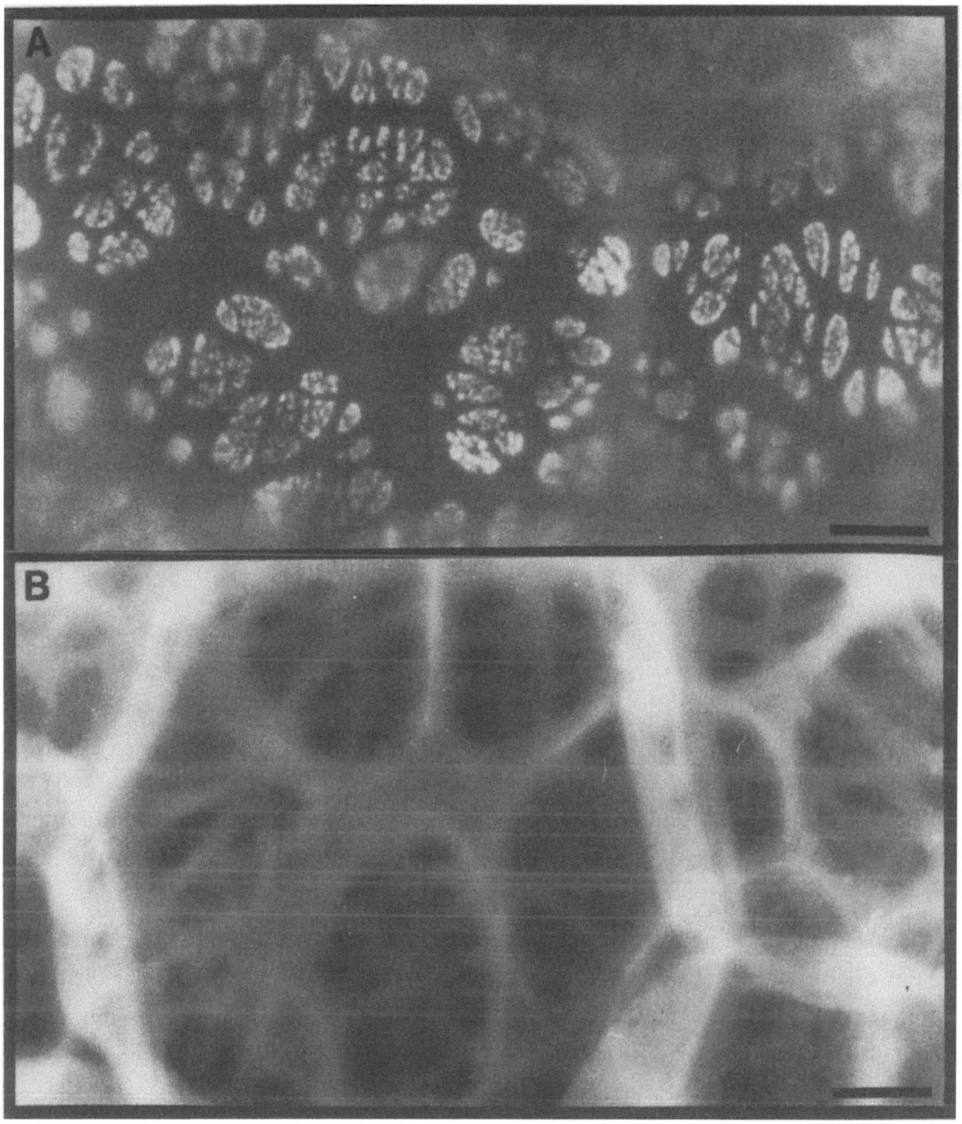

Fig. 6. Fluorescence micrographs of the mesophyll-bundle sheath interface in *Salsola kali* leaves showing face-on views of the pit fields containing abundant plasmodesmata. (Scale markers = 2 μm). Staining with Aniline Blue (A) results in a distinct point fluorescence corresponding to single plasmodesmata, clustered into pit fields, separated by non-staining areas of the cell wall. Staining with Calcofluor White (B) reveals a reverse image with only non-pore areas stained against unstained pit fields.

Physiological Sphincter or Traumatic Response

As is the case in phloem research, functional models for plasmodesmata are substantially based upon anatomical and cytochemical data. With the clear comparison between plasmodesmata and sieve plate pores (Esau and Thorsch 1985) in mind, traumatic responses to physical or chemical pertubations such as cutting and fixation (e.g. Hughes and Gunning 1980) constitute the most serious problem in the interpretation of callose and Aniline Blue fluorescence in terms of dynamics and physiological significance (cf. Eschrich 1975).

Therefore, although it cannot be ruled out that the sphincter fluorescence could represent the deposition of a callose-like substance released in a traumatic wound response, the evidence presented here (although, admittedly, mainly circumstancial in nature) is in favour of the visualised sphincters being involved in a physiological *in vivo* regulation of symplastic transport. It is to be hoped that future research in this area will clarify the issue further which would be of utmost importance to understanding the control of plant growth and development.

A Functional Model

Taken together, both morphological (Olesen 1979 and Overall *et al*, 1982) data and estimates of molecular exclusion limits from dye injection studies (Tucker 1982, Goodwin 1983, Terry and Robards 1987) clearly substantiate the view of Overall *et al.* (1982) that regulation of symplast transport may well take place in the cytoplasmic sleeve of plasmodesmatal neck constrictions. This provides clear analogies to the situation in animal gap junctions.

If the actual functional pores of plasmodesmata are, indeed, as small as those of gap junctions (1-2 nm, cf. Terry and Robards 1987) it follows that even very small dimensional changes of the cytoplasmic sleeve may have highly significant effects on the transport capacity. Olesen (1986) presented a rather coarse model illustrating how dimensional regulation of the cytoplasmic sleeve pathway could take place through the action of an external sphincter. Here it was envisaged that the sphincter complex, with a nine-fold symmetry, might contain polysaccharide synthase molecules involved in the relatively rapid deposition of ß-1, 3-glucans which, by compression against the cell wall, would close an otherwise open pathway. Such a model would neatly fit a primarily traumatic response and would also explain the almost invariable, but not universal, closed appearance of plasmodesmatal neck regions (the neck constriction, Robards 1976, Gunning and Hughes 1976, Olesen 1979). Clearly, such a mechanism could be qualitatively and quantitatively regulated in different tissues, developmental stages, and physiological conditions (such as C_4 photosynthesis). In many respects, this model could be seen as a kind of a mini-phloem system However, the very small dimensional regulations needed, by analogy with gap junctions, would certainly mean that a sphincter mechanism could not be detected from dimensional changes by traditional thin section electron microscopy. If the neck constriction is a rather stable overall configuration and transport regulation takes place between subunits of the cytoplasmic sleeve, rather than by opening and closing a free pathway, dimensional regulation through the action of an external sphincter could still be the operative mechanism. If, as in the model proposed by Olesen (1986), the sphincter contains large particles involved in the reversible synthesis and breakdown of a callose-like compound (callose can bind large quantities of water and Ca^{++}, Eschrich and Eschrich 1964), their activity could cause small variations in pressure against the plasma membrane and the more rigid surrounding cell wall. Such variations in pressure against the plasma membrane could be transduced into small compressions or relaxations of the particle ring against the more rigid desmotubule. As a consequence, the effective pore area between the 5 nm subunits would change, along with differences in their relative packing (Fig. 7).

Focusing on a glucan synthase complex as the molecular candidate for the external sphincter appears obvious for several reasons. Firstly, glucan synthase complexes are found closely associated with the extracellular surface of the plasma membrane and are specifically visualised by tannic acid fixation/staining (Olesen 1980 and references therein). Recently, Northcote (1989) has specifically located callose at plasmodesmata by immunocytochemical methods. Secondly, glucan synthase II, which is responsible for cellulose synthesis, is highly regulated and can rapidly change to the synthesis of the ß-1, 3-glucan callose (Delmer 1987). Such shifts can be mediated by any perturbation of plasma membrane permeability (transmembrane potential changes, polyamines, lysophosphatidylcholine etc, see Delmer 1987 and Kauss 1987). Interestingly, callose synthesis is strongly regulated by the influx of Ca^{++} across the plasma membrane (Kauss 1987), and it has been suggested that the resulting increase in the concentration of cytoplasmic free Ca^{++} is causally involved in triggering the plasma membrane-located ß-1, 3-glucan synthase (Kauss 1987).

If, as in many other biological regulation and signalling mechanisms, sphincter function is regulated by cytosolic Ca^{++} concentration, the source for a strongly localised Ca^{++} influx must be sought in the plasmodesmatal neck constriction complex itself. In fact, evidence has been presented that Ca^{++} may be involved in the regulation of symplastic cell-to-cell transport (Erwee and Goodwin 1983, Baron-Epel *et al.* 1988). A recent paper by Tucker (1988) strongly indicates that the Ca^{++} -phosphoinositide second-messenger system, which many recent data suggest to be active in plants (Poovaiah *et al.* 1987), plays a significant role in intercellular transport between plant cells. Thus, Tucker (1988) obtained evidence that molecules such as inositol bisphosphate and inositol trisphosphate, of the putative signal transduction pathway leading to the Ca^{++} response, inhibit cell-to-cell transport through plasmodesmata.

At least two possible sites can be suggested for the putative Ca^{++} source: either the external callose-containing sphincter structure located in the cell wall; or the desmotubule connected to ER cisternae which are known to be internal Ca^{++} reservoirs (Hepler and Wayne 1985). Since plant cell walls contain high amounts of bound as well as free Ca^{++} (MacRobbie 1989) and the present desmotubule model leaves very little if any space for the accommodation of solutes (Overall *et al.* 1982 and Fig. 3), we would see the external sphincter as the more likely candidate of the two suggestions.

To accommodate available structural and physiological data, we suggest the following functional model for a light-triggered, Ca^{++} activated gating of a sphincter mechanism (Fig. 7): (1) In the open, relaxed state the cytoplasmic sleeve has a relatively high cross sectional area for transport between the 5 nm subunits and a low Ca^{++} concentration. In the external sphincter area the glucan synthase has no associated callose and free Ca^{++} concentration is relatively high. (2) As a result of signal transduction processes (such as propagated changes in the membrane potential of the plasma membrane or other perturbations), Ca^{++} channels in the plasma membrane may be activated, resulting in the net influx of Ca^{++} with the possible involvement of phosphoinositide messenger molecules (Tucker 1988). (3) Triggered by high cytosolic Ca^{++}, callose would be produced locally by the glucan synthase part of the external sphincter and, by compression, the diameter of the cytoplasmic sleeve would be reduced, resulting in a closer packing of its subunit particles. The anticipated decrease in effective pore area between the particles might be caused by their relatively closer packing alone, but morphological changes of the individual particles could also contribute. The synthesis of callose in the outer sphincter would progressively allow the binding of more and more of the available free Ca^{++} so that a steady state situation will eventually develop (4) in which the influx of Ca^{++} is relatively reduced and, eventually, reversed which will in turn reduce callose synthesis so facilitating a relative relaxation of the compression of the cytoplasmic sleeve. The proposed Ca^{++} efflux across the plasma membrane may be mediated by local changes in membrane surface potential caused by the relatively low external concentrations of Ca^{++}. As mentioned earlier, nothing is known of the chemical composition or morphology (such as freeze-fracture data on particle distribution) of the plasma membrane tube of the neck constriction, but the exclusion of a general plasma membrane glycoprotein epitope from this area (Pennell *et al.* 1989) clearly suggests a chemical difference - which might be caused, for example, by a high concentration of ion channels such as found in different junctional complexes (Unwin 1986).

A possible further consequence of local Ca^{++} changes on either side of the plasma membrane would be modulations of membrane rigidity which, again, would facilitate the suggested plasticity of the neck constriction complex needed for even small dimensional modulations by compression/relaxation of the cytoplasmic sleeve.

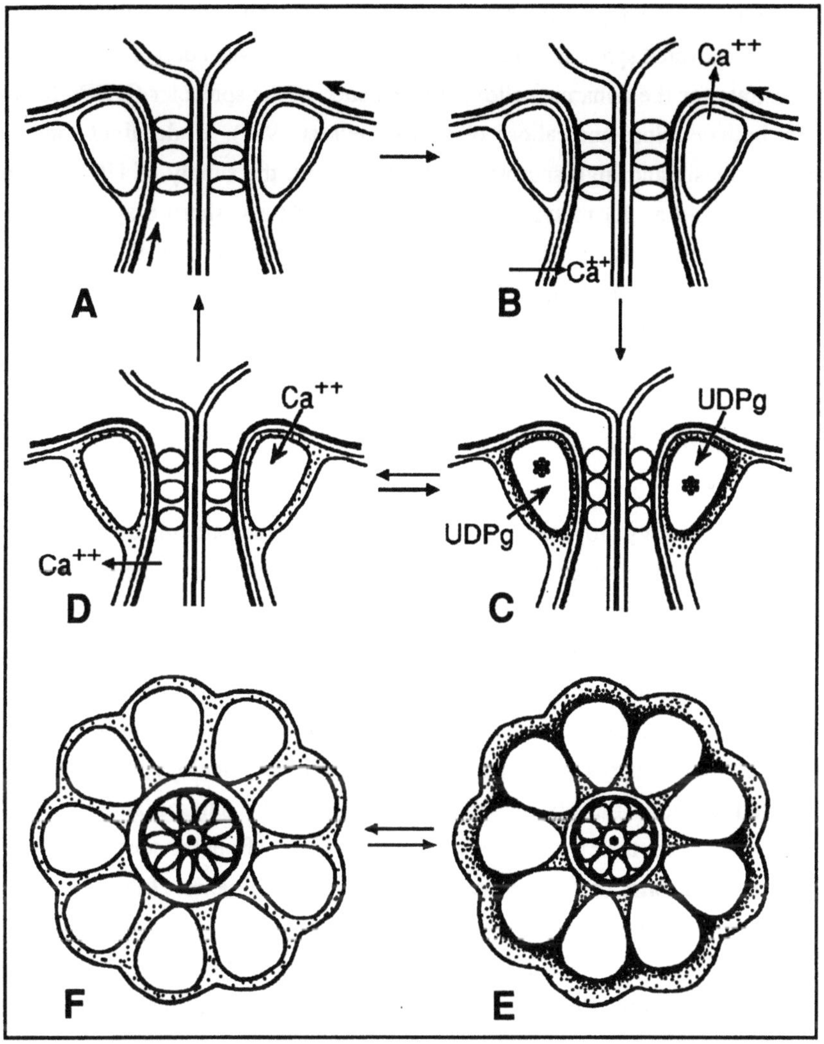

Fig. 7. Diagrams illustrating the proposed functional model for a physiological sphincter activity involved in the modulation of plasmodesmatal transport capacity. A: Basic situation with an open, relaxed structure; the spaces between the subunits of the cytoplasmic sleeve have a relatively high cross sectional area; arrows indicate a "closing signal" or stimulus arriving in the plasmamembrane. B: As a result of signal transduction, Ca++ channels in the plasmamembrane are activated, giving rise to a local increase of cytosolic free Ca++. C: Callose production by a glucan synthase component of the external sphincter triggered by high cytosolic Ca++ and using cytosolic UDP-glucose as substrate causes a relative compression of the cytoplasmic sleeve resulting in a closer packing of its subunits. * = high activity of callose synthesis (glucan synthase). (The concomitant change in subunit structure is drawn for illustrative purposes only and does not necessarily represent actual changes). D: Synthesised callose will progressively bind more and more Ca++, resulting in a kind of steady state situation where the influx of Ca++ is relatively reduced and eventually reversed. E & F: Cross sectional views of the situations depicted in B & C

Further evidence (although still circumstantial) for glucan synthase involvement in the suggested Ca^{++} activated sphincter gating comes from the demonstration of UDP-glucose as an efficient substrate for the dynamic callose fluorescence of the sphincter (UDP-glucose is the substrate for both cellulose and callose synthesis, Delmer 1987). Under light conditions, i.e. active photosynthesis and high sucrose synthase activity, the supply of UDP-Glucose will easily be able to fuel a glucan synthase (Fig. 7). In darkness, sucrose synthase activity is abolished and consequently no UDP-Glucose will be available to fuel the sphincter glucan synthase which, however, can be demonstrated to be restored simply by experimental addition of UDP-Glucose (Table 2).

Structural Analogy with Animal Junctions

In the gap junction analogy proposed by Overall *et al.* (1982) the particulate cytoplasmic sleeve is seen as the structure directly in contact with the two cytosolic compartments - and is thereby analogous to the membrane-spanning connexons of gap junctions. In traditional thin section views of gap junctions (this volume) the juxtaposed connexons of the apposed membranes are visualised as 5 nm subunits or septae occurring at 2-3 nm intervals producing the familiar ladder-like appearance. Structurally, the 5 nm subunits of the cytoplasmic sleeve in plasmodesmatal neck constrictions clearly display a similar morphology, size and distribution but are located between the plasmamembrane *cytoplasmic* surface and the ER-derived desmotubule. Therefore, their localisation differs significantly from the septae of gap junctions placed at the *extraplasmic* surface between two plasma membranes. Interestingly, a recent study (Smereka *et al.* 1988) reported on the presence of a junctional complex formed between invaginated and backfolded plasma membranes in germinating ascospores of the plant pathogenic fungus *Venturia*. Here, the whole structure strongly resembles septate junctions of invertebrate tissues although with a different origin.

Although the structural similarities between the cytoplasmic sleeves of plasmodesmata and the inter-membrane spaces of gap junctions are spectacular indeed, much more work is needed to clarify real analogies in structure as well as function. Some evidence has been reported for the presence of gap junction polypeptides in plant tissue (Meiners and Schindler 1987, Meiners *et al.* 1988, this volume) and, recently, also that a dynamic continuity may exist not only between cytoplasmic but also membrane compartments between plant cells (Baron-Epel *et al.* 1988). This latter observation suggests a significant difference from animal cells where no membrane continuity exists - and, thereby, further substantiates the difference mentioned before between the cytoplasmic sleeve of plasmodesmata and gap junctions with respect to which types of membranes are coupled in the two types of junctional complexes. Taken together, despite the apparent large similarities, there are also important differences between plant and animal junctions.

With such differences in mind, the intracellular nature of the plant junctional complex (plasmodesmatal neck constriction) is possibly more similar to the very complicated junctional connections in the triads of skeletal muscle cells (Franzini-Armstrong and Nunzi 1983, Franzini-Armstrong *et al.* 1987). Here, signal transmission occurs between the cell surface membrane (plasma membrane) invaginated deeply into so-called transverse T-tubules and a system of cytoplasmic membranes, the sarcoplasmic reticulum (SR). The signal for Ca^{++} release in exitation-contraction coupling of the muscle fibre cells is transferred from the T-tubule across the triad junction to terminal cisternae of SR. Thus, connecting a plasma membrane and an ER-like cytoplasmic membrane system, the triad junction may well be a structural as well as functional analogue to plasmodesmatal neck constrictions. As in gap junctions and the cytoplasmic sleeve of neck constrictions, the junctional gap of triads is occupied by evenly spaced densities, called feet. The triad junction gap is significantly larger (10-20 nm) than the similar structures in gap junctions (5 nm) and in neck constrictions (5-6 nm). The triad feet spanning the gap have now been identified as Ca^{++} channels associated with the junctional face membrane of terminal cisternae of the SR and their three dimensional structure has been resolved (Wagenknecht *et al.* 1989). As isolated molecules the feet are flattened structures significantly larger (27 x 27 x 14 nm) than the bridging structures in either gap junctions or plasmodesmata.

The only structural component of plasmodesmatal neck constrictions that may equal the triad feet Ca^{++} channels in size and staining characteristics may be the extracellular structures involved in the external sphincter, where some evidence has been obtained for the presence of nine subunits in apparent radial symmetry, each measuring on average 27 nm in diameter (25-30 nm, Olesen 1979 and Table 1). It is tempting, therefore, to speculate that the external sphincter is involved directly in Ca^{++} release as well as in a glucan synthase reaction - which would fit neatly into the functional model we suggest here (Fig. 7).

Concluding Remarks

This paper has confined itself largely to a discussion of the neck region of plasmodesmata - the region that is now increasingly believed to be responsible for the regulatory activities of plasmodesmata in intercellular transport. If this is so, then more detailed information is urgently needed on the molecular structure of this complex region. There are still large gaps in our knowledge. It has been observed previously that many of the activities of, and resultant problems of studying, plasmodesmata are reminiscent of those applicable to the phloem:

plasmodesmata can, in many respects, be equated to short-distance transport analogues of the long-distance transporting system of the phloem. It is, therefore, surprising that our structural knowledge remains limited to work almost exclusively carried out on fixed and embedded material. Freeze-fracture/etching methods have proved disappointing in the lack of novel information that they have provided but meticulous work using well-frozen and freeze-substituted cells is long overdue.

The extension of the application of cytochemical methods, including immunocytochemical labelling, clearly holds great promise as already hinted at by the work of Schindler and colleagues (this volume). However, such methods will only come into their own when the molecular components of plasmodesmata have been isolated and characterised and, as appropriate, antibodies raised for subsequent use in immunolabelling experiments.

It is tempting to over-emphasise the analogies between gap junctional complexes and plasmodesmata but, as pointed out above, there are inevitably some fundamental points of difference. If there are connexin-like molecules within the cytoplasmic sleeve of plasmodesmata then there still remain a number of potential pathways for transport: between the *outer* interstices of the sleeve subunits and the inner surface of the plasmalemma; between the *inner* interstices of the subunits and the outer surface of the desmotubule; through a connexin-like pore in the molecules - if one exists; and not forgetting the small, but still possibly patent, pore through the centre of the desmotubule itself. In molecular terms, much remains to be done and, as pointed out by Unwin (1986) it is by no means necessary to speculate that the relevant molecules all have a 6-fold symmetry.

The study of plasmodesmata in general, still more their structure, remains a backwater of plant cell biology more than 100 years after they were first described. And yet it cannot be doubted that they are of the utmost importance in the normal functioning of the mature plant as well as in its development. They also act as channels for the dissemination of disease through plants. The furtherance of knowledge, leading to improved understanding, will arise from the integration of studies such as those reported in this volume.

References

Baron-Epel O, Hernandez D, Jiang LW, Meiners S, Schindler M (1988) Dynamic continuity of cytoplasmic and membrane compartments between plant cells. J. Cell Biol. 106: 715-721

Baron-Epel O, Schindler M (1987) Cell to cell communication between soybean cells in culture. Plant Physiol. 83: 42

Botha CEJ, Evert RF (1988) Plasmodesmatal distribution and frequency in vascular bundles and contiguous tissues of the leaf of *Themeda triandra*. Planta 173: 433-441

Burgess J (1971) Observations on structure and differentiation in plasmodesmata. Protopasma 73: 83-95

Currier HB (1957) Callose substances in plant cells. Am. J. Bot. 44: 478-488

Currier HB, Strugger S (1956) Aniline blue and fluoresence. Microscopy of callose in bulb scales of *Allium cepa* L. Protoplasma : 522-559

Delmer DP (1987) Cellulose biosynthesis. Ann. Rev. Plant. Physiol. 38: 259-290

Erwee MG, Goodwin PB (1983) Characterisation of the *Egeria densa* Planch. leaf symplast: inhibition of the intercellular movement of fluorescent probes by group II ions. Planta 158: 320-328

Erwee MG, Goodwin PB (1984) Characterization of the *Egeria densa* leaf symplast: response to plasmolysis, deplasmolysis and to aromatic amino acids. Protoplasma 22: 162-168

Erwee MG, Goodwin PB (1985) Symplastic domains in extrastelar tissues of *Egeria densa* Planch. Planta 163: 9-19

Esau K, Thorsch J (1985) Sieve plate pores and plasmodesmata, the communication channels of the symplast: ultrastructural aspects and developmental relations. Am. J. Bot. 72: 1641-1653

Eschrich W (1975) Bidirectional transport. In: Encyclopaedia of Plant Physiology. Transport in Plants I. Phloem Transport, pp. 245-255, Zimmermann HH, Milburn JA, eds. Springer-Verlag, Berlin/Heidelberg

Eschrich W, Currier HB (1964) Identification of callose by its diachrome and fluorochrome reactions. Stain Technol. 39: 303-307

Evert RF, Eschrich W, Heyser W (1977) Distribution and structure of plasmodesmata in mesophyll and bundle-sheath cells of *Zea mays* L. Planta 136: 77-89

Franzini-Armstrong C, Kenney LJ, Varriano-Marston E (1987) The structure of calsequestrin in triads of vertebrate skeletal muscle: a deep-etch study. J. Cell Biol. 105: 49-56

Franzini-Armstrong C, Nanzi G (1983) Junctional feet and particles in the triads of a fast-twitch muscle fibre. J. Muscle Res. and Cell Motility 4: 233-252

Gunning BES, Hughes JE (1976) Quantitative assessment of symplastic transport of pre-nectar into trichomes of *Abutilon* nectaries. Aust. J. Plant Physiol. 3: 619-637

Gunning BES, Overall RL (1983) Plasmodesmata and cell-to-cell transport in plants. Bioscience 33: 260-265

Gunning BES, Robards AW (1976a) Plasmodesmata: current knowledge and outstanding problems. In: Intercellular Communication in Plants: Studies on Plasmodesmata, pp. 297-311, Gunning BES, Robards AW, eds. Springer-Verlag, Heidelberg

Gunning BES, Robards AW (1976b) Plasmodesmata and symplastic transport. In: Transport and Transfer Processes in Plants, pp. 15-41, Wardlow IF, Passioura JB, eds. Academic Press, New York and London

Gunning BES, Robards AW (1976c) Intercellular Communication in Plants: Studies on Plasmodesmata. Springer-Verlag, Heidelberg

Hawes CR, Juniper BE, Horne JC (1981) Low and high voltage electron microscopy of mitosis and cytokinesis in maize roots. Planta 152: 397-407

Hepler PK (1982) Endoplasmic reticulum in the formation of the cell plate and plasmodesmata. Protoplasma 111: 121-133

Hepler PK, Wayne RO (1985) Calcium and plant development. Ann Rev. Plant Physiol. 36: 397-439

Hughes JE, Gunning BES (1980) Glutaraldehyde induced deposition of callose. Can. J. Bot. 58: 250-258

Jones MGK (1976) The origin and development of plasmodesmata. In: Intercellular Communication in Plants: Studies on Plasmodesmata, pp. 81-105, Gunning BES, Robards AW, eds. Springer-Verlag, Berlin/Heidelberg

Kauss H (1987) Some aspects of calcium-dependent regulation in plant metabolism. Ann. Rev. Plant. Physiol. 38: 47-72

López-Sáez JF, Gimenez-Martin G, Risueno MC (1966) Fine structure of the plasmodesm. Protoplasma 61: 81-84

MacRobbie E (1989) Calcium influx at the plasmalemma of isolated guard cells of *Commelina communis*. Effects of abscisic acid. Planta 178: 231-241

Meiners S, Baron-Epel O, Schindler M (1988) Intercellular communication - filling the gaps. Plant Physiol. 88: 791-793

Meiners S, Schindler M (1987) Immunological evidence for gap junction polypeptide in plant cells. J. Biol. Chem. 262: 951- 953
Mollenhauer H, Morré J (1987) Some unusual staining properties of tannic acid in plants. Histochemistry 88: 17-22
Northcote DH (1989) Use of antisera to localize callose, xylan and arabinogalactan in the cell-plate, primary and secondary walls of plant cells. Planta 178: 353-366
Olesen P (1975) Plasmodesmata between mesophyll and bundle sheath cells in relation to the exchange of C4-acids. Planta 123: 199-202
Olesen P (1979) The neck constriction in plasmodesmata evidence for a peripheral sphincter-like structure revealed by fixation with tannic-acid. Planta 144: 349-358
Olesen P (1980) A model of a possible sphincter associated with plasmodesmatal neck regions. Europ. J. Cell Biol. 22: 250
Olesen P (1986). Interactions between cell wall and plasmodesmata: model of a possible sphincter mechanism. In: Cell Walls '86, Vian B, ed., Paris
Overall RL, Gunning BES (1982) Intercellular communication in *Azolla* roots: II. Electrical coupling. Protoplasma 111: 151-160
Overall RL, Wolfe J, Gunning BES (1982) Intercellular communication in *Azolla* roots. I. Ultrastructure of plasmodesmata. Protoplasma 111: 134-150
Pennell RI, Knox JP, Scofield GN, Selvendran RR, Roberts K (1989) A family of abundant plasma-membrane-associated glycoproteins related to the arabinogalactan proteins is unique to flowering plants. J. Cell Biol. 108: 1967-1977
Pooviah BW, Reddy ASN, McFadden JJ (1987) Calcium messenger system: role of protein phosphorylation and inositol biphospholipids. Physiol. Plantarum 69: 569-573
Robards AW (1968) A new interpretation of plasmodesmatal ultrastructure. Planta 82: 200-210
Robards AW (1971) The ultrastructure of plasmodesmata. Protoplasma 72: 315-323
Robards AW (1976) Plasmodesmata in higher plants. In: Intercellular Communications in Plants: Studies on Plasmodesmata, pp. 15-57, Gunning BES, Robards AW, eds.Springer-Verlag, Heidelberg
Robards AW, Clarkson DT (1984) Effects of chilling temperatures on root cell membranes as viewed by freeze-fracture electron microscopy. Protoplasma 122: 75-85
Smereka KJ, Kausch AP, MacHardy WE (1988) Intracellular junctional structures in germinating ascospores of *Venturia inaequata*. Protoplasma 142: 1-4
Smith MM, McCully ME (1977) Mild temperature "stress" and callose synthesis. Planta 136: 65-70
Stephenson JLM, Hawes CR (1986) Stereology and stereometry of endoplsamic reticulum during differentiation in the maize root cap. Protoplasma 131: 32-46
Tangl E (1879) Ueber offene Communicationen zwischen den Zellen des Endosperms einiger Samen. Jb. wiss Bot. 12: 170-190
Terry BR, Robards AW (1987) Hydrodynamic radius alone governs the mobility of molecules through plasmodesmata. Planta 171: 145-157
Thomson WW, Platt-Aloia K (1985) The ultrastructure of the plasmodesmata of the salt glands of *Tamarix-aphylla* as revealed by transmission electron microscopy and freeze-fracture electron microscopy. Protoplasma 125: 13-23
Tucker EB (1987) Cytoplasmic streaming does not drive intercellular passage in staminal hairs of *Setcreasea-purpurea*. Protoplasma 137: 140-144
Tucker EB (1988) Inositol biphosphate and inositol triphosphate inhibit cell-to-cell passage of carboxyfluorescein in staminal hairs of *Setcreasea purpurea*. Planta 174: 358-363
Unwin N (1986) Is there a common design for cell membrane channels?. Nature 323: 32-33
Wagenknecht T, Grassucci R, Frank J, Saito A, Inui M, Fleischer S (1989) Three-dimensional architecture of the calcium channel/foot structure of sarcoplasmic reticulum. Nature 338:167-170
Willison JHM (1976) Plasmodesmata: a freeze-fracture view. Can. J. Bot. 54: 2842-2847
Zee S-Y (1969) The fine structure of differentiating sieve elements of *Vicia faba*. Aust. J. Bot. 17: 441-456

IMMUNOLOGICAL INVESTIGATIONS OF RELATEDNESS BETWEEN PLANT AND ANIMAL CONNEXINS

A. Xu, S. Meiners, and M. Schindler
Department of Biochemistry
Michigan State University
East Lansing
Michigan 48824
United States of America

INTRODUCTION

Contact and communication between specific cell types are two fundamental properties that shape and control the organization, activity, and development of tissues (Loewenstein, 1979; Ekblom *et al.*, 1986; Carr, 1976; Warner *et al.*, 1984; Sheridan and Atkinson, 1985). Evolution from single to multicellular organisms necessitated the emergence of unique structures to physically link the intracellular environments of individual cells. Such cytoplasmic linkage served to unify tissue response to biochemical triggers and signals and facilitate the coordination of tissue proliferation and metabolism.

In animal tissues, the gap junction has been demonstrated to serve as the transmembrane channel between contacting cells (Loewenstein, 1979; Warner *et al.*, 1984; Unwin and Zampighi, 1980). Low molecular weight metabolic and signaling molecules may diffuse from cell to cell through the transmembrane aqueous channel of the gap junction. Structurally, the gap junction is composed of membrane localized macromolecular structures (connexons) that connect contacting cells. Each connexon is comprised of six identical polypeptide chains (connexins) that form a hexagonally symmetric transbilayer structure (Unwin and Zampighi, 1980).

The distance between plasma membranes imposed by the head-to-head association of connexons is 2 nm. It is, therefore, evident that the gap junction is too short in length to be suitable for cytoplasm-to-cytoplasm connections between

NATO ASI Series, Vol. H 46
Parallels in Cell to Cell Junctions in Plants and Animals
Edited by A. W. Robards et al.

contacting plant cells sharing a combined cell wall thickness of ca. 100-300 nm. The best evidence to date supports the view that, in plants, complex tubular structures through the cell wall, termed plasmodesmata, serve as the functional analogues of gap junctions (Carr, 1976; Robards, 1976). These structures have been intensely studied at the level of transmission electron microscopy and an assortment of models have been presented to explain both their structure and biological functions (Robards, 1976; Olesen, this volume).

Unfortunately, as of yet, plasmodesmata have not been isolated. Furthermore, no set of cellular proteins has been unambiguously demonstrated to be a component of the plasmodesmata. Although this has precluded a direct comparison between plasmodesmata and gap junctions, much existing functional data suggests that the transport and control parameters of these two morphologically distinct structures are remarkably similar (Meiners *et al.*, 1988). In the spirit of the architectural principle that similar structures evolve under the selection pressure of similar problems, our laboratory has undertaken an experimental strategy that was originally suggested by Gunning and Robards (1976): "It should be possible to prepare cell walls, and to isolate plasmodesmata from them ... and using techniques of biochemical fractionation such as in comparable work on isolated animal gap junctions, which has revealed the presence of large amounts of a special protein ... ."

In this belief that analogous mechanisms may use homologous components, albeit in different patterns of organization, we initiated experiments to search for a gap junction-type protein (connexin) in extracts of homogenized cultured plant cells and whole plant organs; e.g., petals, fruits, leaves, and roots (Meiners and Schindler, 1987). The tools for this approach were immunological and, initially, consisted of monospecific antibodies prepared against rat liver connexin and V79 Chinese lung fibroblast connexin. Subsequently, antibody was also prepared against a 29 kD polypeptide from soybean root cells (SB-1 cell line) that demonstrated reactivity with the monospecific antibody raised against the 27 kD connexin of rat liver (Meiners and Schindler, 1987). These reagents demonstrated 1) immunological relatedness between rat liver connexin and a 29 kD polypeptide in soybean root cell, operationally termed soybean connexin; 2) a peripherally localized punctate organization of soybean connexin at contacting regions between soybean root cells (Meiners and Schindler, 1989); and 3) presence of rat liver and soybean connexin cross-reactive polypeptides in extracts from petals, fruits, and leaves from a variety of plants (Meiners and Schindler, 1989).

In a further exploration of the immunological relatedness between animal and a putative plant connexin, we employed an antibody prepared against a peptide loop region of rat liver connexin (amino acid sequence 98-124). Utilizing this immunological probe, putative plant connexin was also observed in *Arabadopsis thaliana*.

IMMUNOLOGICAL RELATEDNESS AS DETERMINED BY A VARIETY OF ANTIBODY PROBES

Reactivity of Antibody to the 27 kD and 29 kD Polypeptides

To further examine the structural relatedness of rat liver and soybean connexin, our laboratory has pursued the use of a series of immunological reagents. Figure 1 demonstrates an immunoblot in which fractions enriched for soybean (lane A) and rat liver connexin (lane B) are probed with monospecific antibody prepared against the 27 kD polypeptide, while lanes C and D are the same fractions, respectively, probed with monospecific antibody prepared against the 29 kD polypeptide (molecular masses for the peptides refer to their migration on reduced SDS-PAGE). Preimmune serum controls for both immunological probes demonstrated no labeling. The use of these antibodies also demonstrated the existence of cross-reactive material in a wide variety of extracts from petals, leaves, and fruits (Figs. 2 and 3).

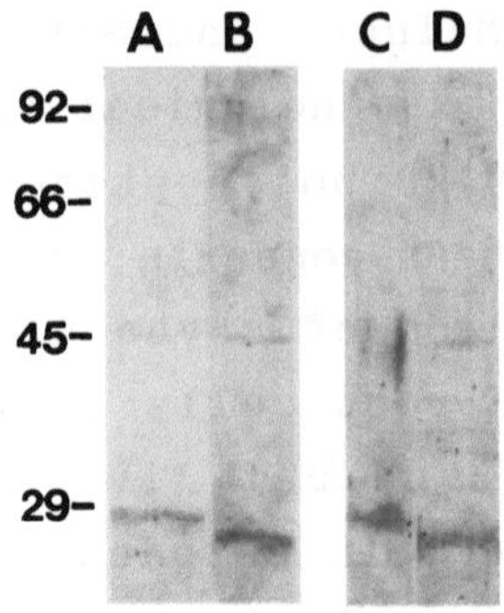

Fig. 1. Immunological characterization of monospecific antibodies prepared from antisera raised against the 27 kD polypeptide (rat liver connexin) and the 29 kD polypeptide (soybean connexin). The antisera were made monospecific against the corresponding antigens by the procedure of Smith and Fisher (1984). Samples were resolved on SDS-PAGE gels (acrylamide composition 10%) and polypeptides transferred to nitrocellulose paper (400 mA, 90 min). Samples were diluted here and elsewhere 1:1 with β-mercaptoethanol sample buffer and maintained for 30 min at room temperature. Samples were not boiled in SDS in an attempt to reduce the observed tendency for self-aggregation (Meiners and Schindler, 1987). Antibody binding was detected here and elsewhere using alkaline phosphatase conjugated goat anti-rabbit IgG (Boehringer Mannheim Biochemicals, FRG) according to the manufacturer's instructions. Lanes A and B, fractions enriched in soybean connexin (A) and rat liver connexin (B), respectively, immunoblotted with monospecific antibody prepared against 27 kD polypeptide from rat liver; lanes C and D, fractions enriched in soybean connexin (C) and rat liver connexin (D), respectively, immunoblotted with monospecific antibody prepared against 29 kD polypeptide from soybean cells. The molecular weight standards were: phosphorylase b, 92,000; bovine serum albumin, 66,000; ovalbumin, 45,000; and carbonic anhydrase, 29,000. Approximately 20 μg of protein was loaded in each lane of the gel and 30 μg/ml of antibody was employed

In an attempt to prepare immunological probes to more defined epitopes, an antibody was prepared against a hydrophilic loop peptide of rat liver connexin (sequence 98-124) (Kumar and Gilula, 1986; Paul, 1986). Peptide was synthesized on an Applied Biosystems Inc. peptide synthesizer. The resultant peptide mixture was chromatographed on a Sephadex G-25 column utilizing as eluant, 25 mM NH_4HCO_3, titrated to pH ~ 3.0 (the peptide is not soluble above pH 3.0). The first

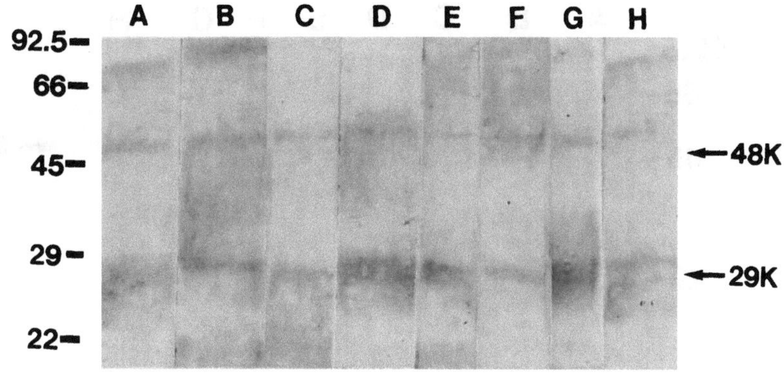

Fig. 2. Immunological analysis of antibody binding to polypeptides from petal extracts resolved on SDS-PAGE gels (acrylamide composition 12.5%) and transferred to nitrocellulose. Lane A, fraction enriched in rat liver connexin; lane B, fraction enriched in soybean connexin; lane C, iris extract; lane D, daisy extract; lane E, rose extract; lane F, lilly extract; lane G, petunia extract; and lane H, African violet extract. Rat liver and soybean connexin fractions in lanes A and B, respectively, were prepared as described previously (Meiners and Schindler, 1987). Petal extracts in lanes C through H were prepared in the following manner: Petals were obtained from a variety of plants and rapidly frozen at -70°C. The frozen material was homogenized with 4 volumes of 1 mM $NaHCO_3$ in a Bellco tissue homogenizer. The homogenate was suspended 1:1 (v/v) in SDS β-mercaptoethanol sample buffer (Laemmli, 1970). The mixture was permitted to remain at 25°C for 30 min and then centrifuged for 1 min in an Eppendorf microfuge at 15,000 x g. The pellet was discarded and the supernatant was subjected to SDS-PAGE, the polypeptides transferred to nitrocellulose and immunoblotted with antibody to the 27 kDa polypeptide from rat liver (30 μg/ml). The molecular weight standards were: phosphorylase b, 92,500; bovine serum albumin, 66,000; ovalbumin, 45,000; carbonic anhydrase, 29,000; and soybean trypsin inhibitor, 22,000. Approximately 10 μg of protein was loaded in lanes A and B of the gel, and approximately 200 μg was loaded in lanes C through H

peak obtained as measured by absorption at 280 nm (~ 22% of total initial load by weight) was utilized for preparing antigen and an immunoaffinity column. Rabbits were immunized with peptide coupled to rabbit serum albumin prepared in the following manner. Peptide was added to rabbit serum albumin (RSA) (Sigma, St. Louis, MO) dissolved in water at a weight

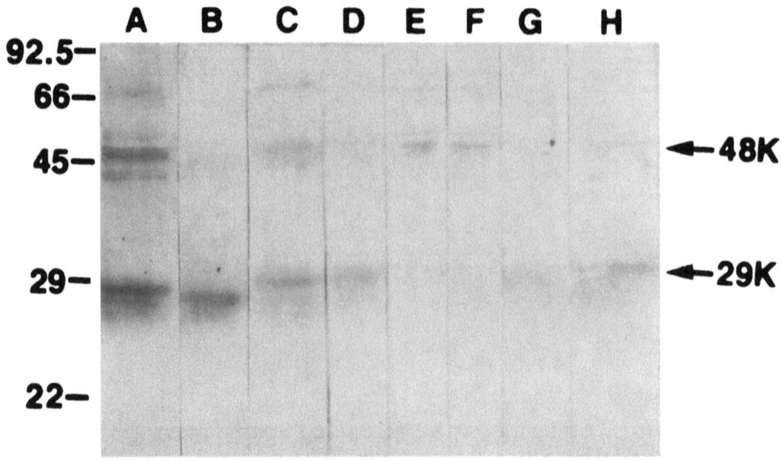

Fig. 3. Immunological analysis of antibody binding to polypeptides from fruit, leaf and vegetative samples resolved by SDS-PAGE and transferred to nitrocellulose. Lane A, fraction enriched in rat liver connexin; lane B, lettuce leaf extract; lane C, tomato extract; lane D, cucumber extract; lane E, green bell pepper extract; lane F, red bell pepper extract; lane G, avocado extract; and lane H, artichoke leaf extract. Extracts were prepared as described in Figure 2 for petal extracts. The molecular weight standards were as described in Figure 2. Approximately 30 µg of protein was loaded in lane A of the gel, and approximately 200 µg was loaded in lanes B through H

ratio of 1:1. Dry 1-ethyl-3(3-dimethyl aminopropyl) carbodiimide hydrochloride (Sigma, St. Louis, MO) was then added in excess (15 x the weight of peptide). The reaction mixture was maintained at room temperature overnight and then dialyzed against phosphate buffered saline (PBS), pH 7.4, for 24 hr with three changes. The resulting peptide/RSA complex was emulsified with an equal volume of complete Freund's adjuvant. Initially, 3 mg of peptide/RSA complex was injected into the toe pads of the rabbit. Booster injections utilized 1 mg of the complex emulsified with an equal volume of incomplete Freund's adjuvant. These injections were started 3 weeks after the primary immunization, followed thereafter every 10 days with an injection of 0.5 mg of the complex. High antibody titers were achieved in 10-15 weeks. Immune sera thus obtained were purified by absorption to a peptide affinity column (Fig. 4).

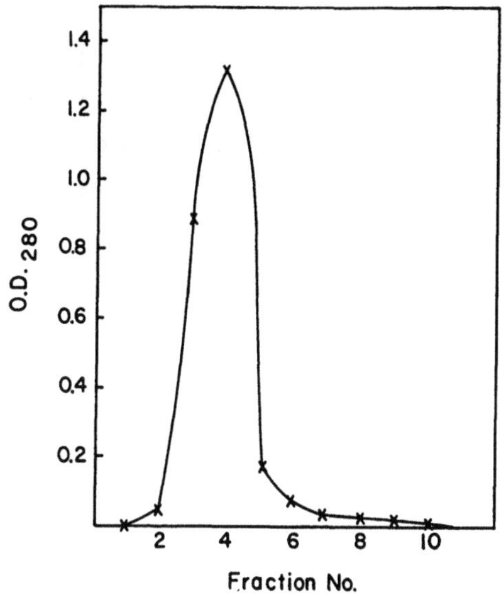

Fig. 4. Affinity purification of antibody raised against a loop peptide prepared against rat liver connexin peptide (amino acid sequence 98-124). Immune sera (10 ml) prepared against the peptide/RSA complex was applied to the column. The column (2-2.5 ml) was washed with 200 ml 0.1 M sodium phosphate buffer, pH 7.0, until the $O.D._{280}$ of the eluant was < 0.01. Bound antibody was then eluted with citric acid buffer, pH 3.0, and the column was then re-equilibrated with phosphate buffer. The column itself was prepared in the following manner. Aldehyde activated agarose (Calbiochem, San Diego, CA) was washed in 10 volumes of coupling buffer (0.1 M NaAc buffer, pH 3.0). The peptide and agarose were dissolved in equal amounts of coupling buffer and combined. Sodium cyanoborohydride ($NaCNBH_3$) was added to the peptide/agarose mixture to a final concentration of 0.1 M and the suspension agitated for 2 hr at room temperature. Following incubation, the gel was washed with 20 volumes of 1 M NaCl. Blocking of unreacted aldehyde groups was achieved by agitating coupled agarose gel in 0.1 M $NaCNBH_3$, pH 3.0, for 2 hr at room temperature in the presence of 0.1 M ethanolamine. The agarose was again washed with 1 M NaCl, followed by 0.1 M $NaPO_4$, pH 7.0, containing 0.01% NaN_3. The coupled agarose was stored at 4°C

Immunoblots of cell fractions enriched in connexin prepared by the Hertzberg procedure (Hertzberg, 1984) from both rat liver and soybean and probed with antiserum prepared against the rat liver connexin loop peptide, demonstrated a similar pattern of labeling in comparison with the labeling of monospecific antibodies prepared against rat liver 27 kD polypeptide and soybean 29 kD polypeptide (Fig. 5). When these fractions were probed with an antibody prepared against an amino acid sequence specific to heart connexin (heart sequence 252-271), no labeling was observed for rat liver or soybean material.

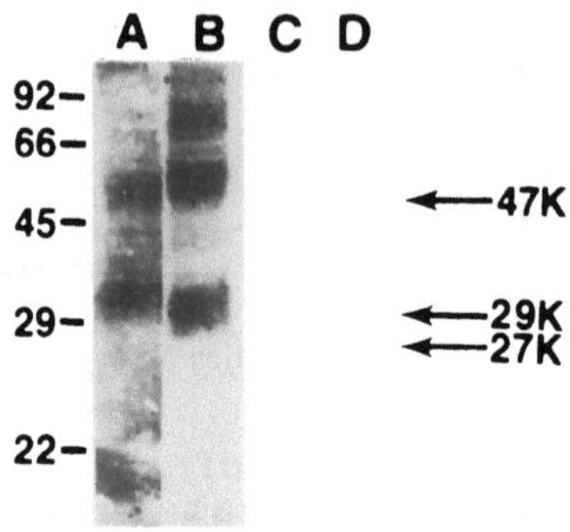

Fig. 5. Immunological analysis of anti-peptide (rat liver connexin sequence 98-124) antiserum binding to polypeptides resolved on SDS-PAGE gels (acrylamide composition 12.5%) and transferred to nitrocellulose. Lanes A and B, fraction enriched in soybean connexin (A) and fraction enriched in rat liver connexin (B) immunoblotted with anti-peptide antiserum; lanes C and D, fraction enriched in soybean (C) and fraction enriched in rat liver (D) immunoblotted with anti-peptide preimmune serum. The molecular weight standards were as described in Figures 2 and 3. Approximately 15 μg of protein was loaded in each lane of the gel

Arabidopsis thaliana has increasingly become the plant material of choice for a variety of molecular biological manipulations. Accordingly, we examined extracts from *Arabidopsis thaliana* leaves kindly supplied by Dr. Chris Somerville, Department of Botany and Plant Pathology, Michigan State University. Immunoblots of extracts of *Arabidopsis thaliana* leaves probed with affinity purified antibodies to the loop peptide of rat liver connexin demonstrated the characteristic 27-29 kD and 48 kD bands observed for soybean and rat liver connexin, respectively (Fig. 6).

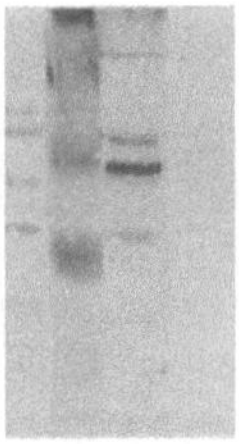

A B C D

Fig. 6. Immunological reactivity of antibody prepared against loop peptide (rat liver connexin sequence 98-124) to extracts of *Arabidopsis thaliana* leaves. *Arabidopsis* leaves (5 g) were homogenized in PBS at 4°C. The homogenate was centrifuged at 14,000 rpm for 10 min and the supernatant electrophoresed on SDS-PAGE gels (acrylamide composition 12.5%). Polypeptides were transferred to nitrocellulose and probed with affinity purified antibody to the rat liver peptide. Lane A contains standards as described in Figures 2 and 3; Lane B, fraction enriched in rat liver connexin; Lane C, *Arabidopsis* PBS extract; Lane D, *Arabidopsis* PBS extract probed with preimmune serum. (Approximately 40 µg of total protein was loaded in lanes B, C, and D of the gel.)

Immunofluorescent Localization Studies

Soybean (*Glycine max* (L.) Merr. cv. Mandarin) root cells (SB-1 cell line) were grown in suspension culture and labeled as previously described (Meiners and Schindler, 1987; Meiners and Schindler, 1989). An important element of the labeling process was that cells had to be mildly treated with pectinase to permit antibody accessibility into and across the cell wall region (Baron-Epel *et al.*, 1988); failure to treat cells resulted in no labeling. Affinity purified antibodies prepared against the 27 kD and 29 kD polypeptides of rat liver and soybean, respectively, were utilized on both soybean cells and mouse liver thin sections. Antibody prepared against both the 27 kD (rat liver) and 29 kD (soybean) polypeptides demonstrated the same pattern of reactivity when utilized on both mouse liver and soybean cells. As shown in Figure 7A,B, fluorescent spots and patches were observed at the cellular periphery between regions in which soybean cells made contact.

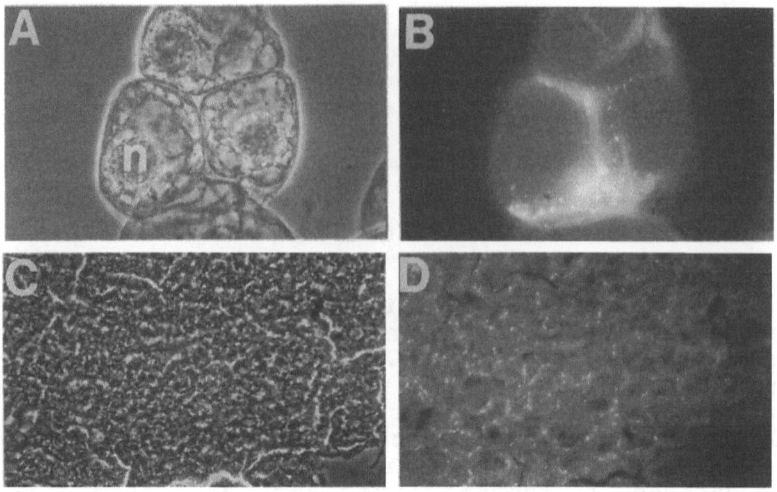

Fig. 7. Immunofluorescence localization of soybean connexin in soybean root cells and rat liver-type connexin in mouse liver sections. Soybean cells were suspended in 1B5C medium containing 0.1 mg/ml pectinase (EC 3.27.15; Sigma, St. Louis, MO) and were digested for 30 min at 25°C. Following pectinase treatment, the SB-1 cells were washed by pelleting and resuspension in fresh 1B5C medium. They were then fixed in 3.7% formaldehyde in 1B5C medium for 30 min, washed in Tris buffered saline (TBS: 20 mM Tris HCl, pH 7.5, 0.5 M NaCl), and incubated for 1 hr in antibody (60 µg/ml) to the 27 kD polypeptide in TBS containing 3% BSA (w/v) (BSA: bovine serum albumin). The cells were washed in TBS/3% BSA and incubated for 1 hr in a 1:30 dilution of rhodamine-conjugated goat anti-rabbit IgG (Boehringer Mannheim) in TBS/3% BSA. The cells were washed again in TBS and mounted on slides in 70% glycerol-phosphate buffered saline (PBS: 0.14 M NaCl, 2.7 mM KCl, 1.5 mM KH_2PO_4, 4.3 mM $NaHPO_4$) containing 5% of the anti-bleaching agent n-propyl gallate (Sigma). Soybean cells were viewed with a Leitz epifluorescence microscope using a 25X objective lens. n represents the nucleus. Soybean cell images are A, phase, and B, fluorescence. Mouse liver sections were prepared as follows: Freshly excised mouse livers were cut into 4 pieces and immersed in -70°C petroleum ether for 2 min. The liver pieces were mounted onto the chuck of a Miles cryostat using "Tissue Tek," an embedding medium for frozen tissue specimens (Miles Laboratories). The liver was sectioned in the -20°C cryostat to a thickness of 8 µm. Liver sections were placed onto coverslips and immersed in -20°C acetone for 10 min. The coverslips were washed with TBS and incubated in primary and secondary antibodies as described for SB-1 cells. The coverslips were mounted on slides in 70% glycerol-PBS containing 5% n-propyl gallate and were viewed with a Leitz epiphase fluorescence microscope using a 40X objective lens. Mouse section images (C,D) are C, phase, and D, fluorescence

Punctate fluorescent patches were observed predominantly at areas of contact between soybean cells and not at non-contacting cell surfaces or within the cytoplasm. This is suggested by the absence of fluorescence in the cytoplasmic volume subtended by the nucleus (Fig. 7A, nucleus (n)) (Fig. 7B). Variations in fluorescent patch size were observed and may be related to difference in the amount of antigen at each site. Such variability has also been observed with respect to the size of pit-fields and the number of plasmodesmata per pit-field (Evert *et al.*, 1977).

In the mouse liver thin sections (Fig. 7C,D), the labeling pattern observed was consistent with previous studies in which punctate cellular labeling was observed at areas of cell-cell contact (Dermietzel *et al.*, 1984; Dermietzel *et al.*, 1987). Reaction of pectinase permeabilized SB-1 cells with preimmune serum prepared for either 27 kD or 29 kD polypeptides (Fig. 8A,B) showed no labeling. Absence of labeling was also observed for the same preimmune sera on mouse liver sections (Fig. 8C,D).

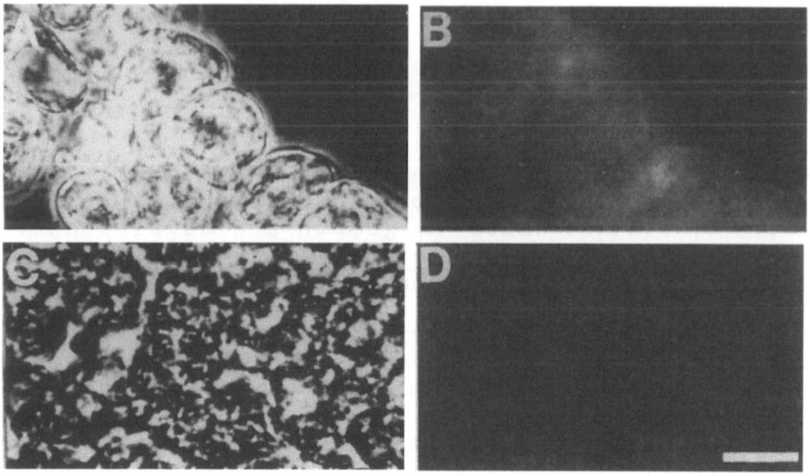

Fig. 8. Immunofluorescence in soybean cells and mouse tissue sections probed with preimmune IgG. Soybean cells (A,B) and mouse liver thin sections (C,D) were prepared for immunofluorescence and examined as described in Figure 7. Preimmune IgG was substituted for antibody to the 27 kD polypeptide

CONCLUSIONS

Our recent immunological studies have demonstrated that a rat liver-type connexin polypeptide may be present in a wide variety of plants. Immunolocalization studies suggest that this polypeptide is localized to the cell periphery and demonstrates a distribution that is similar to the pattern of wall organization observed for pit-fields containing plasmodesmata. Future efforts will utilize an assortment of immunological reagents in precise immuno-gold localization studies to examine the precise intracellular localization.

ACKNOWLEDGMENTS

We wish to thank Dr. Siu-Cheong Ho, Department of Biochemistry, Michigan State University, for his excellent technical advice and assistance. This work was supported by a grant from the U.S.-Israel Binational Agricultural Research and Development Fund (BARD Project No. US-1384-87).

REFERENCES

Baron-Epel O, Garyal PK, Schindler M (1988) Pectins as mediators of wall porosity in soybean cells. Planta 175:389-395

Carr DJ (1976) Plasmodesmata in growth and development. In: Intercellular Communication in Plants: Studies on Plasmodesmata. BES Gunning, AW Robards (eds). Springer-Verlag, Berlin 243-289

Dermietzel RA, Leibstein A, Frixen U, Janssen-Timmen U, Traub O, Willecke K (1984) Gap junctions in several tissues share antigenic determinants with liver gap junctions. EMBO J 3:2261-2270

Dermietzel RB, Yancey B, Janssen-Timmen U, Traub O, Willecke K, Revel JP (1987) Simultaneous light and electron microscopic observation of immunolabeled liver 27 kD gap junction protein on ultra-thin cryosections. J of Histochem and Cytochem 35:387-392

Ekblom P, Vestweber D, Kemler R (1986) Cell-Matrix interactions and cell adhesion during development. Ann Rev Cell Biol 2:27-47

Evert RF, Eschrich W, Heyser W (1977) Distribution and structure of the plasmodesmata in mesophyll and bundle-sheath cells of *Zea mays* L. Planta 136:77-89
Gunning BES, Robards AW (1976) Plasmodesmata: Current knowledge and outstanding problems. In: Intercellular Communication in Plants: Studies on Plasmodesmata. BES Gunning, AW Robards (eds). Springer-Verlag, Berlin 306
Hertzberg EL (1984) A detergent-dependent procedure for the isolation of gap junctions from rat liver. J Biol Chem 259:9936-9943
Kumar NM, Gilula NB (1986) Cloning and characterization of human and rat liver cDNAs coding for a gap junction protein. J Cell Biol 103:767-776
Laemmli UK (1970) Cleavage of structural proteins during the assembly of the head of bacteriophage T4. Nature 227:680-685
Lowenstein WR (1979) Junctional intercellular communication and the control of growth. Biochim Biophys Acta 560:1-65
Meiners S, Baron-Epel O, Schindler M (1988) Intercellular communication - Filling in the gaps. Plant Physiol 88:791-793
Meiners S, Schindler M (1987) Immunological evidence for gap junction polypeptide in plant cells. J Biol Chem 262:951-953
Meiners S, Schindler M (1989) Characterization of a connexin homologue in cultured soybean cells and diverse plant organs. Planta, in press
Olesen, P (1989) The neck region of plasmodesmata: general architecture and some functional aspects. In: Parallels in Cell to Cell Communication in Plants and Animals. AW Robards, HJ Jongsma, WJ Lucas, JD Pitts, DC Spray (eds). Springer-Verlag, Berlin XX-YY
Paul DL (1986) Molecular cloning of cDNA for rat liver gap junction protein. J Cell Biol 103:123-134
Robards AW (1976) Plasmodesmata in higher plants. In: Intercellular Communication in Plants: Studies on Plasmodesmata. BES Gunning, AW Robards (eds). Springer-Verlag, Berlin 15-57
Sheridan JD, Atkinson MM (1985) Physiological roles of permeable junctions: some possibilities. Annu Rev Physiol 47:337-353
Smith DE, Fisher PA (1984) Identification, developmental regulation and response to heat shock of two antigenically related forms of a major nuclear envelope protein in *Drosophila* embryos: Application of an improved method for affinity purification of antibodies using polypeptides immobilized on nitrocellulose blots. J Cell Biol 99:20-28
Unwin PNT, Zamphighi G (1980) Structure of the junction between communicating cells. Nature 283:545-549
Warner AE, Guthrie SC, Gilula NB (1984) Antibodies to gap-junctional protein selectively disrupt junctional communication in the early amphibian embryo. Nature 311:127-131

SECONDARY FORMATION OF PLASMODESMATA IN CULTURED CELLS - STRUCTURAL AND FUNCTIONAL ASPECTS

J Monzer*
Botanisches Institut
Universität Kiel
Olshausenstr 40
2300 Kiel 1
Federal Republic of Germany

Introduction

Almost immediately after the first description of protoplasmic connections between plant cells (Tangl 1879) the question arose, whether these connections are of cytokinetic origin or if they also could be a result of secondary formation. Strasburger (1901) first described secondary plasmodesmata between partners of a graft. Since then, they have been reported in chimeras (Buder 1911, Hume, 1913, Binding *et al.* 1987), between parasitic higher plants and their hosts (Dörr 1968, 1969, Tainter 1971, Dell *et al.* 1982), as well as in grafts (Funck 1929, Jeffree and Yeoman 1983, Kollmann *et al.* 1985), including the final ultrastructural proof by Kollmann and Glockmann (1985). On the other hand, secondary plasmodesmata have also been proposed to be present in the intact, regularly growing plant. Secondary cell contacts have been reported for the postgenital fusion of carpels (Boeke 1971) and between pollen mother cells (Cheng *et al.* 1987). Schnepf and Sych (1983), as well as Seagull (1983) observed an increase of plasmodesmata during cell elongation. Closely related to the problem of secondary cell contacts, half plasmodesmata have been described in several studies (Dörr 1968, Boeke 1971, Burgess 1972, Kollmann *et al.* 1985); presumably these structures are also of secondary origin (cf. Jones 1976, Kollmann and Glockmann 1989). As a major point of interest, the mechanism of secondary plasmodesmata formation has also been discussed over a long period (Strasburger 1901). The course of regular plasmodesmata formation during cell division has been demonstrated to be closely related to cell plate formation (Hepler 1982, cf. Jones 1976). However, for secondary plasmodesmata formation, the situation is less clear. Almost hypothetically, a mechanism of directed enzymic wall degradation has been postulated (Jones 1976, cf. Cheng *et al.* 1987).

---

* *Present address:*

Botanisches Institut
Technische Universität München
D-8050 Freising-Weihenstephan

NATO ASI Series, Vol. H 46
Parallels in Cell to Cell Junctions in Plants and Animals
Edited by A. W. Robards et al.

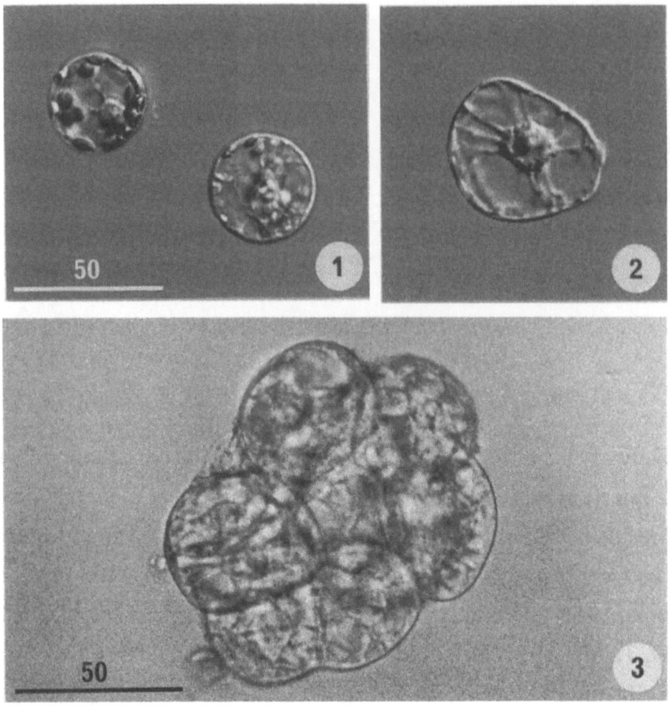

Labeling of bars = μm.
Figs. 1-3. Dedifferentiation and regeneration of cultured Solanum nigrum-protoplasts.
Fig. 1. Cells 2 days in culture. Differentiated cell (left) with large plastids and peripheral nucleus; dedifferentiated cell (right) with reduced plastid size and central nucleus.
Fig. 2. Dedifferentiated cell with central nucleus, numerous cytoplasmic strands and irregular shape.
Fig. 3. Microcallus, 8 days in culture.

In the present study, both ultrastructural results and indirect proof by fluorochrome microinjection are presented for the formation of secondary plasmodesmata during the early stages of regenerating *Solanum nigrum* protoplast cultures. These results allow new suggestions on at least one mechanism for secondary plasmodesmata formation.

Regeneration of Cultured Cells

The course of regeneration was followed in isolated *Solanum nigrum* protoplasts. Isolation and embedding in agarose gel droplets covered by liquid culture medium followed the usual procedures (see Binding and Mordhorst 1984, Binding *et al.* 1988). The protoplasts must dedifferentiate to regain division capability. This process can be characterized by plastid size

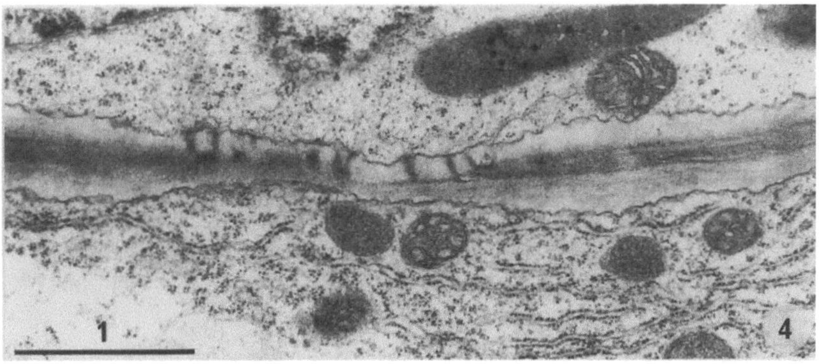

Fig. 4. Branched half plasmodesmata in callus, inserted in the cell wall only as far as a densely contrasted median line.

reduction (Fig. 1) and the central localization of the nucleus within the cell, numerous cytoplasmic strands maintaining the connection with the peripheral cytoplasm (Fig. 2). Following dedifferentiation, the first division occurs within 2-3 days of culture. The division wall is inserted in the cell form; the daughter cells are not substantially rounded from each other. Usually, less than 30% of the cells undergo dedifferentiation and first division. These cells form microcallus within 8-12 days of culture (Fig. 3). Callus formation and subsequent shoot regeneration on agar media occurs within 40-50 days.

Half Plasmodesmata in Regenerating Protoplast Cultures

Half plasmodesmata have been suggested to be structures of secondary origin (Jones 1976, Kollmann and Glockmann 1989). In the present work, half plasmodesmata and probable precursor structures could be identified and traced during the course of regeneration.

In callus material, half plasmodesmata were frequently observed structures. They were usually branched and appeared as multiple arrays of protoplasmic strands linked with each other in one plane within the wall (Fig. 4). In cells cultured 4-8 days, early stages of half plasmodesmata were observed. They appeared as arrays of tubular, membrane-surrounded protoplasmic strands of about 40 nm diameter, featuring a central membranous tubule and a connection with endoplasmic reticulum (Figs. 5a-c). Usually, they showed branches oriented in a plane parallel to the cell membrane and were closely aligned to the cell wall. The walls of regenerating cells were thin and of low contrast. Below the arrays, the plasma membrane seemed to be locally retracted, featuring deposition of contrasted, flocculent material between the protoplasmic structures outside the cell. Reconstructions of serial sections demonstrate the three-dimensional

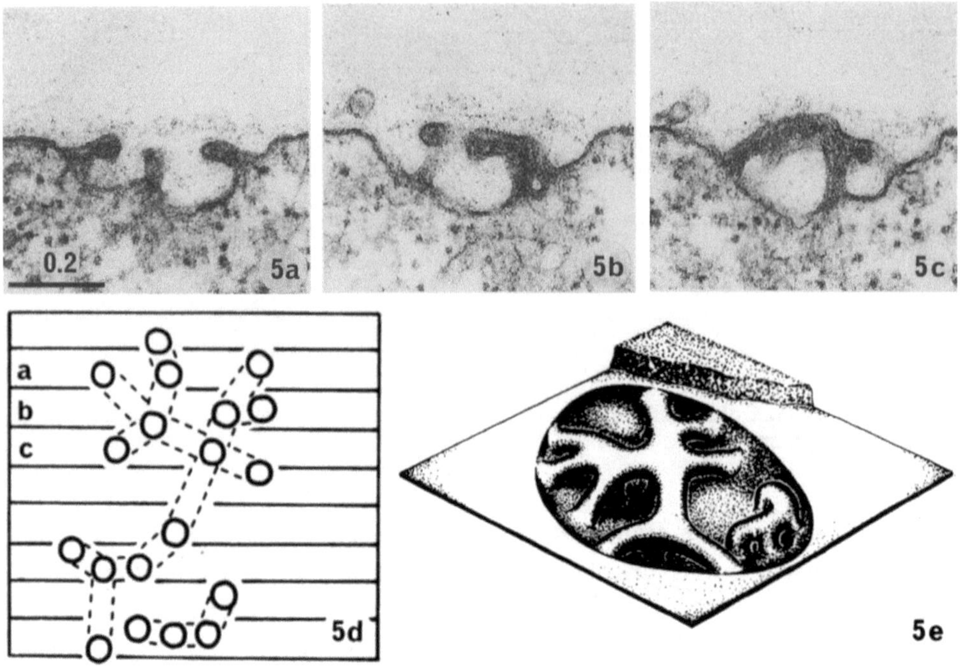

Fig. 5. Early stages of half plasmodesmata. Cells 7 days in culture.
Figs. 5a-c. Serial sections of half plasmodesmata array. Local retraction of plasma membrane; flocculent, contrasted material around protoplasmic structures. Loose cell wall with low contrast.
Fig. 5d. Reconstruction of above serial sections, presented sections indicated. Branched (broken lines) array of protoplasmic strands, insertion of protoplasmic strands in cytoplasm (circles).
Fig. 5e. Perspective image of reconstructed array; view of plasma membrane from outside the cell, wall removed except to a rest (top). Multiple, branched plasmatic strands.

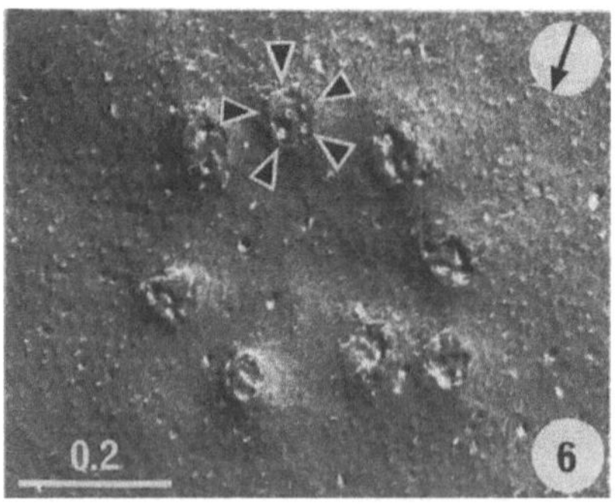

Fig. 6. Freeze-fracture of half plasmodesmata, cells 4 days in culture. Evaporation direction indicated in the upper right (arrow). Exoplasmic fracture face. Half plasmodesmata-array. 40 nm-structures with particular substructures (arrow-heads) in circular arrangement around a central particle.

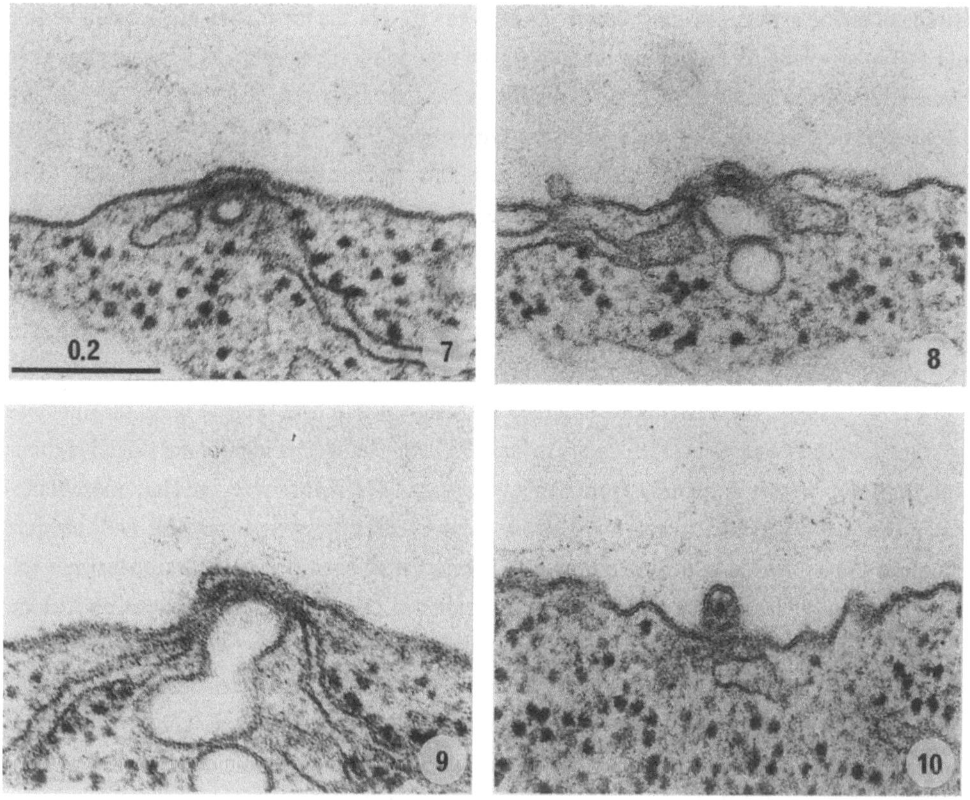

Figs. 7-10. Early stages of half plasmodesmata-precursors. Cells 2 days in culture.
Fig. 7. Constricted cisterna closely aligned with plasma membrane and a small vesicle. Longitudinal section along constriction.
Fig. 8. Vesicle aggregation below precursor.
Fig. 9. Vesicle fusion below precursor. Elongated protoplasmic strand.
Fig. 10. Cross-section of precursor excluded from cytoplasm. Surrounding membrane and central tubule.

arrangement of branched protoplasmic strands with a total diameter of about 0.5 μm (Figs. 5d,e). These precursors generally showed plasmodesmatal structure and dimensions (cf. Robards 1976). Half plasmodesmata precursors were also detected in freeze-fractured material. Arrays of cross-fractured structures, about 0.5 μm in diameter, were visible best in the exoplasmic fracture face of the cell membrane. They consisted of circular structures with about 40 nm diameter (Fig. 6). Several particles could be seen surrounding a central one. In general, they could be compared with cross-fractured plasmodesmata as described by Thomson and Platt-Aloia (1985).

The earliest precursors were identified in cells two days in culture. At least at this stage, they appeared to be restricted to non-differentiated cells only. They consisted of a tubular, constricted cisterna of endoplasmic reticulum, located in between the plasma membrane and a small vesicle below (Fig. 7; plane of sectioning parallel to the constriction). Similar, but larger structures showed elongation of the constricted cisternae as well as larger vesicles (Fig. 8). The vesicles appeared to be enlarged by means of fusion with others (Fig. 9). Cross-sections (Fig. 10) of these protoplasmic strands showed the tubular, membrane-surrounded structure. Again, these structures resembled plasmodesmata. Besides these small precursors, consisting of one protoplasmic strand, branched precursors were also present in this stage.

These results allow a new interpretation of half plasmodesmata formation (Fig. 11). Following dedifferentiation, the process may be initiated by an assembly of constricted, tubular cisternae of endoplasmic reticulum closely aligned to the plasma membrane together with small vesicles below (Figs. 11a,b). The vesicles are enlarged by fusion with others (Figs. 11c,e), they enclose the constricted cisternae (Fig. 11d) and fuse laterally with the plasma membrane (Fig. 11f), thus excluding the whole structures from the cytoplasm. The results are tubular, membrane-surrounded protoplasmic structures, linked on two sides to cytoplasm and endoplasmic reticulum (Figs. 11g,h). In a similar way, the formation of branched half plasmodesmata (cf. Fig. 5) may be explained by the exclusion of branched cisternae of endoplasmic reticulum. Following the initial formation of an half plasmodesmata precursor array in the plane of the cell membrane (Fig. 12a), the structure is overlayered with wall material, the connecting strands between branches and cytoplasm become increasingly elongated (Figs. 12b,c). The result resembles the usual view of branched half plasmodesmata.Dedifferentiated cells, presumed to have become associated during culture, were analyzed for secondary contacts by electron microscopy. Furthermore, microinjection experiments with the fluorescent dye Lucifer Yellow (cf. Steinbiss and Stabel 1983, Terry and Robards 1987) were made in order to find *in vivo* evidence for secondarily formed symplastic connections between regenerating protoplasts.

Fig. 11. Schematic drawing. Half plasmodesmata formation. Details of cross-sectioned cells (frames). Longitudinal sections (left row) of tubular cisternae of endoplasmic reticulum (dotted); perpendicular sections (right row; plane of sectioning indicated in Fig. 11a by broken line). Cell wall (irregularly shaded area; top), plasma membrane (irregular line and cytoplasm (pale). Movement directions indicated with arrows.

Fig. 11a. Alignment of constricted, tubular cisterna and vesicle below plasma membrane (cf. Fig. 7).

Fig. 11b. Same situation, cross-section.

Fig. 11c. Vesicle aggregation (cf. Fig. 8).

Fig. 11d. Enclosing of constricted cisterna between plasma membrane and vesicle.

Fig. 11e. Vesicle fusion below protoplasmic strand (cf. Fig. 9).

Fig. 11f. Exclusion of protoplasmic structure by membrane fusion of vesicle and plasma membrane. The protoplasmic structure is surrounded with membrane material.

Fig. 11 g. Longitudinal view of excluded protoplasmic strand.

Fig. 11h. Cross-section of same stage (cf. Fig. 10).

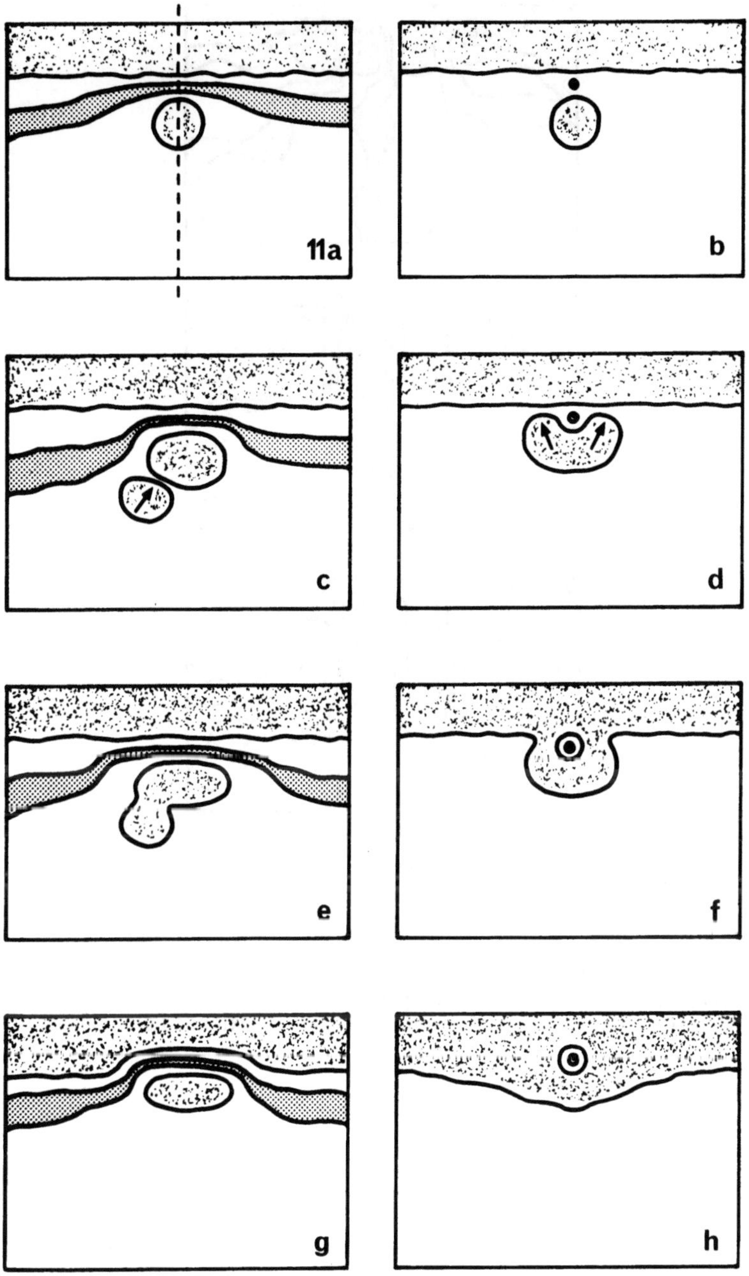
11a
b
c
d
e
f
g
h

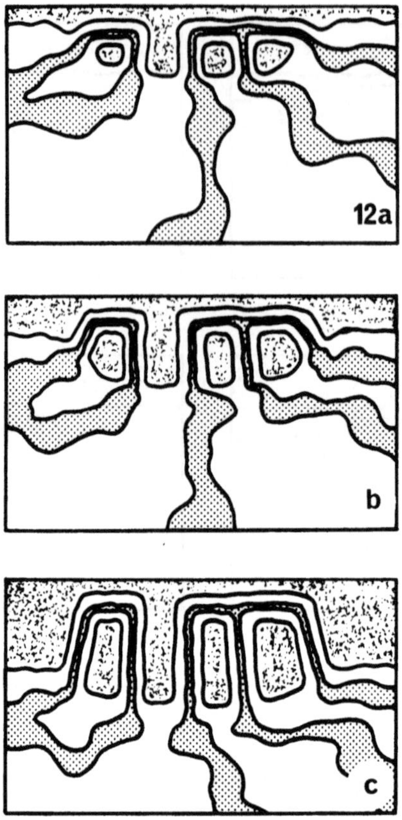

Fig. 12. Schematic drawing. Further development of half plasmodesmata-precursors. Details of cross sectioned cells (frames). Cell wall (irregularly shaded are; top); endoplasmic reticulum (dotted), cell membrane (irregular line).
Fig. 12a. Branched half plasmodesmata precursor, early stage (cf. Fig. 5).
Fig. 12b. Early wall regeneration. The plane of branches is overlayered with wall material, connecting strands to the cytoplasm are elongated.
Fig. 12c. Later stage. Increasing wall deposition leads to extended elongation of connecting strands (cf. Fig. 4)

Secondary Plasmodesmata in Protoplast Cultures

Cell contacts to be analyzed in serial ultrathin sections were selected from specimens about 2 days in culture according to the cytological criteria described for the course of dedifferentiation.

In contrast to divided cells, isolated cells located together during preparation exhibited a nearly spherical shape (Figs. 13, 14). In a small area, where the cells were flattened against each other, had contacts been established. These cells were interconnected by a characteristic type of

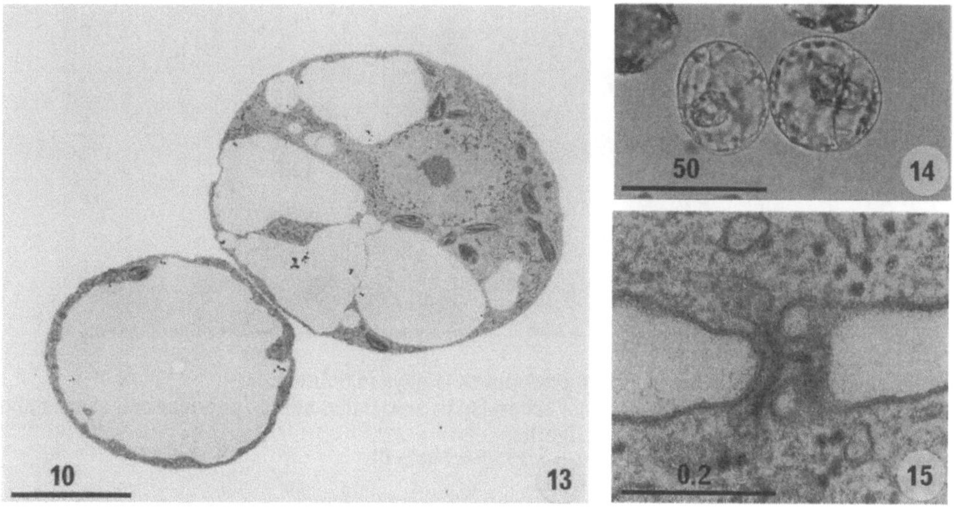

Figs. 13-15. Plasmodesmata between cells associated secondarily, 2 days in culture.
Fig. 13. Section of cells with spherical shape, flattened contact area in between. Nucleus of right cell not oriented to contact area.
Fig. 14. Light micrograph. Small contact area between cells. Nuclei not oriented towards contact area.
Fig. 15. Branched plasmodesma between associated cells. Serial section of Fig. 13.

branched plasmodesma, usually featuring a joint plane of branches parallel to the cell membranes (Fig. 15).

For microinjection experiments, regenerating cells had to be reliably attached to culture vessel surfaces. For this purpose, a commercially available adhesive preparation was used. "Cell-Tak" (BioPolymers, Farmington CT, USA), based on the byssus protein of the marine mussel *Mytilus edulis* (cf. Waite and Tanzer 1981) has so far been in use for the attachment of animal cells. Attached *Solanum* protoplasts (for technical details, see Monzer 1989) did show a regeneration behaviour similar to agarose embedded material. Cell divisions were observed from the second day of culture, half plasmodesmataprecursors could be seen as well as regeneration of callus and shoots (data not shown). Neighbouring cells (Fig. 16a) were selected according to the dedifferentiation criteria mentioned above. After injection of one cell, the fluorescence in some cases also appeared to arise in the adjacent cell within several seconds. It was then presumed that a symplastic contact had been made (Fig. 16b). Nevertheless, in the majority of injection experiments, fluorescence did not pass to neighbouring cells. Using the above precautions, 11 presumptive secondary contacts were detected from 261 injections.

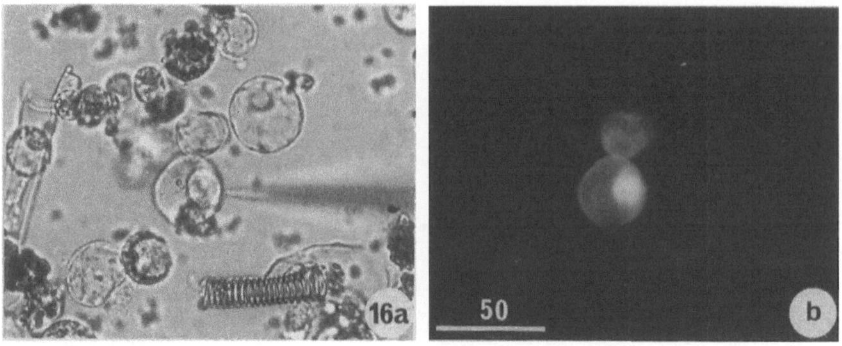

Fig. 16. Microinjection of regenerating protoplasts, 3 days in culture.
Fig. 16a. Brightfield image. Injection of secondarily associated cells; spherical cell shapes and small contact area. Injection capillary (right).
Fig. 16b. Fluorescence present in injected and adjacent cell.

Conclusions

In this study, a model for the secondary establishment of plasmodesmata has been presented. Tubular cisternae of endoplasmic reticulum are constricted to the size of a desmotubule. Golgi vesicles seem to be involved in enclosing and exclusion of protoplasmic strands containing the constricted cisternae. At this stage, striking coincidences with the regular formation of plasmodesmata become apparent. According to Hepler (1982), the cell plate is formed from masses of Golgi vesicles. In between the vesicles, cisternae of endoplasmic reticulum are constricted. These tubuli are surrounded by membrane material during vesicle fusion. This process may be understood in exactly the same way as the half plasmodesmata formation described here. The only difference is the fact that, during cell plate formation, the cytoplasm is divided and connecting strands are retained, whereas half plasmodesmata formation excludes groups of protoplasmic strands, internally linked by branches. The mechanism, indeed, may be the same.

This mechanism of half and continuous secondary plasmodesmata formation remains the central question. The hypothesis discussed so far has been based on the local enzymic digestion of an existing wall and the insertion of plasmodesmatal structures (cf. Jones 1976). In fact, that hypothesis features several outstanding disadvantages. So far direct proof for the enzymes needed for wall lysis has not been obtained (cf. Jones 1976, Cheng *et al.* 1987). Branching of plasmodesmata is not explained sufficiently. It would need repeated changes of direction during wall penetration. The formation of half plasmodesmata raises even more problems in explanation. To remain in accord with the penetration hypothesis, half plasmodesmata formation in fused walls, for example between partner cells in a graft, would also demand a recognition of "own" and "foreign" cell wall.

As a remarkable fact, the necessity to penetrate a wall is virtually absent from the cells analyzed in this study. Thus, secondary plasmodesmata described in this system might be explained by a completely different mechanism than by penetration of an existing wall. In contrast to the previous hypothesis, cell contacts could be established by fusion of pairs of half plasmodesmata between closely aligned, cells. Such a contact would exactly resemble the structure of the secondary plasmodesmata shown here (Fig. 15). During wall regeneration, the contact as a whole may again be overlayered with wall material, leading to a branched cell contact inserted in the wall, the common view of branched secondary plasmodesmata. The initial penetration of a cell wall here seems to be a marginal problem, if necessary at all. Indeed, a quite similar explanation has already been suggested by Strasburger (1901). There, he proposed an extreme thinning of the walls of outgrowing partner cells preceding the establishment of cellular contacts in a graft, followed by subsequent elongation of the secondary contacts during wall thickening.

The mechanism proposed here coincides with details from several other reports. Secondary plasmodesmata formation in general seems to be related to a nondifferentiated state of the cells. All cells involved in the establishment of secondary contacts within a graft are dedifferentiated, growing cells (cf. Stoddard and McCully 1979, Jeffree and Yeoman 1983, Kollmann *et al.* 1985). Secondary plasmodesmata between *Cuscuta* and its host frequently appear at the hyphal tip, the youngest part of growing searchinghyphae of the parasite (Dörr 1968). The apical meristem of a *Solanum tuberosum* (+) *Solanum nigrum* chimera (cf. Binding *et al.* 1987) features numerous secondary plasmodesmata (Monzer 1989). A similar situation is present in all published cases of secondary plasmodesmata formation within intact plants. Schnepf and Sych (1983) and also Seagull (1983) have described the secondary increase of plasmodesmata during elongation, i.e. during differentiation. Nematode-induced giant cells (Jones and Dropkin 1976, cf. Jones 1976) appear to be metabolic highly active, dedifferentiated syncytia.

Whether the model for secondary plasmodesmata formation presented here proves valid for other systems, is still to be examined. There are, in fact, many questions still to be answered. Secondary cell contacts have been characterized here as branched structures. On the other hand, secondary plasmodesmata are known as both single and branched structures (Kollmann and Glockmann 1985). Within intact plants, areas of probable secondary plasmodesmata formation do not show an increased incidence of branched plasmodesmata (Schnepf and Sych 1983, Seagull 1983). Thus, the question remains as to whether secondary plasmodesmata may also arise from other mechanisms.

Acknowledgements

The author wishes to thank R. Kollmann, H. Binding and Mrs. C. Glockmann for helpful discussion and suggestions on the manuscript. This work was supported by Deutsche Forschungsgemeinschaft (Ko 368/7-1).

References

Binding H, Mordhorst G (1984) Haploid *Solanum dulcamara* L.: Shoot culture and plant regeneration from isolated protoplasts. Plant Sci Lett 35: 77-79

Binding H, Witt D, Monzer J, Mordhorst G, Kollmann R (1987) Plant cell graft chimeras obtained by co-culture of isolated protoplasts. Protoplasma 141: 64-73

Binding H, Görschen E, Jörgensen J, Krumbiegel-Schroeren G, Ling HQ, Rudnick J, Sauer A, Zuba M, Mordhorst G (1988) Proto-plast culture in agarose media with particular emphasis to streaky culture lenses. Botanica Acta 101: 233-239

Boeke JH (1971) Location of the postgenital fusion in the gyno-ecium of *Capsella bursa-pastoris* L.. Med Acta Bot Neerl 20: 570-576

Buder J (1911) Studien an *Laburnum adami*. Z Vererbungsl 5: 209-284

Burgess J (1972) The occurrence of plasmodesmata-like struc-tures in a non-division wall. Protoplasma 74: 449-458

Cheng KC, Nie XW, Chen SW, Jian LC, Sun LH, Sun DL (1987) Studies on the secondary formation of plasmodesmata between the pollen mother cells of lily before cytomixis. Acta Biol Exper Sinica 20: 1-11

Dell B, Kuo J, Burbidge HH (1982) Anatomy of *Pilostyles hamil-tonii* (Rafflesiaceae) on stems of *Daviesia*. Austr J Bot 30: 1-9

Dörr I (1968) Plasmatische Verbindungen zwischen artfremden Zellen. Naturwissenschaften 55: 396

Dörr I (1969) Feinstruktur intrazellulär wachsender *Cuscuta*- Hyphen. Protoplasma 67: 123-137

Funck R (1929) Untersuchungen über heteroplastische Transplantationen an Solanaceen und Cactaceen. Beitr Biol Pflanzen 17: 404-468

Hepler PK (1982) Endoplasmic reticulum in the formation of the cell plate and plasmodesmata.Protoplasma 111: 121-133

Hume M (1913) On the presence of connecting threads in graft hybrids. New Phytol 12: 216-220

Jeffree CE, Yeoman MM (1983) Development of intercellular connections between opposing cells in a graft union.New Phytol 93: 491-509

Jones MGK (1976) The origin and development of plasmodesmata. In: Gunning BES, Robards AW (eds) Intercellular communication in plants: Studies on plasmodesmata. Springer, Berlin, 81-105

Jones MGK, Dropkin VH (1976) Scanning electron microscopy of nematode-induced giant transfer cells. Cytobios 15: 149-161

Kollmann R, Dörr I (1987) Parasitische Blütenpflanzen. Naturwissenschaften 74: 12-21

Kollmann R, Glockmann C (1985) Studies on graft unions I. Plasmodesmata between cells of plants belonging to different unrelated taxa. Protoplasma 124: 224-235

Kollmann R, Glockmann C (1989) Sieve elements in graft unions. In: Behnke HD, Sjolund RD (eds) Sieve elements - Comparative structure and development. Springer, Berlin, in press

Kollmann R, Yang S, Glockmann C (1985) Studies on graft unions II. Continuous and half plasmodesmata in different regions of the graft interface. Protoplasma 126: 19-29

Monzer J (1989) Strukturelle und funktionelle Aspekte der sekundären Entstehung von Plasmodesmen in Protoplasten-Regeneraten. Thesis, Kiel

Robards AW (1976) Plasmodesmata in higher plants. In: Gunning BES, Robards AW (eds) Intercellular Communication in Plants: Studies on Plasmodesmata. Springer, Berlin, 15-57
Schnepf E, Sych A (1983) Distribution of plasmodesmata in developing *Sphagnum* leaflets. Protoplasma 116: 51-56
Seagull RW (1983) Differences in the frequency and disposition of plasmodesmata resulting from root cell elongation. Planta 159: 497-504
Steinbiss HH, Stabel P (1983) Protoplast derived cells can survive capillary microinjection of the fluorescent dye Lucifer Yellow. Protoplasma 116: 223-227
Stoddard FL, McCully ME (1979) Histology of the development of the graft union in pea roots. Can J Bot 57: 1486-1501
Strasburger E (1901) Über Plasmaverbindungen pflanzlicher Zellen. Jb wiss Bot 36: 493-610
Tainter FH (1971) The ultrastructure of *Arceuthobium pusillum*. Can J Bot 49: 1615-1622
Tangl E (1879) Ueber offene Communicationen zwischen den Zellen des Endosperms einiger Samen. Jb wiss Bot 12: 170-190
Terry BR, Robards AW (1987) Hydrodynamic radius alone governs the mobility of molecules through plasmodesmata.Planta 171:145-157
Thomson WW, Platt-Aloia K (1985) The ultrastructure of the plasmodesmata of the salt glands of *Tamarix* as revealed by transmission and freeze-fracture electron microscopy.Protoplasma 125: 13-23
Waite JH, Tanzer ML (1981) Polyphenolic substance of *Mytilus edulis*. Science 212: 1038-1040

1983

DISTRIBUTION OF PLASMODESMATA IN LEAVES. A COMPARISON OF CANANGA ODORATA WITH OTHER SPECIES USING DIFFERENT MEASURES OF PLASMODESMATAL FREQUENCY

David G. Fisher
Department of Botany
University of Hawaii - Manoa
Honolulu, HI 96822
USA

Introduction

The functioning of plasmodesmata, the cytoplasmic channels connecting adjacent plant cells, may be considered at two levels of organization: the individual and the collective. Of course the first level involves studies of plasmodesmata as single units, and is treated elsewhere in this volume. The second involves plasmodesmatal frequency, or the numbers of plasmodesmata in relation to some physical parameter of the interconnected cells or surrounding cells and tissues. It is this collective level of organization which will be discussed here.

Plasmodesmata have been found in some algae and fungi, in bryophytes, pteridophytes, and all gymnosperms and angiosperms that have been examined. They are not, however, present between all contiguous living cells at maturity, nor are they equally distributed over the walls where they do appear. Such discontinuous distribution has led to the accumulation of comparative plasmodesmatal frequencies for various cell combinations both within a species and among different species. Robards (1976) has compiled a list of species from the above groups which indicates a range of frequency from 0.01 to more than 50 plasmodesmata per square micrometre over whole wall surfaces. Most of the subsequent studies using this measure of frequency report values well within this range (e.g., Offler and Patrick 1984, Fisher 1986, Kronestedt *et al.* 1986).

One of the areas in which plasmodesmatal frequency has become increasingly important concerns the functioning of angiosperm leaves. For years there has been a lively debate about how photosynthetic assimilates produced in the mesophyll are transported to the smallest (or minor) veins and "loaded" into the long-distance conducting elements of those veins, the sieve elements. Do the sugars traverse the distance in the

NATO ASI Series, Vol. H 46
Parallels in Cell to Cell Junctions in Plants and Animals
Edited by A. W. Robards et al.

extra-protoplastic continuum, the apoplast, or do they travel symplastically from cell to cell via plasmodesmata? Or is the pathway some combination of apoplast and symplast? Similarly, by which pathway(s) are imported assimilates from mature leaves "unloaded" from sieve elements in developing leaves and distributed amongst the immature tissues? The structural parameter most pertinent to answering these questions is the plasmodesmatal frequency at the various cell-to-cell interfaces along the assimilate route. The assumption here is that the greater the numbers of plasmodesmata at a given interface, the greater the potential for symplastic transport through it. Unfortunately, different measures of plasmodesmatal frequency have been used in different studies, so that only some overlap is available from one species to the next. Some of the more common or recently introduced measures are: 1) plasmodesmata/μm interface/thin section (Fisher and Evert 1982, Russin and Evert 1985, Evert and Mierzwa 1986, Fisher 1986); 2) plasmodesmata/μm^2 of interface (Kuo *et al.* 1974, Gamalei and Pakhomova 1981, Gamalei 1985, Fisher 1986); 3) percent of the total number of plasmodesmata that appear at each interface (Gunning *et al.* 1974, Kaneko *et al.* 1980, Botha and Evert 1988); 4) plasmodesmata at each interface/mm of vein (van Bel, *et al.* 1988); and 5) the surface to volume ratio of the cells in a given compartment multiplied by the number of plasmodesmata/μm which connect it to the next compartment (Ding *et al.* 1988). Various of these measures have been used to summarize plasmodesmatal frequencies diagrammatically, e.g., the "plasmodesmograms" of van Bel *et al.* (1988) and the pictorial classification of minor veins into three types based on wall ingrowths and plasmodesmatal frequencies (Gamalei 1985). Also, Russin and Evert (1985) used a diagram to designate frequencies numerically.

The purpose of this paper is threefold: first, to present the ultrastructure and plasmodesmatal distribution of the *Cananga odorata* leaf, second, to assess the usefulness of the various measures of plasmodesmatal frequencies listed above, and third, to compare the distribution of plasmodesmata in *C. odorata* with those of *Amaranthus retroflexus* (Fisher and Evert 1982), *Coleus blumei* (Fisher 1986) and other species, using the most appropriate measures of frequency. The three named species are especially appropriate for comparison because each has distinctive characteristics relating to the importance of plasmodesmata. *Amaranthus* is an NAD - malic enzyme C_4 plant with extraordinarily large numbers of plasmodesmata at the mesophyll - bundle sheath interface, the site of large fluxes of alanine and aspartate (Fisher and Evert 1982). *Coleus* is unusual in having apparently two separate systems for sieve element loading, one apoplastic and the other symplastic (Fisher 1986). And *Cananga* is unique (among species studied so far) in lacking sieve elements in about 60% of its minor veins. This dearth probably places added importance on plasmodesmata for transport of assimilate to the 40% of the veins which do have sieve elements (Fisher, 1989).

Materials and Methods

Sun leaves of *Cananga odorata* (Lam.) Hook. f. and Thoms. were collected from a tree in Lyon Arboretum, Honolulu, HI. Strips measuring 1 x 5 mm were cut from intercostal regions near the center of leaves which were the 10th - 12th oldest ≥ 1 cm long and which were located at the periphery of the tree. (The youngest fully expanded leaves were the fifth oldest ≥ 1 cm long.) The samples were fixed in 4% glutaraldehyde, post-fixed in 2% OsO_4, dehydrated, and embedded in Spurr's (1969) epoxy resin. Thin (gold) sections were cut with a diamond knife on a Porter-Blum MT-2 ultramicrotome (RMC, Tucson, AZ, USA), stained with uranyl acetate and lead citrate and examined and photographed with a Hitachi HS-8-11 electron microscope. The number of plasmodesmata/μm interface and μm interface/transverse vein section were measured as in a previous study (Fisher 1986). Plasmodesmata/μm vein was calculated by multiplying plasmodesmata/μm^2 interface (determined by the method of Gunning in Robards 1976) by μm interface/transverse vein section. Twenty minor veins were sampled, all of which exhibited "type II" anatomy (Fisher 1989). Raw data from previous studies (Fisher and Evert 1982, Fisher 1986), also based on 20 minor veins each, were used in the manner described above to determine the same types of parameters for *Amaranthus retroflexus* and *Coleus blumei*.

Results

Organization of Leaf Tissues

The overall anatomy of the minor veins and surrounding tissues has already been described in detail (Fisher, 1989) and therefore will be recounted here only briefly, for context. In transverse view, the mesophyll consists of two layers of highly branched palisade parenchyma cells and 5 layers of horizontally-oriented spongy parenchyma cells (Fig. 1). Embedded in this photosynthetic tissue are minor veins of four anatomically distinct but interconnected types. In order of largest to smallest, they contain the following cell types: Type I (Fig. 2) - tracheary elements, vascular parenchyma cells, companion cells, sieve elements most of which have nacreous walls, and fibers; type II (Figs.

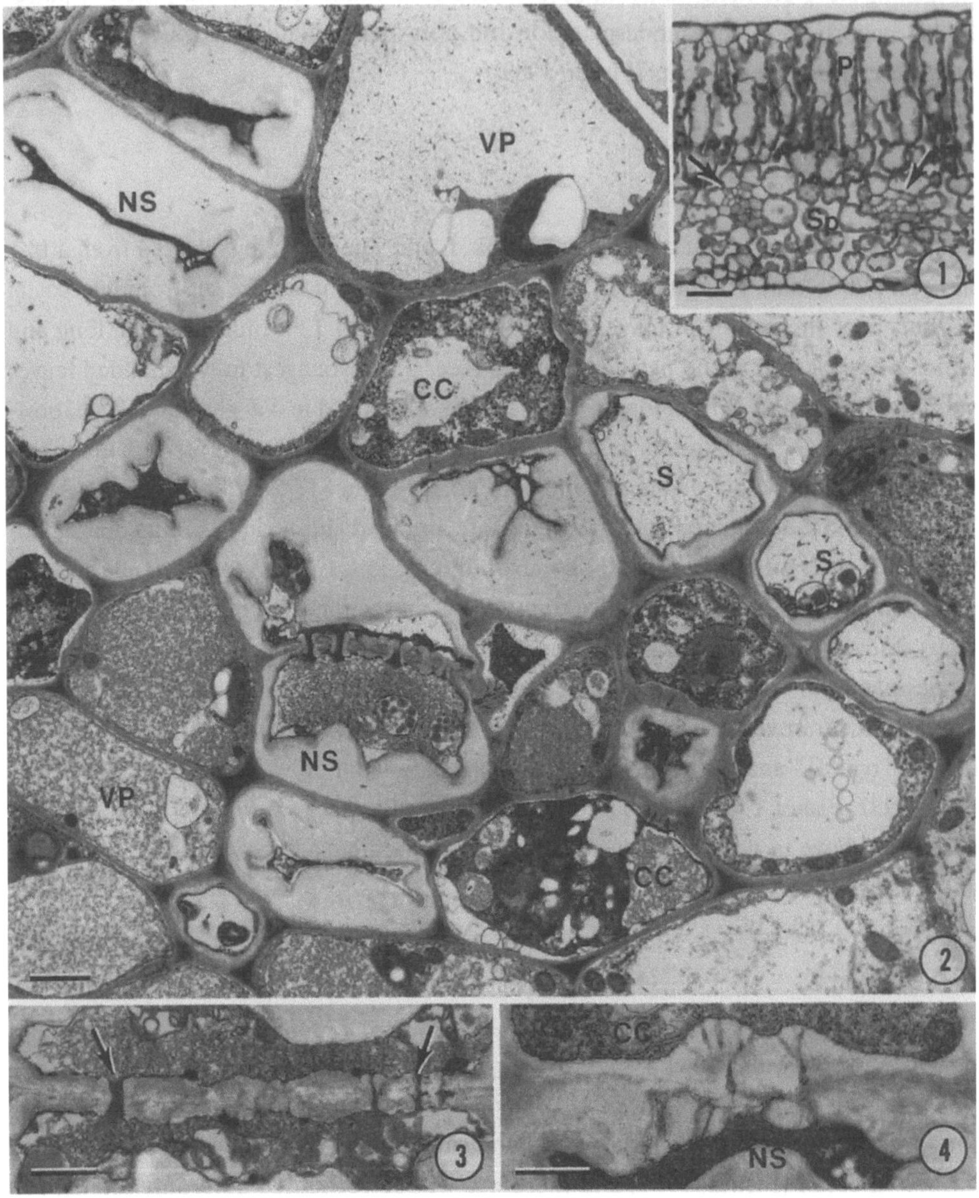

Figs. 1-4. Fig. 1. Cross section of lamina. P=palisade parenchyma; Sp=spongy parenchyma. Arrows point to minor veins. Bar=20μm. Fig. 2. Portion of transverse section of type I vein. CC=companion cell; NS=nacreous-walled sieve element; S=sieve element with little or no nacreous wall; VP=vascular parenchyma cell. Bar=2μm. Fig. 3. Sieve area between two sieve elements, showing variation in pore diameter. Arrows point to P-protein plugging pores. Bar=1μm. Fig. 4. Branched plasmodesmata-pore connections between companion cell (CC) and nacreous-walled sieve elements (NS). Bar=1μm.

5-7) - same cell types as I, but most of the sieve elements lack nacreous walls; type III (Fig. 11) - tracheary elements and vascular parenchyma cells; type IV (vein endings; Figs. 12, 13) - tracheary elements only. All of the minor veins are surrounded by chlorenchymatic bundle-sheath cells which, with decreasing vein size, extend farther and farther out into the mesophyll by means of branching arms.

Mesophyll and Bundle-Sheath Cells

Although the mesophyll and bundle-sheath cells vary considerably in size and shape, their fine structure is quite similar. All tend to have one or more large vacuoles, with a fairly dense parietal cytoplasm containing chloroplasts (see bundle-sheath cells in Figs. 5, 11-13) and a prominent nucleus (Fig. 12). The chloroplasts have well-developed grana, prominent starch grains, plastoglobuli, and are sometimes associated with microbodies. In bundle-sheath cells, the chloroplasts are more or less centrifugally arranged, whereas in palisade and spongy parenchyma cells the chloroplasts occur all around the periphery. Mitochondria are common, with endoplasmic reticulum, dictyosomes, and small lipid droplets less so.

Sieve Elements

As mentioned above, most of the sieve elements in veins of type I anatomy have walls with nacreous thickenings (Fig. 2). Typically, the thickenings are much more electron lucent than peripheral portions of the sieve element wall. Although they tend to occlude most of the cell lumen, they usually are absent or greatly reduced at sieve areas (Fig. 3) or at the sites of pore-plasmodesmata connections to companion cells (Fig. 4). Possibly as a result of the restricted lumens, the contents of these cells are often quite dense, making identification of individual components difficult (Figs. 3,4). Plastids and mitochondria can sometimes be discerned, however, embedded in a granular or fibrous substance which may be in part P-protein. The occasional membranous components may be smooth ER. In the few cases where the nacreous walls are relatively thin (Fig. 2, S), the lumens of the sieve elements are much clearer. No attempt was made to compare quantitatively the pores in lateral and end-wall sieve areas, but a marked range in pore diameter within sieve areas was noted (Fig. 3). Such pores are typically filled with fibrous P-protein. Connections to companion cells are typical of most angiosperms in consisting of several plasmodesmata leading away from each pore (Fig. 4). Sieve ele-

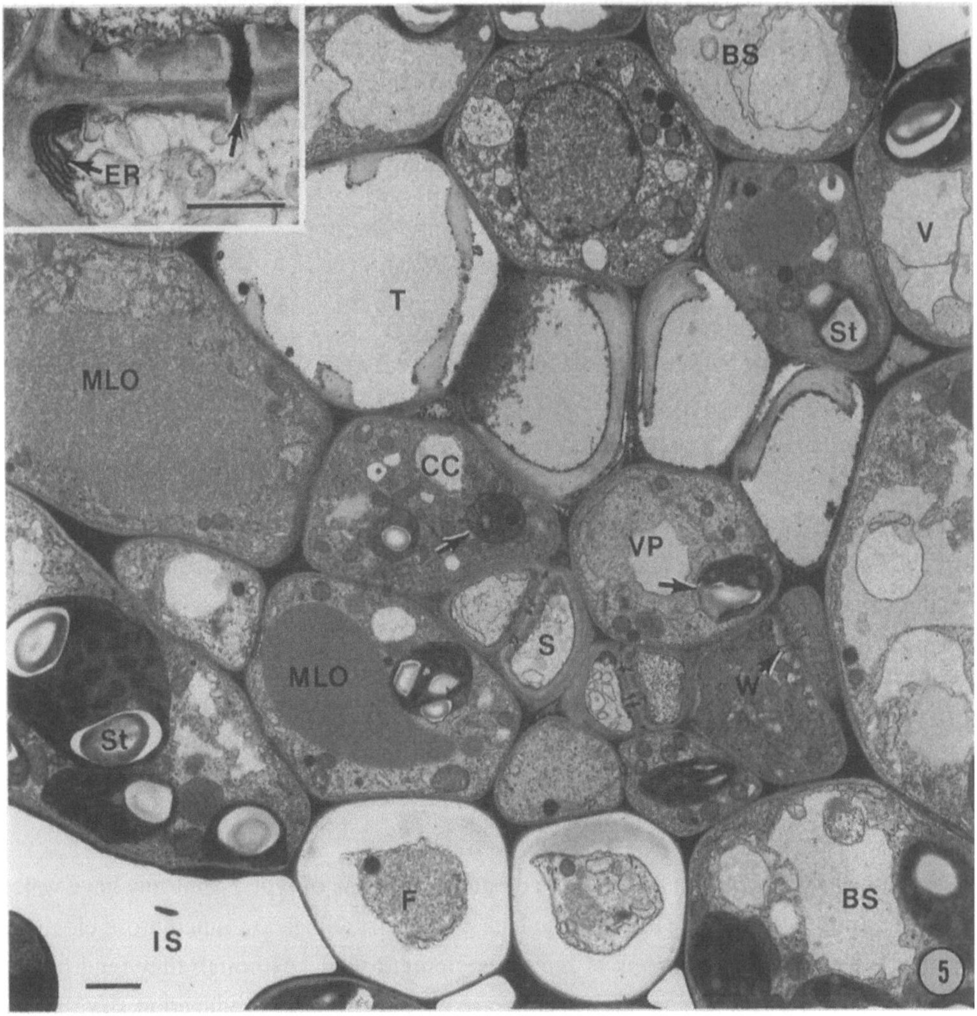

Fig. 5. Transverse section of type II vein. BS=bundle sheath cell; CC=companion cell, with arrow pointing to plastid; F=fiber; IS=intercellular air space; MLO=microbody-like organelle in vascular parenchyma cell; S=sieve element; St=starch grain in chloroplast, T=tracheary element; V=vacuole, VP=vascular parenchyma cell, W=wall thickening traversed by plasmodesmata. Bar=2μm. Insert: portion of sieve element with stacked endoplasmic reticulum (ER); unlabeled arrow points to P-protein in pore. Bar=1μm.

ments in the abaxial-most portion of the veins are invariably crushed, thus delimiting the protophloem.

In minor veins of type II anatomy, most of the sieve elements lack nacreous walls entirely or have only very reduced nacreous thickenings (Figs. 5,6). A few of the larger type II veins have at least some sieve elements with thick nacreous walls (Fig. 7), but these never add up to more than half of the total number of sieve elements (otherwise, the vein would be classified as having type I anatomy). Occasionally the more electron-dense portion of the sieve element wall is greatly thickened, either uniformly (Fig. 7, upper left; Fig. 8, right center), or only around part of the cell periphery (Fig. 8, left center). However, even when the cell lumen is reduced in volume, it is still clear, in contrast to that of most nacreous-walled sieve elements. Thinner-walled sieve elements may also appear relatively clear, depending on the plane of section, or they may exhibit various types of inclusions, such as vesicular membranous components, stacked ER (Fig. 5, insert), mitochondria, plastids, or fibrous P-protein. As with nacreous-walled sieve elements, the pores in sieve areas vary in width (Fig. 5) and are filled with P-protein (Fig. 5, insert). Pore-plasmodesmata connections to companion cells (Fig. 9) are also similar to those associated with nacreous-walled sieve elements. Although most minor veins with type II anatomy are too small to have protophloem, a few of the larger ones do (Fig. 7). In contrast to obviously crushed sieve elements of the larger veins, however, some medium-sized type II veins have sieve elements with either concave walls or protoplasts which have disintegrated to various degrees (Fig. 6, arrow; Fig. 8, arrow). In some cases it was obvious from the degenerated contents that these elements were not functional at the time of sampling and thus were not included in quantitative data. But a nonfunctional status could not be assigned with reasonable certainty as long as the plasmalemma was intact and plasmolysis was slight. Sieve elements in this condition were therefore included in quantitative data.

Parenchymatic Elements

Of the two types of parenchymatic element, vascular parenchyma cells occur in veins with types I, II, and III anatomies, but companion cells occur only in types I and II. It is sometimes difficult to distinguish between the two cell types when key characteristics do not appear in the plane of section, especially in type I veins. The cytoplasmic density of both varies from quite diffuse to relatively dense, and the degree of vacuolation can range from nil in one cell section to half or more of the cell lumen in another. More over, mitochondria range from scarce to abundant, and dictyosomes, lipid droplets, ER, and a nucleus can be seen commonly in both cell types. However, the plastids in vascular parenchyma cells have well-developed grana and starch grains (Figs. 5-7), whereas the plastids in companion cells have few internal membranes and rarely contain grana or starch grains (Figs. 5,7). In addition, a prominent feature of vascular parenchyma cells which is lacking in companion cells is what might be called microbody-like organelles (MLO's). Though relatively large, MLO's are like microbodies in having granular contents, a single delimiting membrane, and no internal membranes, (Fig. 10).

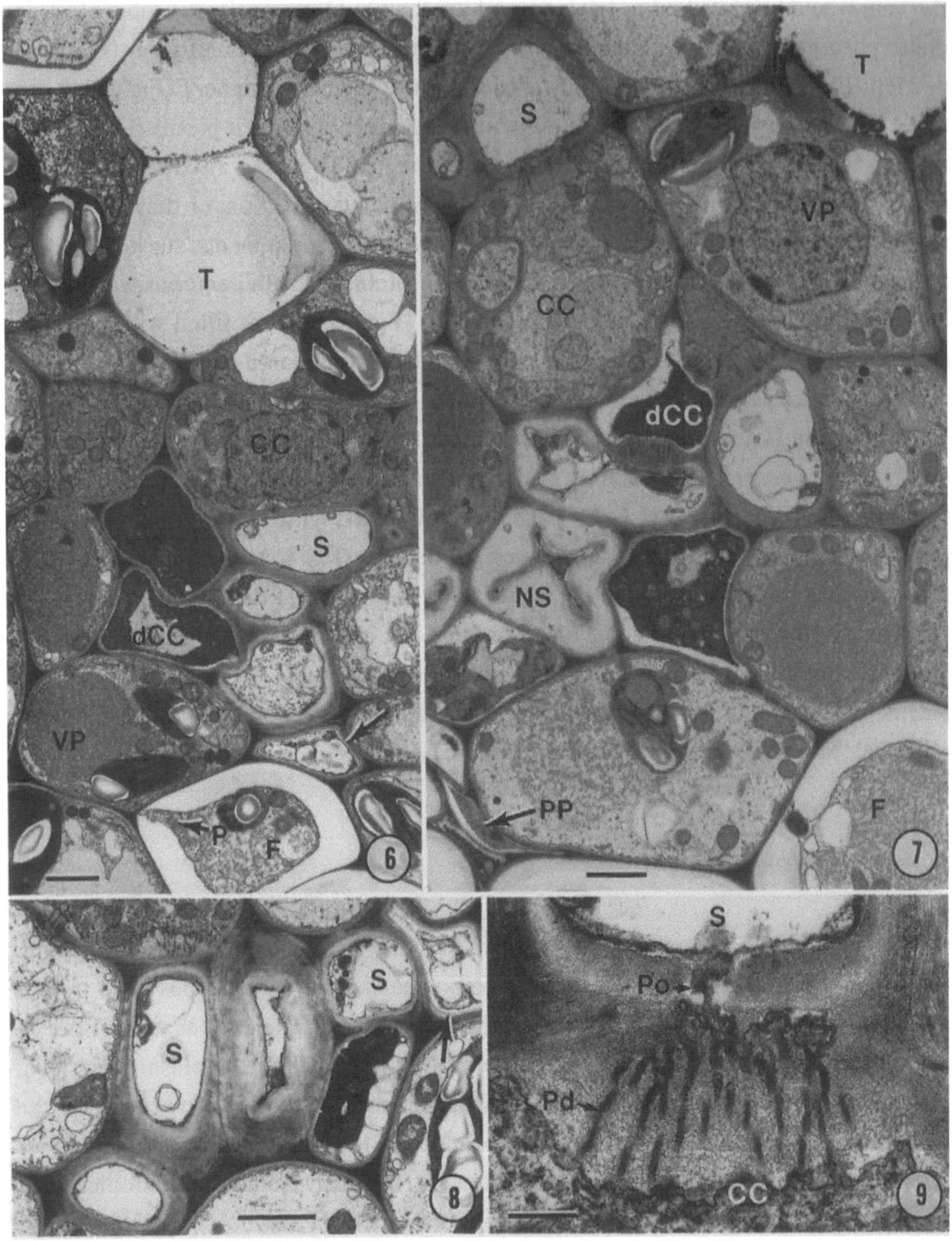

Figs. 6-9. Portions of transverse sections of type II veins. CC=ordinary companion cell; dCC=very densely cytoplasmic companion cell; F=fiber; P=pit; Pd=plasmodesma; Po=pore; PP=protophloem; NS=nacreous-walled sieve element; S=sieve element with ordinary walls; T=tracheary element; VP=vascular parenchyma cell; unlabeled arrows (figs. 6,8) point to presumably nonfunctional sieve elements. Figs. 6-8: Bars=2μm. Fig. 9: Bar=0.5μm.

The granular contents range in appearance from almost flocculent to quite dense (Figs. 5-7,11), differing strikingly from vacuoles in the same cell when the two occur together (e.g., Fig. 5, lower left). Vascular parenchyma cells are further distinguished from companion cells by an almost total lack of symplastic connections with sieve elements. Connections between companion cells and sieve elements typically occur at wall thickenings on the companion cell side (Figs. 4, 9). Similar thickenings may also occur at sites of plasmodesmata among the parenchymatic elements of the vein (e.g., Fig. 5,W). Not all companion cells are associated with sieve elements, nor are all sieve elements, especially those in type I veins, associated with companion cells. In larger minor veins companion cells are usually separated from the bundle sheath by vascular parenchyma cells, but in the smaller ones they sometimes contact the sheath cells.

Companion cells vary in cytoplasmic density much more than vascular parenchyma cells. Some, usually in type II veins, have extremely dense contents. When present, these cells tend to be more abaxial than ordinary companion cells (Figs. 6, 7). They have relatively few plasmodesmata-pore connections with ordinary sieve elements and often have somewhat concave walls. In addition, they are frequently associated with partly crushed or nacreous-walled sieve elements. When this is the case, they are usually plasmolyzed, but when they are associated with non-plasmolyzed sieve elements, they are not plasmolyzed. Of the 20 minor veins examined, 10 had 1-3 companion cells comparable in density to the very dense ones in Figure 6, and another 4 had companion cells which were intermediate in density between the lighter and darker ones in Figure 6. Only when a companion cell was associated with an obviously nonfunctional sieve element, using the criteria listed above, was it excluded from the quantitative data.

Fibers

Fibers are typically found at the periphery of the bundle, though sometimes they appear to be part of the bundle-sheath (Fig. 5) and at other times part of the vascular tissue (Fig. 6). In type I bundles there may be 2-3 layers of fibers on both the abaxial and adaxial sides of the bundle, but in type II bundles they are usually abaxial and few in number. With advancing leaf age there are more fibers and their walls tend to become thicker, so at least some of them are probably sclerified vascular parenchyma cells. Their contents may in fact be identical to those of vascular parenchyma cells (Figs. 5, 6) or they may have only a very thin peripheral layer of cytoplasm surrounding a large central vacuole. In any event, they are always elongate and have plasmodesmatal connections to vascular parenchyma and/or bundle sheath cells via simple pits (Fig. 6,P).

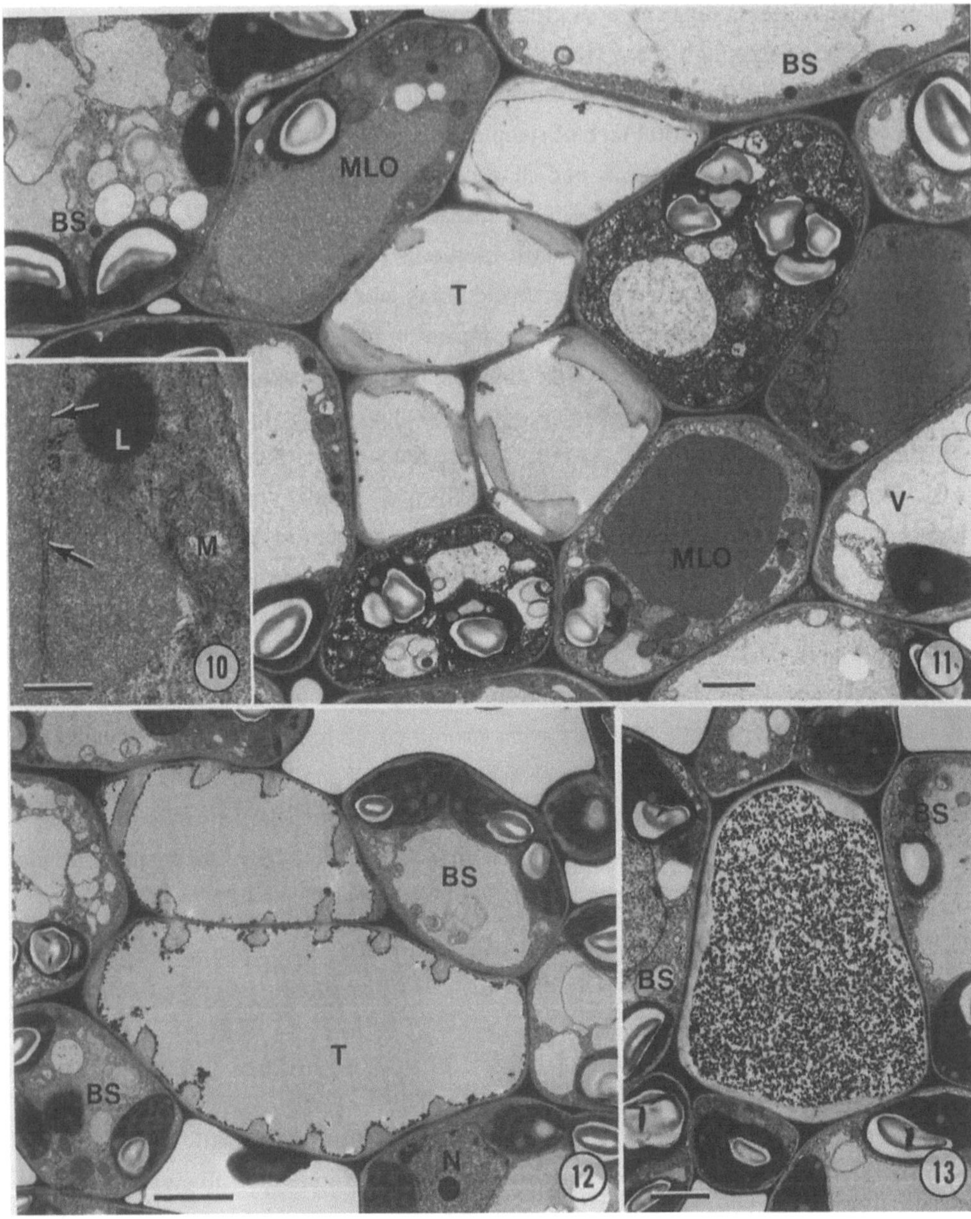

Figs. 10-13. Fig. 10. Arrows point to enclosing membrane of microbody-like organelles. L=lipid droplet; M=mitochondrion. Bar=0.5μm. Figs. 11-13. Transverse sections of type III vein (fig. 11) and type IV veins (figs. 12,13). BS=bundle sheath cell; N=nucleus in bundle sheath cell; MLO=microbody-like organelle in vascular parenchyma cell; T=tracheary element; V=vacuole. Fig. 11: Bar=2μm. Fig. 12: Bar=5μm. Fig. 13: Bar=3μm.

Tracheary Elements

Although virtually all of the tracheary elements appear to have spiral secondary wall thickenings, they vary considerably in size, contents, and degree of contact with intercellular air space. The largest ones (in diameter) occur in type IV bundles at the very tips of the vein endings. Here they may have considerable contact with air space (Fig. 12) or none at all (Fig. 13), depending on the site of the section. Only rarely does air space come in direct contact with tracheary elements in types I, II, and III veins. Except in type I veins, the lumens have varying amounts of granular contents, either dispersed evenly or along the periphery (Figs. 5-7, 11-13). These granules tend to be largest and most noticeable at the vein tips (Fig. 13).

The distribution of plasmodesmata is given in Tables 1 and 2, and will be considered in the context of a comparison with other species.

Discussion

Anatomy and Ultrastructure of the *Cananga odorata* Leaf

The *Cananga* leaf is fairly typical for a dicotyledon in having a mesophyll that consists of two layers of vertically-oriented palisade parenchyma cells and five layers of horizontally-oriented spongy parenchyma cells. The minor veins embedded in this mesophyll, however, are unusual in consisting of four distinct but interconnected types. Type I, the largest, consists of tracheary elements, vascular parenchyma cells, fibers, companion cells, and sieve elements most of which have nacreous walls. Type II, the next largest, is made up of the same cell types as I but most of the sieve elements lack nacreous walls. Type III, the third largest, contains only tracheary elements and vascular parenchyma cells, while type IV, typical of vein endings and the smallest, consists of tracheary elements only. The proportions of total vein length made up of each are: I - 15.1%, II - 27.2%, III - 24.4%, and IV - 33.3% (Fisher 1989).

The vascular parenchyma cells can be distinguished from companion cells in part by their microbody-like organelles, the MLO's. These organelles, which appear to be quite distinct from vacuoles in the same cells, resemble microbodies in having a single delimiting membrane, granular contents of varying density, and no internal membranes. However, microbodies are defined as ranging in diameter from 0.2 to 1.7μm (Huang, *et al.* 1983), while MLO's are usually much larger -- up to several μm in diameter. In any case, MLO's do not occur in companion cells. Vascular parenchyma cells can also be distinguished from companion cells by their plastids, which contain well-developed grana and starch grains, and an almost complete lack of plasmodesmata-pore connections to sieve elements. Companion cells have plastids with no grana and very little starch, and their symplastic connections to sieve elements are relatively numerous. The extreme density of the cytoplasm in some companion cells apparently indicates that they are no longer functional, since such cells are often plasmolyzed, have partially collapsed walls, and are often associated with nonfunctional sieve elements.

Fibers are evidently peripheral vascular parenchyma cells or bundle sheath cells which simply develop thick secondary walls and become extremely vacuolate. Tracheary elements, the only cell type found in all four types of veins, gradually decrease in number and increase in diameter with decreasing vein size. The nature of their granular contents, most noticeable at the vein endings, is unknown.

As mentioned previously (Fisher 1989), nacreous-walled sieve elements have been reported in angiosperm leaf minor veins only once before -- for *Zostera nana* (Pate and Gunning 1969) -- though they have been found in stems and/or petioles of *C. odorata* (Behnke 1971, 1988) as well as other species (e.g., Dute 1983, Behnke 1988). Nacreous walls have also been reported for leaf blade sieve elements of seagrasses (Kuo 1983), but it is not clear from the evidence presented that the sieve elements were in minor viens. It has been argued that nacreous walls are an artifact (Fisher 1975), but the presence in *Cananga* of sieve elements with and without such walls in the same mature leaf tissue -- in some cases side by side in the same veins -- indicates otherwise. Both types of sieve elements have intact plasmalemmas and a few plastids and mitochondria in a thin layer of parietal cytoplasm which also contains ER and P-protein. Both also have intact pore-plasmodesmata connections to companion cells. Whether the nacreous walls enhance, inhibit, or have no effect on the loading of assimilate is an open question. Based on the percentages of vein length occupied by types I and II veins, vein surface-to-volume ratio, and the degree of direct contact between adjacent bundle sheath cells and the mesophyll, it seems likely that type II veins load significantly more assimilate than type I (Fisher 1989).

Comparison of Different Measures of Plasmodesmatal Frequency and of Different Species

<u>Interfaces within the Mesophyll and at the Mesophyll-Bundle Sheath Boundary</u> A valid measure of the symplastic transport capacity of the mesophyll and bundle-sheath cells would be complex because of the variation in cell shape and arrangement and the presence of large air spaces, which limit cell-to-cell contact. Such a measure would have to consider the number of plasmodesmata per unit of wall, the amount of wall contact,

the number of cell walls along the assimilate pathway, and the size, shape, and orientation of all the cells. In addition, the relative amounts of photosynthate produced by the various layers of palisade and spongy parenchyma cells should be known, as well as the preferential assimilate route to the nearest phloem-bearing minor vein. Naturally, it is assumed that for comparative purposes the ultrastructure of the cells, especially that of plasmodesmata, does not vary significantly in relation to symplastic transport (Gunning 1976).

Based on experiments of the rate of transport of fluorescent dyes along filaments one cell wide, it appears that the single most limiting factor in symplastic transport is the resistance offered by plasmodesmata. The diffusion of uranin in staminal hairs of *Tradescantia*, for example, is limited by the number of plasmodesmata-containing cell walls it has to pass through, not the cell length or total distance covered (Tyree and Tammes 1975). A similar conclusion was reached by Barclay, Peterson, and Tyree (1982) for the movement of uranin in *Lycopersicon* trichomes. It is equally clear that the rate of transport across a cell wall is proportional to the number of plasmodesmata present in it (Gunning 1976, and references cited therein). If these same principles apply to assimilate transport in the mesophyll, it probably means that the preferred route is the one that supplies the least number of cell walls to traverse and the greatest number of plasmodesmata at those walls.

As far as mesophyll tissues are concerned, there has been no attempt yet to weave all of the above aspects into a measure of plasmodesmatal frequency that could be used to compare symplastic transport capacity among different species. The best measure available, and that for only a few species, is plasmodesmata/μm or /μm^2 for mesophyll - mesophyll interfaces, and plasmodesmata/μm vein for the mesophyll - bundle sheath interface. A comparison of *Amaranthus*, *Coleus*, and *Cananga* will serve to illustrate some of the inadequacies of these measures.

To begin with, *Cananga* has 2 layers of palisade and 5 layers of spongy parenchyma cells as compared to 1 and 3 layers, respectively, for *Coleus* (Fisher 1985). Consequently, the average number of horizontal cell walls that assimilate must pass through is very likely to be greater for *Cananga* than for *Coleus*. This in turn would dictate slower symplastic transport in *Cananga* if all other factors were equal, including plasmodesmata/μm at the various interfaces. As it is, *Coleus* averages 0.1 - 0.26 (Fisher 1986) and *Cananga* 0.18 - 0.62 plasmodesmata/μm (Table 1). (It is interesting that in *Cananga* the vertical interfaces consistently have higher frequencies than the horizontal ones, which would favor lateral transport of assimilate even in the palisade mesophyll.)

But there are other differences which a simple measure of plasmodesmata/μm does not take into account. Since about 60% of the total minor vein length in *Cananga* lacks sieve elements, assimilate in mesophyll cells nearest these veins will be forced to travel through even more cell walls to reach sieve elements. This will be true whether the assimilate route is mostly through the mesophyll or via mesophyll, then phloemless veins to the nearest sieve elements. By contrast, *Coleus* has phloem in all of its veins (Fisher 1985), so that assimilate may take a more direct route - and probably cross fewer cell walls -- to reach the sieve elements. These facts alone may account for the relatively large interveinal distance in *Coleus* (499 μm; Fisher 1985) as compared to 329 μm between phloem-bearing veins of *Cananga* (Fisher 1989).

Table 1. Plasmodesmatal frequencies within the mesophyll

	Average plasmodesmata/μm interface						
	Pl-Pl	Sp-Sp	Pl-Sp	Pl-Pvm	Sp-Pvm	Pvm-Pvm	Ms-Ms
Coleus[1]	0.19	0.26	0.10	-	-	-	-
Cananga	0.37[v] 0.18[h]	0.62[v] 0.46[h]	0.35	-	-	-	-
Populus[2]	0.09[v] 0.13[v] 0.20[h]	0.16[h]	-	0.28	0.13	0.34	-
Beta[3]	-	-	-	-	-	-	0.09

[v]vertical interfaces
[h]horizontal interfaces

[1](Fisher 1986)
[2](Russin and Evert 1985)
[3](Evert and Mierzwa 1986)

Pl = palisade parenchyma cell
Sp = spongy parenchyma cell
Pvm = parveinal mesophyll cell
Ms = generalized mesophyll cell
- = not applicable

Many other mesophyll cell arrangements occur which would also alter the number of cell walls and plasmodesmata that photosynthate would have to pass through. Leaves containing paradermal mesophyll are just one example (Kevekordes *et al.* 1988). In *Populus deltoides*, the number of plasmodesmata/μm interface is higher at the walls between paradermal mesophyll cells (0.34) than at other interfaces within the mesophyll (0.09-0.28) (Russin and Evert 1985). Assuming that transport is symplastic, this would be consistent with Russin and Evert's belief that the bulk of photosynthate traverses the paraveinal mesophyll enroute to the phloem. *Amaranthus*, because it is a C_4 plant, is basically exempt from considerations of mesophyll-mesophyll interfaces since all the photosynthetic cells are in direct contact with bundle-sheath cells. However, Table 2 shows that there are far more plasmodesmata/μm vein at the mesophyll - bundle sheath interface than in *Coleus* or *Cananga*. This is most likely because of the two-way transport of aspartate and alanine at this interface as well as high rates of photosynthesis and assimilate export. In this case the biochemical requirements of C_4 photosynthesis evidently make demands on plasmodesmata and symplastic transport that do not apply to C_3 plants (Olesen 1975; Weiner *et al.* 1988).

The Bundle Sheath - Vein Boundary and the Vascular Tissues The situation at the vein boundary and within is somewhat simpler, since the vascular cells are usually long and cylindrical or polyhedral and there are typically few if any intercellular spaces. This means that plasmodesmata can be distributed over any part of the walls of most cells. For the most part, the pertinent cells are sieve elements, companion cells, and vascular parenchyma cells, though some authors divide the latter into phloem and xylem parenchyma cells based on position. With these basics in mind, the various measures of plasmodesmatal frequency can be explored.

Plasmodesmatal/μm Interface and Plasmodesmata/μm^2 Interface Both of these measures give some idea of the capacity for symplastic transport. However, neither gives a reliable indication of the actual or even relative numbers of plasmodesmata at different interfaces, because they do not take into account the amount of contact at those interfaces. A pair of cell types which have only a little contact with one another may have more plasmodesmata/μm^2 wall than a second pair of cell types which have far more symplastic connections, but fewer plasmodesmata/μm^2 wall because they have much more cell-to-cell contact than the first pair. So plasmodesmata/μm^2, though essential for calculating actual flux rates between two given cells, *by itself* can be very misleading when comparing symplastic transport capacities at the various minor vein interfaces. The same can be said for plasmodesmata/μm interface. For example, in *Amaranthus*, plasmodesmata/μm wall at the companion cell - sieve element interface is 0.13, and that at the vascular parenchyma cell - sieve element interface is 0.05 (Table 2B), for an apparent ratio of 2.6 : 1. (A similar ratio holds for plasmodesmata/μm^2, not shown.) However, the sieve elements have much more interface with companion cells than with vascular parenchyma cells (Table 2A). As a consequence, the actual ratio of plasmodesmata is 17.0 : 1 (Table 2C), indicating that plasmodesmata/μm greatly overestimates symplastic continuity at the sieve element - vascular parenchyma interface in relation to the sieve element - companion cell interface. A comparison of Table 2B and C reveals similar differences in plasmodesmatal ratio between other pairs of interfaces, the only exceptions being where the amount of interface is about the same (e.g., vascular parenchyma - companion cell and companion cell - sieve element). This problem can be partly solved by using plasmodesmata/vein/section (Fisher and Evert 1982, Fisher 1986, Bota and Evert 1988), but since section thickness varies among studies in different laboratories, it is probably not especially useful for comparative purposes.

Table 2. Parameters concerning the mesophyll - bundle sheath interface and within

	MS-BS	BS-VP	BS-CC	VP-CC	CC-SE	VP-SE	
	A. Average μm interface/transverse section						
Amaranthus	228	48	24	49	46	8	
Coleus	131	49	39^i	12^i	8^i	8	
			5^c	17^c	15^c		
Cananga	64	49	1	25	12	7	
	B. Average plasmodesmata/μm interface						
Amaranthus[1]	1.09	0.53	0.11	0.25	0.13	0.05	
Coleus[2]	0.17	0.07	1.33^i	0.27^i	0.88^i	0	
			0.01^c	0.10^c	0.21^c		
Cananga	0.60	0.18	0.00	0.38	0.85	0	
	C. Average plasmodesmata/μm vein						Total
Amaranthus	2,114	216	23	104	51	3	2,511
Coleus	259	40	610^i	37^i	78^i	0	
			0^c	20^c	37^c	0	1,081
Cananga	326	75	0	81	87	0	569
Commelina[3]	52.5	42.4	0.1	0.8	0.3^m	13.7^m	
					3.7^p	$-^p$	113.5

-^{i}(intermediary cells)
-^{c}(ordinary companion cells)
-^{m}(metaphloem)
-^{p}(protophloem)
-[1](Fisher and Evert 1982)
-[2](Fisher 1986)
-[3](van Bel *et al.* 1988)

MS = mesophyll cell
BS = bundle sheath cell
CC = companion cell
SE = sieve element
VP = vascular parenchyma cell
\- = less than 0.1

Percent of the Total Number of Plasmodesmata that Appears at each Interface This measure is useful only for comparing the relative numbers of plasmodesmata within a species, since the average total number of plasmodesmata per cross section or unit of vein length can vary widely from one species to the next (Gunning *et al.* 1974). As a result, percent of total can be very misleading when used to compare different species. Suppose, for instance, that for the four species in Table 2C the proportion of total plasmodesmata present at each type of interface happened to be the same from one species to the next. That is, the mesophyll-bundle sheath interface might have 22% of the total for each species, the companion cell - sieve element interface might have 15% for each, and so on. Plasmodesmatal frequencies based on percent of total would thus appear to be the same, even though *Amaranthus* would average 22.1, *Coleus* 9.5, and *Canaga* 5.0 times as many plasmodesmata at each interface as *Commelina*. The reason is that the total number of plasmodesmata differ by these factors in Table 2C.

Van Bel *et al* (1988) used percent of total plasmodesmata to compare frequencies in 9 different species by means of "plasmodesmograms," diagrams in which plasmodesmata were represented by short lines at right angles to the cell walls of the veins, bundle sheath and mesophyll. These diagrams, however, were constructed from data which ranged from percent of total to plasmodesmata/μm interface to estimates based on other quantititative data or published electron micrographs. One of the problems arising from the use of original data other than percent of total for this type of plasmodesmogram can be illustrated by *Beta vulgaris, Populus deltoides* and *Amaranthus retroflexus*. These plasmodesmograms (below) were taken from van Bel et al (1988), who based them on plasmodesmata/μm interface (Evert and Mierzwa 1986; Russin and Evert 1985; Fisher and Evert 1982). *Beta* appears to have more plasmodesmata at most interfaces

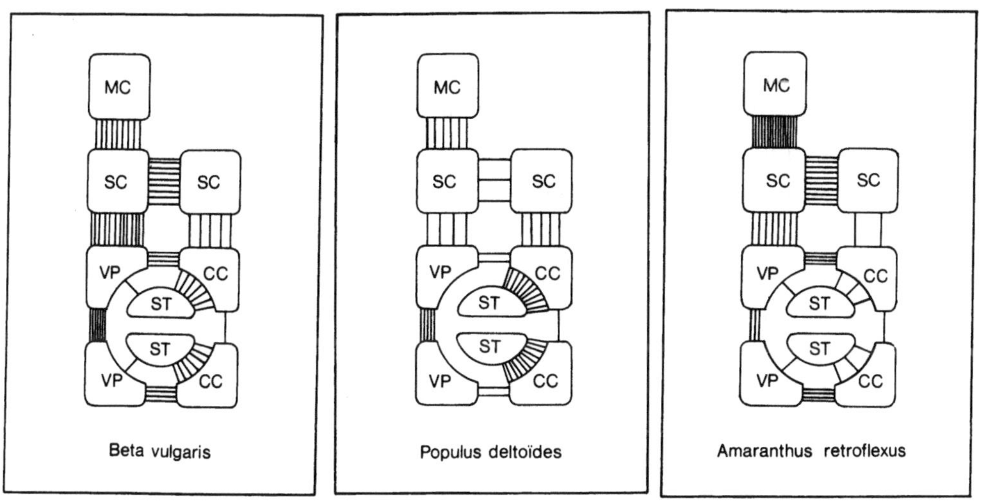

than *Populus*, yet the numbers of plasmodesmata/μm at these interfaces in *Polulus* are 3.3 to 12.1 times as high as in *Beta* (calculated from Table 1, Evert and Mierzwa 1986). Unless *Beta* has far more cell-cell contact than *Populus* at each type of interface, *Populus* must average many more plasmodesmata per interface than *Beta*, contrary to the plasmodesmogram. Similar problems can arise within a species. For *Amaranthus*, the ratio of plasmodesmata at the mesophyll - bundle sheath to the bundle sheath - vascular parenchyma interface is 2.1 (17 : 8) in the plasmodesmogram, but 9.8 (2,114 : 216) when the number of plasmodesmate/μm of vein are considered (Table 2C). Thus the plasmodesmogram greatly underestimates the number of plasmodesmata at the mesophyll - bundle sheath border in relation to those at other interfaces.

The plasmodesmograms of van Bel *et al.* (1988) do have the advantage of more or less instant visualization of relative plasmodesmatal frequencies. If a measure of frequency that included all plasmodesmata were used, such as plasmodesmata/μm vein, the plasmodesmograms would be truly representative. However, the range of numbers for only 4 species is too great to be contained within a small diagram, as Table 2C shows. The same disadvantage would apply to Gamalei's (1985) diagrams if actual, rather than broadly summarized, frequencies had been used. Gamalei's diagrams have the further disadvantage of depicting frequencies only at the bundle sheath - companion cell and companion cell - sieve element interfaces. As is evident from Table 2 and a number of ultrastructural studies of minor veins (e.g., Gunning *et al.* 1974, Evert *et al.* 1978, Russin and Evert 1985, Evert and Mierzwa 1986), this disregards alternate symplastic pathways through the vascular parenchyma cells, a significant omission for many species. The best means of presenting frequencies in diagrammatic form is probably that of Russin and Evert (1985), though for reasons given above plasmodesmata/unit of vein length would be a better indicator of relative transport capacity than plasmodesmata/μm interface.

N, the Surface to Volume Ratio of the Cells in a Given Compartment Multiplied by the Plasmodesmata/μm Which Connect it to the Next Compartment This measure has been used to demonstrate that the numbers of plasmodesmata at successively greater distances from the sieve element-companion cell complex of "Class III" veins in developing tobacco leaves is consistent with diffusion of imported assimilate away from those complexes (Ding *et al.* 1988). Since N relates the numbers of plasmodesmata to the amount of interface between compartments as well as the cross-sectional area of those compartments, it has the potential to be the most realistic measure of plasmodesmatal frequency yet. However, before it can be applied to minor veins of exporting leaves for comparison of different species, several problems must be solved. First, the various types of phloem cell of most smaller minor veins are not arranged in concentric layers, unlike the relatively large minor veins to which N was applied in tobacco. Instead, small minor veins tend to have not only vascular parenchyma but also companion cells in contact with the bundle-sheath. In addition, both vascular parenchyma and companion cells are often isolated from other cells of their respective types. This would greatly complicate the task of treating groups of cells of the same type as single compartments. A corollary of this problem is the treatment of sieve elements and companion cells as only one compartment. Since in exporting leaves the sieve elements are the cells into which most of the assimilate ultimately is loaded, it would be more realistic to treat them as a separate cell type. Another complicating factor for some species is the presence of two distinct types of sieve elements, each associated with its own equally distinct type of parenchyma or companion cell (Evert, *et al.* 1978, Fisher 1986, van Bel *et al.* 1988). Typically, one type of sieve element has symplastic continuity to the bundle sheath via parenchymatic elements while the other has symplastic continuity only as far as the companion cells. (*Cananga* also has two types of sieve elements, but they are usually in different types of minor veins.) Another structural variation which would cause compartmentation problems is transfer cells, with up to 3 different types present in the minor veins of some species (Pate and Gunning 1969).

It is possible that some of the smaller minor veins in some species could be found in which companion cell - sieve element complexes, vascular parenchyma cells, and bundle sheath cells form concentric layers. For these individual veins it should be possible to determine N for the various compartments, though the problem of treating companion cells and sieve elements as one cell type would remain. However, for the purposes of comparing different species, it is difficult to see how N could be applied to all types of minor veins for a wide variety of species.

Plasmodemata/mm Vein Recently proposed by van Bel *et al.* (1988), this is the only measure that gives all the plasmodesmata available for transport at each type of interface. (Because of the numbers it generates, the unit of vein length has been changed to μm in Table 2C.) It is the most useful means yet for comparing minor vein frequencies at different interfaces and in different species, because 1) it does not skew the numbers of plasmodesmata due to differences in the amount of contact at the various interfaces or variations in the total number of plasmodesmata, and 2) it can be easily applied to minor veins of all sizes and all types. It does not, however, indicate directly the amount of interface for the various combinations of cell types or the cross-sectional areas or volumes of the cells. As such, it probably cannot be used to predict actual rates of symplastic transport from one cell type to the next. Also, like most other measures its usefulness rests largely on the assumptions that the transport capacity of all plasmodesmata is roughly the same and that numbers of plasmodesmata alone yield a valid estimate of relative transport capacity.

Using plasmodesmata/μm vein, the 4 species in Table 2C can now be compared. First, it is clear that the 3 dicotyledonous species have considerably more plasmodesmata at almost every interface than *Commelina*. This may be due in part to the fact that the studies on dicot species included minor veins (i.e. veins surrounded by mesophyll) of all sizes, whereas the *Commelina* study apparently included only the smallest minor veins. Excluding the large difference in numbers of plasmodesmata at the mesophyll-bundle sheath interface of *Amaranthus* (as opposed to C_3 species, which has already been discussed), the next distinction involves symplastic connections at the vein boundary. In *Amaranthus*, *Cananga* and *Commelina*, the major symplastic entryway into the veins is evidently through the vascular parenchyma cells. In *Coleus*, the major pathway is via the intermediary cells (a specialized type of companion cell), though a not inconsequential number of plasmodesmata also occur at the vascular parenchyma cell boundary. An important factor with *Coleus* is that the plasmodesmata connecting intermediary cells to bundle-sheath and vascular parenchyma cells are structurally quite different from other plasmodesmata in the vein. This difference may be related to the much higher solute concentration in the intermediary cells than in bundle sheath and vascular parenchyma cells (Fisher 1986). In fact, the sieve element - companion cell complexes generally have higher solute concentrations than adjacent cell types (Geiger, *et al.* 1973, Evert *et al.* 1978, Fisher and Evert 1982, Russin and Evert 1985), which means that any plasmodesmata moving assimilates into them must be capable of active transport. If they are, they may be physiologically distinct enough from plasmodesmata lacking that capability to render comparisons based on frequency alone spurious.

Since the species in Table 2 have only rare plasmodesmata-pore connections between vascular parenchyma cells and sieve elements, virtually the only symplastic entryway for those elements is via the relatively numerous connections with companion cells. In *Amaranthus*, *Cananga* and *Commelina*, this requires adequate numbers of plasmodemata between vascular parenchyma and companion cells. What constitutes adequacy is not yet known, but *Amaranthus* and *Cananga* have far more connections at this interface than *Commelina*. However, *Commelina*, like *Coleus*, apparently has one type of sieve tube-companion cell complex consistent with loading from the apoplast and one consistent with symplastic loading, as mentioned earlier. Thus vascular parenchyma - companion cell connections may not be as important as in *Amaranthus* and *Cananga*.

When the assimilate pathway is summarized as vascular parenchyma - companion cell - sieve element, it can be seen that there is a steady drop in numbers of plasmodesmata at each interface for *Amaranthus* and little change for *Cananga*. Assuming that individual plasmodesmata along this pathway all have the same transport capacity (i.e., within each species), and that cell volumes are largely irrelevant, these numbers are consistent with assimilate leakage into the apoplast in *Amaranthus* but retainment in the symplast in *Cananga*. A similar analysis may not be possible for the bundle sheath - intermediary cell - sieve element pathway in *Coleus* because of the previously-mentioned distinctive plasmodesmata at the interface between the first two cell types. And compared with all three of these species, the numbers of plasmodesmata connecting both types of companion cell-sieve element complexes with surrounding cell types in *Commelina* are quite low. What this means in terms of apoplastic vs. symplastic loading remains to be seen.

In summary, an accurate and reliable means of comparing symplastic transport capacity from mesophyll to sieve elements of minor veins in different species will only be possible when plant biologists all begin to use the same means of comparison. At present, the simplest and probably the best measures are the number of plasmodesmata/unit of vein length for interfaces at the mesophyll - bundle sheath boundary and within, and plasmodesmata/μm or /μm^2 for interfaces in the mesophyll.

Acknowledgements

Many thanks are due Ms. Gail Nakamoto for typing the original manuscript and Ms. Marilyn Chung for assistance in revising it.

Literature Cited

Barclay, G.F., Peterson, C.A., and Tyree, M.T. (1982) Transport of fluorescein in trichomes of *Lycopersicon esculentum*. Can. J. Bot. 60: 397-402.

Behnke, H.-D. (1971) Über den Feinbau verdickler (nacre') Wände und der Plastiden in den Siebröhren von *Annona* und *Myristica*. Protoplasma 72: 69-78.

Behnke, H.-D. (1988) Sieve element plastids, phloem protein, and evolution of flowering plants. III. Magnoliidae. Taxon 37: 699-732.

Botha, C.E.J. and Evert, R.F. (1988) Plasmodesmatal distribution and frequency in vascular bundles and contiguous tissues in the leaf of *Themeda triandra*. Planta 173: 433-441.

Ding, B., Parthasarathy, M.V., Niklas, K., and Turgeon, R. (1988) A morphometric analysis of the phloem-unloading pathway in developing tobacco leaves. Planta 176: 307-318.

Dute, R.R. (1983) Phloem of primitive angiosperms. I. Sieve-element ontogeny in the petiole of *Liriodendron tulipifera* L. (Magnoliaceae). Am. J. Bot. 70: 64-73.

Evert, R.F., Eschrich, W. and Heyser, W. (1978) Leaf structure in relation to solute transport and phloem loading in *Zea mays* L. Planta 138: 279-294.

Evert, R.F. and Mierzwa, R.J. (1986) Pathway(s) of assimilate movement from mesophyll cells to sieve tubes in the *Beta vulgaris* leaf. In: Phloem Transport, pp. 419-432, Cronshaw, J., Lucas, W. and Giaquinta, R.T., eds. Alan R. Liss, New York.

Fisher, D.B. (1975) Structure of functional soybean sieve elements. Plant Physiol. 56: 555-569.

Fisher, D.G. (1985) Morphology and anatomy of the leaf of *Coleus blumei* (Lamiaceae).

Am. J. Bot. 72: 392-406.
Fisher, D.G. (1986) Ultrastructure, plasmodesmatal frequency, and solute concentration in green areas of variegated *Coleus blumei* Benth. leaves. Planta 169: 141-152.
Fisher, D.G. (1989) Leaf structure of *Cananga odorata* (Annonaceae) in relation to the collection of photosynthate and phloem loading. Morphology and Anatomy. Can. J. Bot. (in press).
Fisher, D.G. and Evert, R.F. (1982) Studies on the leaf of *Amaranthus retroflexus* (Amaranthaceae): ultrastructure, plasmodesmatal frequency, and solute concentration in relation to phloem loading. Planta 155: 377-387.
Gamalei, Yu.V. (1985) Characteristics of phloem loading in woody and herbaceous plants. Sov. Plant Physiol. 32: 656-665.
Gamalei, Yu. V. and Pakhomova, M.V. (1981) Distribution of plasmodesmata and parenchyma transport of assimilates in the leaves of several dicots. Sov. Plant Physiol. 28: 649-661.
Geiger, D.R., Giaquinta, R.T., Sovonick, S.A. and Fellows, R.J. (1973) Solute distribution in sugar beet leaves in relation to phloem loading and translocation. Plant Physiol. 52: 585-589.
Gunning, B.E.S. (1976) The role of plasmodesmata in short distance transport to and from the phloem. In: Intercellular communication in plants: studies on plasmodesmata, pp. 203-227, Gunning, B.E.S. and Robards, A.W., eds. Springer. Berlin, Heidelberg, New York.
Gunning, B.E.S., Pate, J.S., Minchin, F.R. and Marks, I. (1974) Quantitative aspects of transfer cell structure in relation to vein loading in leaves and solute transport in legume nodules. Symp. Soc. Exp. Biol. 28: 87-126.
Huang, A.H.C., Trelease, R.N. and Moore, T.S. (1983) Plant Peroxisomes. Academic Press. New York, London.
Kaneko, M., Chonan, N., Matsuda, T. and Kawahara, H. (1980) Ultrastructure of the small vascular bundles and transfer pathways for photosynthate in the leaves of rice plant. Jap. J. Crop. Sci. 49: 42-50.
Kevekordes, K.G., McCully, M.E., and Canny, M.J. (1988) The occurrence of an extended bundle sheath system (paraveinal mesophyll) in the legumes. Can. J. Bot. 66: 94-100.
Kronestedt, E. C., Robards, A. W., Stark, M. and Olesen, P. (1986) Development of trichomes in the *Abutilon* nectary gland. Nord. J. Bot. 6: 627-639.
Kuo, J. (1983) The nacreous walls of sieve elements in seagrasses. Am. J. Bot. 70: 159-164.
Kuo, J., O'Brien, T.P., and Canny, M.J. (1974) Pit-field distribution, plasmodesmatal frequency, and assimilate flux in the mestome sheath cells of wheat leaves. Planta 121: 97-118.
Offler, C. E. and Patrick, J. W. (1984) Cellular structures, plasma membrane surface areas and plasmodesmatal frequencies of seed coats of *Phaseolus vulgaris* L. in relation to photosynthate transfer. Aust. J. Plant Physiol. 11: 79-99.
Olesen, P. (1975) Plasmodesmata between mesophyll and bundle sheath cells in relation to the exchange of C_4-acids. Planta 123: 199-202.
Pate, J.S. and Gunning, B.E.S. (1969) Vascular transfer cells in angiosperm leaves. A taxonomic and morphological survey. Protoplasma 68: 135-156.
Robards, A.W. (1976) Plasmodesmata in higher plants. In: Intercellular communication in plants: studies on plasmomdesmata, pp. 15-57, Gunning, BES and Robards, A.W., eds. Springer-Verlag, Berlin, Heidelberg, New York.
Russin, W.A. and Evert, R.F. (1985) Studies on the leaf of *Populus deltoides* (Salicaceae): Ultrastructure, plasmodesmatal frequency, and solute concentrations. Am. J. Bot. 72: 1232-1247.
Spurr, A.R. (1969) A low-viscosity epoxy resin embedding medium for electron microscopy. J. ultrastruct. Res. 26: 31-43.
Tyree, M.T. and Tammes, P.M.L. (1975) Translocation of uranin in the symplasm of staminal hairs of *Tradescantia*. Can. J. Bot. 53: 2038-2046.

van Bel, A.J.E., van Kesteren, W.J.P. and Papenhuijzen, C. (1988) Ultrastructural indications for coexistence of symplastic and apoplastic phloem loading in *Commelina benghalensis* leaves. Planta 176: 159-172.

Weiner, H., Burnell, J. N., Woodrow, I. E., Heldt, H. W., and Hatch, M. D. (1988) Metabolite diffusion into bundle sheath cells from C_4 plants. Relation to C_4 photosynthesis and plasmodesmatal function. Plant Physiol. 88: 815-822.

PLASMODESMATAL STRUCTURE AND FUNCTION IN NECTARIES

E P Eleftheriou
Department of Botany
University of Thessaloniki
GR 54006 Thessaloniki
Greece

Introduction

Elucidation of how water and solutes move from cell to cell and how stimuli are transmitted through plants constitutes a major question in plant cell biology. In the multicellular bodies of higher plants, cells are enclosed by a rigid cell wall which does not readily permit changes in shape, neither does it favour the exploitation of mechanical pumping mechanisms to face nutritional requirements, such as pinocytosis and reverse pinocytosis common in animal cells. Facing this inherent restriction, plant cells have instead evolved and utilised effectively alternative structures for their intercellular communication, the plasmodesmata. Plasmodesmata are fine strands of cytoplasm bounded by the plasma membrane that connect the living cells with their living neighbours by penetrating through perforations of the intervening cell walls. The presence of plasmodesmata subdivides the plant body into two major compartments, the symplast, comprising the interconnected protoplasts bounded by the continuous plasma membrane, and the apoplast, consisting of the non-living compartment outside the plasma membrane (for details see Gunning and Steer 1975, Gunning 1976).

Plasmodesmata have been extensively investigated and are still being studied both structurally and functionally for their indisputably important role in cell to cell transport of water and solutes. For the study of a given biological question it is imperative to use a suitable model system. Plant glands, and especially nectaries, have so far proved to be one of the most suitable and convenient experimental "tools" for the study of many biological questions, among which are those relating to plasmodesmatal structure and function.

Nectaries are glands secreting sugars in the form of nectar and are encountered in flowers (floral nectaries) and other aerial vegetative organs (extrafloral nectaries). They display a broad range of anatomical and presumably physiological diversity (Fahn 1979a), those developing trichomatous secretory structures being experimentally the most amenable systems for the study of short distance transport. Some nectaries, such as the floral nectaries of *Abutilon* and *Hibiscus*, combine anatomical simplicity with secretion of copious quantities of nectar providing extremely elegant systems for the quantitative assessment of symplastic transport.

NATO ASI Series, Vol. H 46
Parallels in Cell to Cell Junctions in Plants and Animals
Edited by A. W. Robards et al.

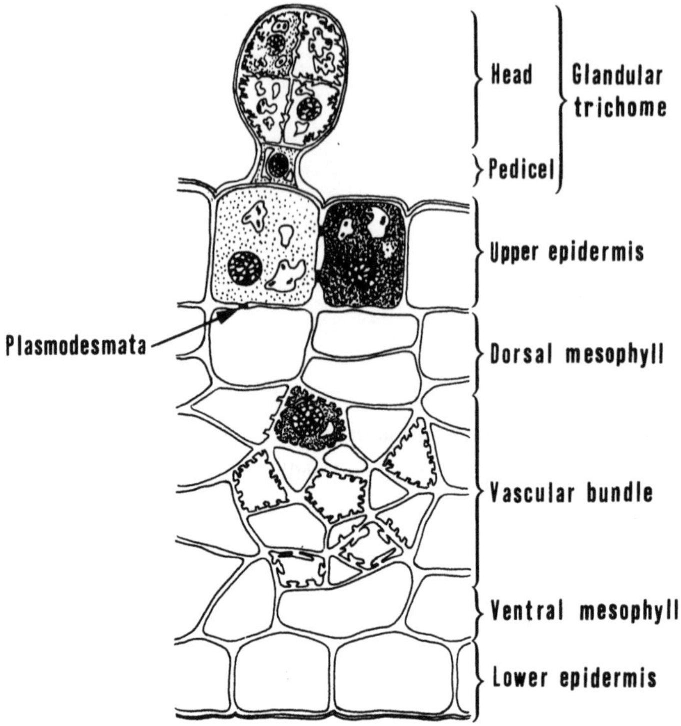

Fig. 1. Transverse section of the stipule of *Vicia faba* at the level of the nectary (re-drawn from Figier 1971).

Models of Nectaries

In the extrafloral nectaries of *Vicia faba*, nectar is secreted by secretory hairs which consist of a four-celled head and a pedicellar cell resting on an epidermal cell (Fig. 1) (Figier 1971). Many well developed plasmodesmata occur in the proximal and the distal wall of the pedicellar cell, as well as between the basal cell and the subglandular mesophyll cells. Endoplasmic reticulum (ER) cisternae were seen clearly associated with the plasmodesmata. The cells of the head and the companion and vascular parenchyma cells of the subjacent conducting bundle develop numerous transfer-cell-type wall protuberances (Fig. 1). Plasmodesmata could rarely be found in these cells; instead, acid phosphatase activity was localised around and in the plasma membrane of these transfer-cell-type cells (Figier 1968). From the ultrastructural and cytochemical evidence, Figier (1971) suggested that the pedicellar and the basal cells seem to represent a less elaborate zone where symplastic transport should take place through plasmodesmata as mediated by ER.

In other places, such as the sieve element-companion cell complexes, sugars are actively transported across the plasma membranes by a series of phosphorylations and dephosphorylations of the sugar molecules (Figier 1968).

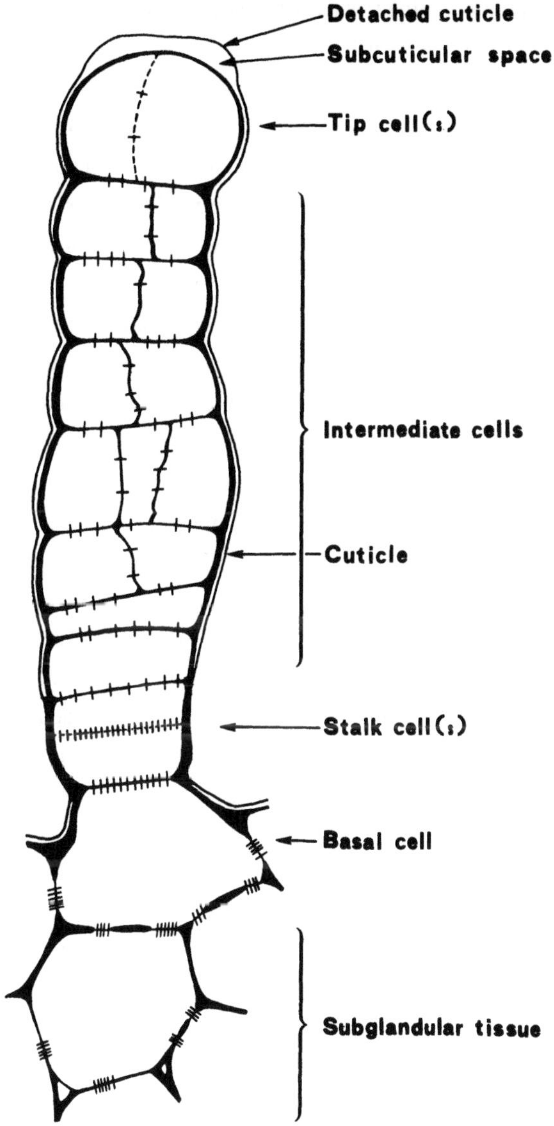

Fig.2. Diagram of a secretory hair of *Hibiscus rosa-sinensis* illustrating the plasmodesmatal frequencies and their pattern of distribution (adapted from Sawidis *et al.* 1987b).

In *Lonicera japonica* the nectar is secreted by single-celled secretory hairs which, similarly to *Vicia faba*, have profuse wall protuberances covered with the plasma membrane (Fahn and Rachmilevitz 1970). Fahn (1979a, b) proposed a model for *Lonicera japonica* in which prenectar is envisaged to be transported from the phloem tissue to the secretory hairs symplastically via the plasmodesmata of the intervening tissue, while nectar is eliminated by vesicles originating from the ER and fusing with the plasma membrane.

In the foliar nectaries of cotton (*Gossypium hirsutum*) nectar is secreted by numerous papillae closely packed in pyriform or sagittate depressions of the midvein on the lower surface of the leaf. The secretory papillae are multicellular structures consisting of many tip cells, considered to be the secretory cells, three to four stories of intermediate cells and a stalk cell resting on an epidermal cell (Eleftheriou and Hall 1983a). The cell walls of the papillae are traversed by numerous plasmodesmata. Structural and cytochemical evidence support the view that in cotton foliar nectaries prenectar flows symplastically from the phloem to the secretory cells, while the localisation of ATPase activity at the plasma membrane of the sieve tube companion cells and of the secretory cells indicate an active transport taking place at these sites (Eleftheriou and Hall 1983a, b).

Two of the most intensively studied nectaries are those of *Abutilon* (Findlay and Mercer 1971a, b, Findlay *et al.* 1971, Gunning and Hughes 1976, Hughes 1977, Kronestedt *et al.* 1986, Terry and Robards 1987, Robards and Stark 1988) and, to a lesser extent, of *Hibiscus rosa-sinensis* (Sawidis *et al.* 1987a, b, 1989). The floral nectaries of these closely related plants occur on the inner side of the sepals and consist of numerous secretory hairs. Each hair is a multicellular trichome. Although there exist some differences between the two species concerning the number and arrangement of cells within the hair, their gross anatomy is in general the same and is shown in Figure 2 which also illustrates the frequencies and the pattern of distribution of plasmodesmata in all cell walls of a secretory hair, including the subglandular cells.

The cowpea *(Vigna unguiculata)* bears two distinctive types of extrafloral nectary, the stipel nectary and that of the inflorescence stalk (Kuo and Pate 1985). In the stipel nectaries nectar is secreted by widely-spaced trichomes (papillae). Among other features, Kuo and Pate (1985) consider the frequencies of plasmodesmatal connections at certain key boundaries to be of special interest for the presumed flow pathway of prenectar. Based on structural information, Pate *et al*, (1985) proposed models for prenectar transport and nectar secretion for both types of extrafloral nectary of cowpea. In the stipel nectaries "passage of prenectar through the trichome might be symplastic via plasmodesmata or apoplastic via plasma membranes of adjacent cells or longitudinally between cells". Nectar is secreted through the readily detached cuticular layer of the transfer-cell-type apical cells. In the nectaries of the inflorescence stalk nectar is secreted mainly in a "hydathode-like" fashion by the pressure flow of the phloem sap, but a symplastic

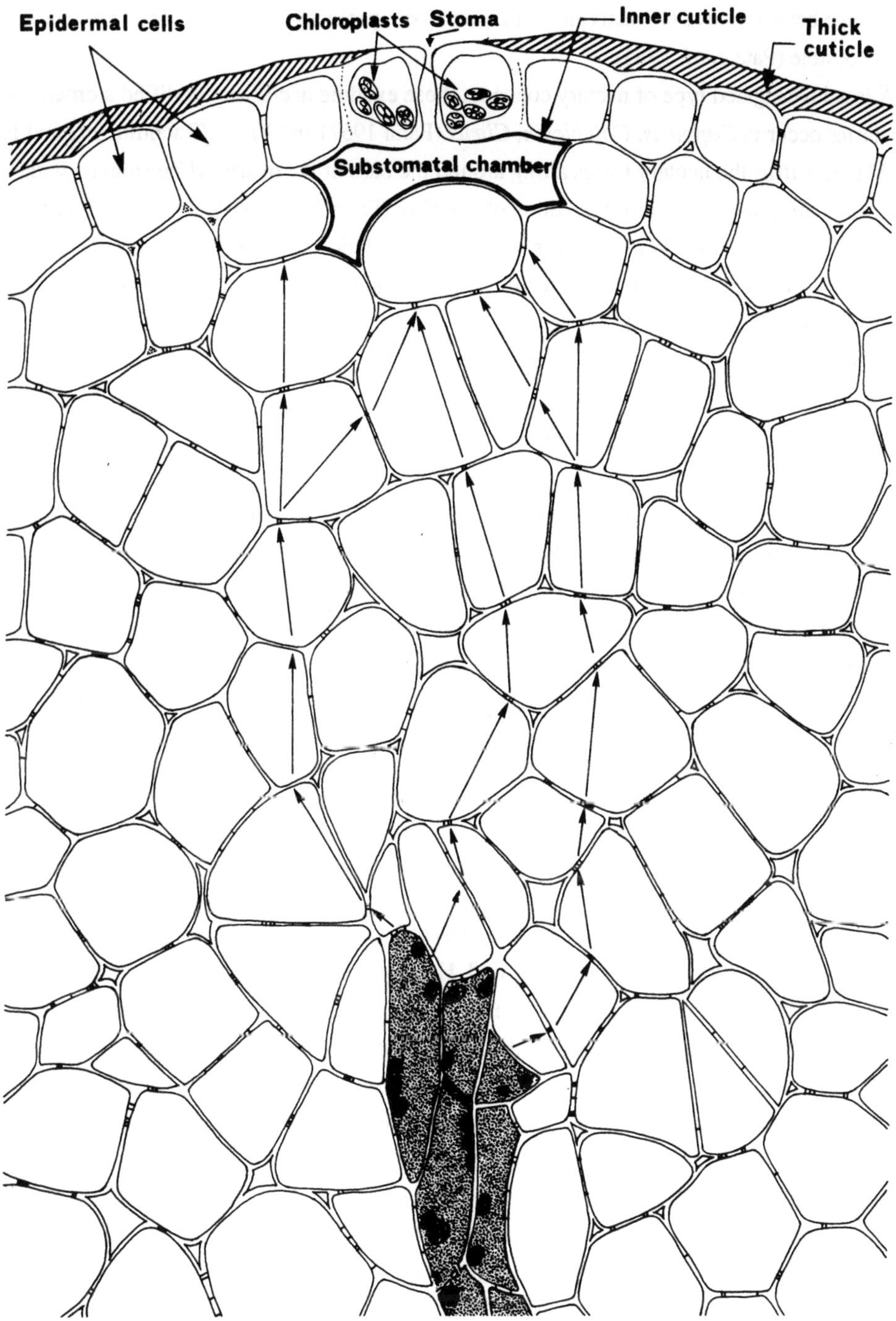

Fig. 3. Model of the floral nectary of *Citrus sinensis* showing some of the possible symplastic routes for prenectar from the sieve elements (stippled) through the nectariferous parenchyma via plasmodesmata to the cells abutting the substomatal chamber (from E P Eleftheriou, unpublished).

exchange between sieve elements and adjacent phloem parenchyma and attached secretory cells is also possible (Pate *et al.* 1985).

A less investigated type of nectary concerns those exuding nectar via modified stomata. Such nectaries occur in *Capparis*, *Colchicum*, *Citrus* (Fahn 1952) and *Vinca* (Rachmilevitz and Fahn 1973). In *Citrus* the nectary forms a ring around the base of the ovary, while stomata with wide apertures are present on raised portions of the ring (Fahn 1952, 1979a, 1982). According to Rachmilevitz and Fahn (1973) sugar is secreted as a solution by means of vesicles derived from the ER. Nectar is secreted to the intercellular spaces and the substomatal chambers from which it is exuded via the stomata. However, further investigation is needed to elucidate functional parameters such as the route of prenectar transport from the sieve elements to the cells abutting the substomatal chambers and the mode whereby nectar is secreted from the symplast to the apoplast. A model is proposed here for the floral nectaries of *Citrus sinensis* (Fig. 3). The secretory tissue consists of compact cells with small intercellular spaces. The epidermis is covered by a thick cuticle and bears stomata with wide subglandular chambers. The cells abutting the chambers are covered by a thin inner cuticle (Fig. 3). An inner cuticle was also observed in the nectaries of *Vinca* (Rachmilevitz and Fahn 1973). Sieve elements of phloem strands, encountered about 7-8 cells deep from the epidermis, vascularise in abundance the nectary. The walls of the nectary cells are perforated by numerous plasmodesmata. Prenectar probably moves from the sieve elements to the cells abutting the substomatal chambers symplastically via the plasmodesmata of the intervening tissue (Fig. 3).

Plasmodesmatal Fine Structure

Only a few studies have been conducted on plasmodesmatal fine structure in nectaries, the most through being in *Abutilon* (Gunning and Hughes 1976, Terry and Robards 1987) and *Gossypium hirsutum* (Wergin *et al.* 1975, Eleftheriou and Hall 1983a).

Most plasmodesmata appear to pierce the walls at right angles (Fig. 4) and their length is about equal to the wall thickness (Figs. 8-10). When they occur in pit fields their length is equal to the wall thickness within the field, which is usually thinner than the nearby wall (Fig. 2, and Wergin *et al.* 1975).

In the nectary cells of *Abutilon*, *Hibiscus*, *Gossypium* and *Citrus* most plasmodesmata occur individually and display a rather simple structure. The plasma membrane is continuous from one cell to the next through the plasmodesmatal canal. Depending upon the section, a distinct or vague desmotubule (Robards 1975, 1976) with uniform thickness is consistently seen passing axially along the plasma membrane-lined plasmodesmatal canal. The space between the plasma membrane and the desmotubule is known as the cytoplasmic sleeve (previously referred to as the cytoplasmic annulus, but see Olesen and Robards, this volume - Figs. 4, 8, 9, 11-13). At the orifices the plasmodesmatal canal is narrower so that it constricts the cytoplasmic sleeve forming two neck regions where the plasma membrane appears to be in close contact with the desmotubule (Figs. 4, 8, 9). Frequently, membranes of ER are seen close to or oriented towards the plasmodesmata (Figs. 4, 5, 8-10). Favourable images give clear evidence for continuity of ER membranes with the desmotubule (Figs. 5, 8) or even luminal continuity between the two structures (Eleftheriou and Hall 1983a).

In *Abutilon*, however, there are no constrictions in about 80% of the cases so that the plasma membrane and the desmotubule are approximately parallel along the entire plasmodesmata

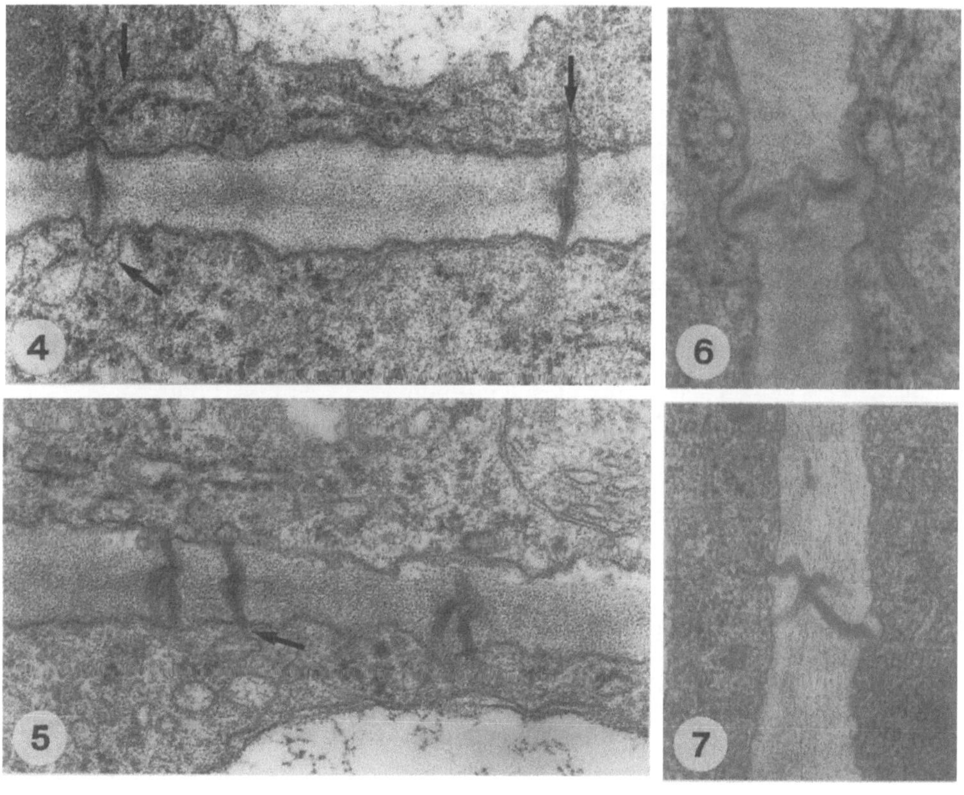

Figs. 4-7. Plasmodesmata of the nectary tissue of *Citrus sinensis*
Fig. 4. Single plasmodesmata in an anticlinal wall just beneath the epidermis. Arrows point to ER cisternae close to the plasmodesmatal canals. X 53.000
Fig. 5. One-sided branched plasmodesmata in a periclinal wall. Arrow points to an ER-plasmodesmatal relationship. X 53.000
Fig.6. A plasmodesma with a median cavity interconnecting it with a neighbouring plasmodesma, the latter being out of section. X 56.000
Fig.7. Two plasmodesmata having an X configuration. X 39.000

(Gunning and Hughes 1976, Kronestedt *et al.* 1986). Based on ultrathin transverse sections Terry and Robards (1987) proposed a model for a plasmodesma of *Abutilon*. In the model the cytoplasmic sleeve has a width of 6.0 nm and the desmotubule is depicted as a tightly rolled cylinder of membrane with a virtually closed central channel. Within the cytoplasmic sleeve there are nine major particles each about 5 nm diameter alternating with discrete cylindrical channels of about 3 nm diameter each. The plasmodesmatal model for *Abutilon* nectaries is in accord with that proposed for *Azolla* root primordia (Overall *et al.* 1982).

In some wall types plasmodesmata have a more complicated structure. In the proximal wall of the stalk cell of cotton they have larger median cavities, which are variously shaped and situated closer to the stalk cell than to the contiguous epidermal cell (Fig. 10). Often, two or more

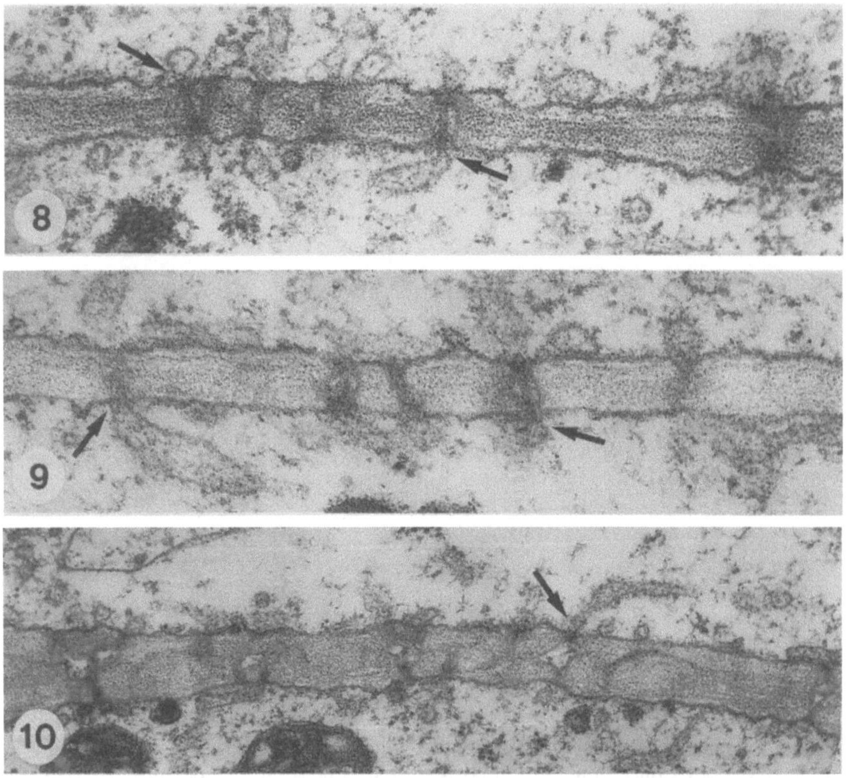

Figs. 8, 9, 10. Longitudinal views of plasmodesmata of the secretory papillae of *Gossypium hirsutum*. In all figures arrows point to ER-plasmodesmatal continuities (from Eleftheriou and Hall 1983a).
Fig. 8. Plasmodesmata in a periclinal wall of intermediate cells. X 65.000
Fig. 9. Plasmodesmata in the distal wall of a stalk cell. X 65.000
Fig. 10. Portion of the proximal wall of a stalk cell illustrating plasmodesmata with a more complex structure. X 32.000

plasmodesmata anastomose at the median cavity or have multiple connections on one side leading to a single channel on the other. Interconnected plasmodesmata also occur in the subglandular cells of cotton (Wergin *et al.* 1975) and *Citrus sinensis* (Fig. 6). In the latter species branched plasmodesmata on one side (Fig. 5) or on both sides (Fig. 7) are encountered. The central cavities of the pairs may be connected at the mid-line of the cell wall (Fig.6).

Measurements of plasmodesmatal dimensions have been carried out in the distal wall of the stalk cell of Abutilon (Gunning and Hughes 1976) and the distal and proximal wall of the stalk cell of cotton nectaries (Eleftheriou and Hall 1983a). The length of the plasmodesmata canals in *Abutilon* was found to be 87 ± 17 nm (in longitudinal sections), while the inner and outer faces of the plasma membrane had diameters 43.6 ± 4.6 and 28 ± 2.9 nm, respectively (in transverse sections). The diameter of the outer face of desmotubule was 16.0 ± 2.1 nm. Results for plasmodesmata dimensions of cotton are listed in Table 1. Comparison of the values between the two species indicates somewhat greater figures for cotton nectaries.

Table 1. Plasmodesmatal dimensions in cotton nectaries from transverse sections to the cell walls (after Eleftheriou and Hall 1983a)

Cell wall type	n	Average diameter at the narrowest position (nm)	Average diameter of the median cavity (nm)	Average plasmo-desmatal length (nm)	Average desmotubule diameter (nm)
Distal wall	25	42.4	75.5	143	21.0
Proximal wall	22	40.9	152.8	235	-

n = sample size (number of plasmodesmata measured)

Plasmodesmatal Frequencies and Distribution Patterns

Although plasmodesmata occur in all walls of the nectary cells, their frequency and distribution pattern vary considerably among different walls (Table 2). In the nectary hair cells the plasmodesmata appear to be evenly distributed throughout a given wall (Figs. 2, 10). However, micrographs from freeze-fracture replicas of *Abutilon* have recently shown plasmodesmata to be mainly arranged in pit fields (Robards and Stark 1988), which however, were not very crowded. Obliquely sectioned walls in the nectary tissue of *Citrus sinensis* reveal plasmodesmata occurring either randomly distributed (Fig.11) or in pairs (Fig. 12). It seems that single plasmodesmata prevail in the anticlinal walls, while the periclinal and oblique walls bear a greater percentage of plasmodesmata in pairs or in pit fields. Densely accumulated plasmodesmata in typical pit fields occur in the basal cells and the subglandular tissue of *Abutilon* (Findlay and Mercer 1971b), *Vicia faba* (Figier 1971), *Gossypium hirsutum* (Wergin *et al.* 1975, Eleftheriou and Hall 1983a) and *Hibiscus* (Sawidis *et al.* 1987b). Within the pit fields plasmodesmata are by far more crowded per unit area than when they are evenly distributed (Figs. 13, 14, Table 2).

In *Abutilon* (Kronestedt *et al.* 1986) and *Hibiscus* (Sawidis *et al.* 1987b) plasmodesmatal frequency was found to decline from the stalk cell towards the tip, a situation considered to be significant from the physiological viewpoint. Evidence has been accumulating that in such long secretory structures nectar is secreted not only from the tip cell but also from all cells distal to the stalk cell (Kronestedt *et al.* 1986, Sawidis *et al.* 1989). Prenectar is actively loaded into a "secretory reticulum" and then unloaded into an extracellular space provided by the lateral cell walls of all trichome cells through sphinctered junctions (Robards and Stark 1988). Should that hypothesis prove to be correct, there would be a reducing need for symplastic transport between the more distal cells of the trichome. It appears that the plant is "programmed" to provide variation in plasmodesmatal number along the trichome and that this variation correlates with the anticipated requirements for transport (Kronestedt *et al.* 1986).

Table 2. Plasmodesmatal frequencies per μm^2 of total wall area in some trichomatous nectaries

Cell wall type	*Abutilon* Floral nectary (Gunning and Hughes 1976)	*Gossypium* Extrafloral nectary (Eleftheriou and Hall 1983a)	*Hibiscus* Floral nectary (Sawidis *et al.* 1987b)
Anticlinal walls of the tip cells	-	-	3.8
Anticlinal walls of the intermediate cells	-	5.7	7.2
Periclinal walls of the intermediate cells	4-5[1]	9.1	10.3
Distal wall of the stalk cell	12.5	15.4	11.8
Intermediate wall of the stalk cell	-	-	19.9
Proximal wall of the stalk cell	-	12.9	20.9
Anticlinal walls of basal cells	-	1.6	14.1
Periclinal wall of basal cells	-	14.1 (50.2*)	15.5
Subglandular tissue	-	40*[2]	10.6 (45*)

*Frequencies within the pit fields alone
[1]In the cells close to the tip (Kronestedt *et al.* 1986)
[2]Estimated by Wergin *et al.* (1975)

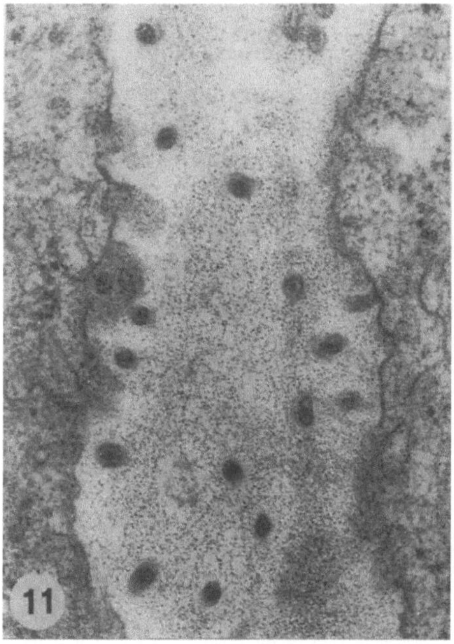

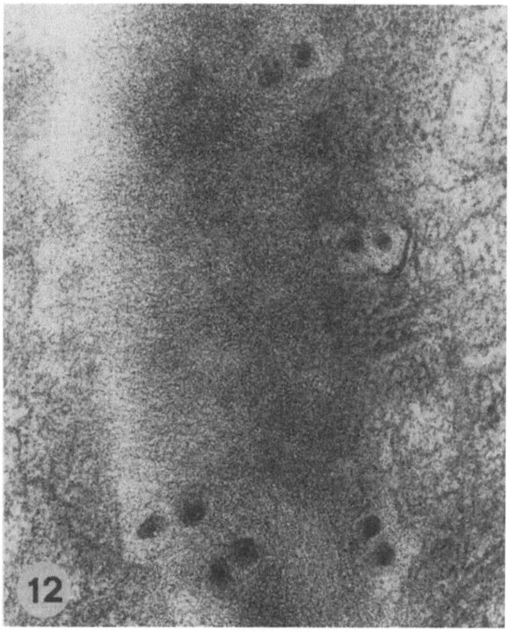

Figs. 11, 12. Obliquely sectioned walls of the nectary tissue of *Citrus sinensis*
Fig. 11. Randomly distributed plasmodesmata between an epidermal and a subepidermal cell. X 64.000
Fig. 12. Plasmodesmata in pairs in a wall between two subepidermal cells. X 61.000

Plasmodesmatal Dynamics

One of the main questions of nectary function is the way by which the prenectar reaches the secretory cells. According to Gunning and Hughes (1976), in the trichomatous nectaries of *Abutilon* there are three possible routes by which prenectar could pass beyond the stalk cell towards the apex of the secretory hair: 1. via plasmodesmata 2. across the successive plasma membranes and the intervening walls of the stalk cell and 3. by the apoplastic route around the stalk cell protoplast.

The apoplastic route was structurally and experimentally shown to be inaccessible to both solutes and solvents (Findlay and Mercer 1971a, Gunning and Hughes 1976, Eleftheriou and Hall 1986a, Sawidis *et al.* 1987a, Terry and Robards 1987); thus, the apoplastic route, if open at all, contributes but slightly to the overall volume flow. Of the other two routes, the plasma membrane might have a large enough hydraulic conductivity to cope with the requirements of high rates of water flow, but membrane permeability is not enough in terms of sugar transport (Gunning and Hughes 1976).

By contrast, the plasmodesmatal pathway offers a much more feasible route for flow of both the solute and solvent components of the prenectar. The current model of plasmodesmatal ultrastructure (Overall *et al.* 1982, Terry and Robards 1987) offers two possible pathways for the cell to cell transport; the cytoplasmic sleeve and the desmotubule itself.

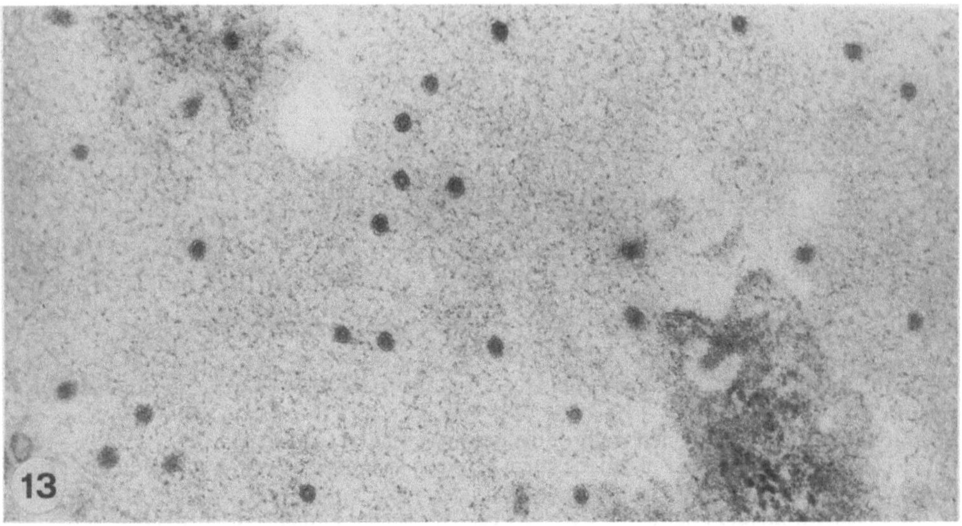

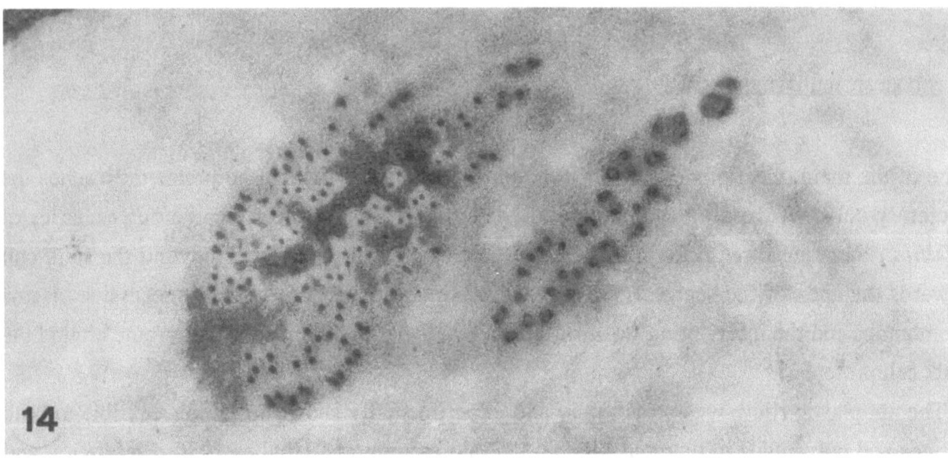

Figs. 13, 14. Tangential sections through different walls of the nectary trichomes of *Hibiscus rosa-sinensis* illustrating two contrasting distribution patterns.
Fig. 13. Evenly distributed plasmodesmata over the whole wall area in a cross wall of intermediate cells. X 64.500
Fig.14. High concentration of plasmodesmata within the pit fields in a wall of a basal cell. There are no plasmodesmata outside the pit fields. X 22.000

The volume flow rates of the cytoplasmic sleeve for *Abutilon* was estimated to be 2.1 x 10^{-2} $\mu m^3 s^{-1}$ per plasmodesma (Gunning and Hughes 1976). Although the cytoplasmic annuli occupy a very small proportion of the total wall area, they are some 250 times more efficient than the membranes as regards water transport. Furthermore, the same authors have calculated that the solute and the solvent will be almost, but not quite, equally well carried by the cytoplasmic annuli, in contrast to the transmembrane pathway. This is because the transport of the solute molecules will be hindered by interactions with the walls of the cytoplasmic sleeve, rendering the efficiency of transport for sucrose slightly less than for the solvent, thus introducing a degree of ultrafiltration by the plasmodesmata (Gunning and Hughes 1976).

Recently, the conductivity of the plasmodesmata in the nectary trichomes of *Abutilon* was ascertained by using various fluorescent molecular probes (Terry and Robards 1987). This conductivity was found to be slightly greater than any other plant tissue and different in several respects from the conductivities reported for the plasmodesmata of other plant species. Furthermore, the same authors have proposed that in *Abutilon* the mobility of the probe is determined solely by the effective Stokes radius of the molecule. The discrete cylindrical channels within the plasmodesmatal sleeve are the most likely route for the diffusion of probes.

Transport through the desmotubule is questionable. Although there is good evidence for continuity of desmotubule and ER (Figs. 8, 9, and Gunning and Hughes 1976, Hepler 1982, Overall *et al.* 1982, Thomson and Platt-Aloia 1985), there is no convincing basis to believe that desmotubule offers a feasible pathway. In *Azolla* roots, a convincing argument has been forwarded that the desmotubule is a virtually closed cylinder providing little or no open pathway for transport (Overall *et al.* 1982). A functionally similar model has been suggested for the root tips of *Lactuca sativa* (Hepler 1982), the salt glands of *Tamarix* (Thomson and Platt-Aloia 1985) and the nectary trichomes of *Abutilon* (Terry and Robards 1987).

Consequently, communication between cells can only be via the cytoplasmic sleeve and its regulation might be brought about by neck constrictions (Olesen 1975) or by the specialised particle complexes between the plasma membrane and the desmotubule (Willison 1976, Overall *et al.* 1982, Hepler 1982, Terry and Robards 1987).

Concluding Remarks

Although some plant glands and especially nectaries have proved suitable experimental systems for the study of cell to cell communication in plants, relatively few attempts have been made to combine and correlate structure and physiological function. In all nectary models the subglandular cells were reported to contain plasmodesmata, the frequency, distribution and fine structure of which vary from one model to the other, these features presumably reflecting local flux rates of symplastic transport. Nevertheless, detailed investigations of the transport capacity of plasmodesmata that will contribute towards extending our relatively sparse knowledge of the physiological function of these intercellular structures are few and are mainly represented by studies carried out in *Abutilon* (Gunning and Hughes 1976, Terry and Robards 1987).

The symplastic permeability through walls with greater or smaller plasmodesmatal frequencies or the permeability of the cytoplasmic annuli in plasmodesmata having neck constrictions, might well be different than those of *Abutilon*. Further experimental research work, presumably by application of similar or even more specialised methods to any other suitable system could provide a clearer picture of the cell to cell transfer mechanisms in plants.

References

Eleftheriou E P, Hall J L (1983a) The extrafloral nectaries of cotton. I. Fine structure of the secretory papillae. J Exp Bot 34: 103-119

Eleftheriou E P, Hall J L (1983b) The extrafloral nectaries of cotton. II ATPase activity, Ca^{2+} -binding sites and selective osmium impregnation. J Exp Bot 34: 1066-1079

Fahn A (1952) On the structure of floral nectaries. Bot Gaz 113: 464-470

Fahn A (1979a) Secretory Tissues in Plants. 302 pp. Academic Press, New York, San Francisco, London

Fahn A (1979b) Ultrastructure of nectaries in relation to nectar secretion. Amer J Bot 66: 977-985

Fahn A (1982) Plant Anatomy, 3rd edition. Pergamon Press, Oxford

Fahn A, Rachmilevitz T (1970) Ultrastructure and nectar secretion in *Lonicera japonica*. In: "New Research in Plant Anatomy" (N K B Robson, D F Cutler and M Gregory, eds). Academic Press, London and New York, and Bot J Linn Soc 63, Suppl 1: 51-56

Figier J (1968) Localisation infrastructurale de la phosphomono-estérase acide dans la stipule de *Vicia faba* L. au niveau de nectaire. Rôles possibles de cet enzyme dans les mechanismes de la sécrétion. Planta 83: 60-79

Figier J (1971) Etude infrastructurale de la stipule de *Vicia faba* L. au niveau du nectaire. Planta 98: 31-49

Findlay N, Mercer F V (1971a) Nectar production in *Abutilon* I. Movement of nectar through the cuticle. Aust J Biol Sci 24: 647-656

Findlay N, Mercer F V (1971b) Nectar production in *Abutilon*. II. Submicroscopic structure of the nectary. Aust J Biol Sci 24: 657-664

Gunning B E S (1976) Introduction to plasmodesmata. In: Intercellular Communication in Plants: Studies on Plasmodesmata, Eds B E S Gunning and A W Robards. Springer-Verlag Berlin, Heidelberg, New York 1976

Gunning B E S, Hughes J E (1976) Quantitative assessment of symplastic transport of pre-nectar into the trichomes of *Abutilon* nectaries. Aust J Plant Physiol 3: 619-637

Gunning B E S, Steer M W (1975) Ultrastructure and the Biology of Plant Cells. London, Edward Arnold

Hepler P K (1982) Endoplasmic reticulum in the formation of the cell plate and the plasmodesmata. Protoplasma 111: 121-133

Hughes J E (1977) Aspects of ultrastructure and function in *Abutilon* nectaries. M. Sc. Thesis, Australian National University, Canberra (Unpubl.)

Kronestedt E C, Robards A W, Stark M, Olesen P (1986) Development of trichomes in the *Abutilon* nectary gland. Nord J Bot 6: 627-639

Kuo J, Pate J S (1985) The extrafloral nectaries of cowpea (*Vigna Unguiculata* (L) Walp): I. Morphology, anatomy and fine structure. Planta 166: 15-27

Olesen P (1975) Plasmodesmata between mesophyll and bundle sheath cells in relation to the exchange of C_4 acids. Planta 123: 199-202

Overall R L, Wolfe J, Gunning B E S (1982) Intercellular communication in *Azolla* roots: I. Ultrastructure of plasmodesmata. Protoplasma 111: 134-150

Pate J S, Peoples M B, Storer P J, Atkins C A (1985) The extrafloral nectaries of cowpea (*Vigna Unguiculata* (L.) Walp): II. Nectar composition, origin of nectar solutes, and nectary functioning. Planta 166: 28-38

Rachmilevitz T, Fahn A (1973) Ultrastructure of nectaries of *Vinca rosea* L., *Vinca major* L. and *Citrus sinensis* Osbeck cv. *Valencia* and its relation to the mechanism of nectar secretion. Ann Bot 37: 1-9

Robards A W (1975) Plasmodesmata. Annu Rev Plant Physiol 26: 13-29

Robards A W (1976) Plasmodesmata in higher plants. In: Intercellular Communication in Plants: Studies on Plasmodesmata, pp. 15-53, eds B E S Gunning, and A W Robards. Springer-Verlag, Berlin, Heidelberg, New York

Robards A W, Stark M (1988) Nectar secretion in *Abutilon*: a new model. Protoplasma 142: 79-91

Sawidis T, Eleftheriou E P, Tsekos I (1987a) The floral nectaries of *Hibiscus rosa-sinensis* L. I. Development of secretory hairs. Ann Bot 59: 643-652

Sawidis T, Eleftheriou E P, Tsekos I (1987b) The floral nectaries of *Hibiscus-rosa-sinensis* L. II. Plasmodesmatal frequencies. Phyton 27: 155-164

Sawidis T, Eleftheriou E P, Tsekos I (1989) The floral nectaries of *Hibiscus rosa-sinensis*. III. A morphometric and ultrastructural approach. Nord J Bot 9 (in press)

Terry B R, Robards A W (1987) Hydrodynamic radius alone governs the mobility of molecules through plasmodesmata. Planta 171: 145-157

Thomson W W, Platt-Aloia K (1985) The ulstrastructure of the plasmodesmata of the salt glands of *Tamarix* as revealed by transmission and freeze-fracture electron microscopy. Protoplasma 125: 13-23

Wergin W P, Elmore C D, Hanny B W, Inger B F (1975) Ultrastructure of the subglandular cells from the foliar nectaries of cotton in relation to the distribution of plasmodesmata and the symplastic transport of nectar. Amer J Bot 62: 842-849

Willison J H M (1976) Plasmodesmata: a freeze-fracture view. Can J Bot 54: 2842-2847

ANALYTICAL STUDIES OF DYE-COUPLING BETWEEN PLANT CELLS

Edward B Tucker
Department of Natural Sciences
Baruch College
City University of New York
17 Lexington Avenue
New York
NY 10010
USA

Introduction

Cell-to-cell communication, defined as the regulated diffusion of ions, metabolites, and other informational molecules through the plasmodesmata, was first suggested by Tangl in 1879. The process allows for the distribution of metabolites throughout a plant and is required for the proper growth and development of plants. Renewed interest in this topic - 100 years after its proposal - has resulted in several reviews (Spanswick, 1975; Gunning and Robards, 1976; Gunning and Overall, 1983; Goodwin and Erwee, 1985). To understand this phenomenon our approach has been (1) to study the ultrastructure of plasmodesma; (2) to determine which plant cells are coupled by using dye passage studies; (3) to analyze the kinetics of cell-to-cell diffusion; (4) to examine the regulation of cell-to-cell communication; and (5) to observe changes in whole plant growth and development when cell-to-cell coupling has been altered. The tissue we use to study the kinetics and regulation of cell-to-cell transport is staminal hairs of Setcreasea purpurea, because such tissue provide a system in which symplastic cell-to-cell passage can be quantitatively measured. The tissue used to study growth and development is Onoclea sensibilis (fern) prothallia.

Structure of S. purpurea staminal hair plasmodesmata

Although the general structure of plasmodesmata is covered by other authors in this volume, I will review the plasmodesmata of S. purpurea staminal hairs. These plasmodesmata are intercellular channels which are lined with the plasma membrane and traversed by a centrally located tubule, the desmotubule (Tucker 1982). They are cylinders 37.6 ± 10 nm in diameter by 288 ± 100 nm long, surrounded by a cylinder of material 29.8 ± 15 nm in thickness which stains less densely than cell wall material. The plasmodesmata contain a cytoplasmic annulus

NATO ASI Series, Vol. H 46
Parallels in Cell to Cell Junctions in Plants and Animals
Edited by A. W. Robards et al.

which is partially occluded with electron-opaque material. The occlusion appears to be ubiquitous in plants as densely staining material can be observed in the cytoplasmic annulus of plasmodesmata from all plant tissues examined: ray cells of willow, Robards, 1968; mesophyll and bundle-sheath cells of Zea mays, Evert et al., 1977; seedlings of Phalaris canariensis, Lopez-Saez et al., 1965; root cells of Azolla, Overall et al., 1982; leaf cells of Salsola kali and root cells of Epilobium hirsutum, Olesen, 1979; leaf cells of Themeda triandra, Botha and Evert, 1988; salt glands of Tamarix, Thompson and Platt-Aloia, 1985; root cells of Lactura sativa, Hepler, 1982; leaf cells of Egeria densa, Erwee and Goodwin, 1985.

Dye coupling

It generally is assumed that if plasmodesmata are observed in tissue, then the cells are physiologically coupled. Conclusions about the degree of cell-to-cell coupling often are based on the number and density of plasmodesmata in cell walls. Now, with the development of plant microinjection techniques, the plant researcher can include dye coupling experiments when making conclusions about plasmodesmata function. Fluorescent tracer dyes, such as fluorescein, Lucifer Yellow, or rhodamine, can be microinjected directly into the cytoplasm and the symplastic transport of these dyes can be followed by fluorescent microscopy. The methodology may be qualitative (one counts the number of cells away from the microinjected cell that the dye has permeated) or quantitative (coefficients of fluorescent dye transport can be calculated from the kinetics of the fluorescent dye cell-to-cell passage). In either case, it is imperative that the probe material be microinjected into the cytoplasm in a manner that will result in the least amount of mechanical damage. We use a very fine micromanipulator and the microscope is placed on a vibration-free table. Dye material should not be microinjected into the vacuole. The researcher can keep chemical toxicity and radiation damage to a minimum by microinjecting only small amounts of the fluorescent probes and by using neutral density filters to remove some of the UV radiation. Weak light emission from the cells can then be enhanced using a SIT television camera or other image-enhancing system. The fluorescent dyes will enter the cytoplasm from the pipette tip simply by diffusion, or they can be delivered by applying either a polarizing current or hydrostatic pressure. To avoid damage, the polarizing current must be small and applied in pulses. It is assumed that these different methods give the same results, but a comparative study should be undertaken. We microinject dyes into the cytoplasm by simple diffusion (Tucker and Spanswick, 1985) or by iontophoresis (Tucker et al., 1989). The diffusion method is very dependent upon the size of the pipette tip, which must be small enough to prevent capillary uptake of the bathing fluid and large enough to allow dye to diffuse into the cytoplasm at a good rate. The iontophoretic method is easier (not so dependent upon the size of the pipette tip) and allows the dye to be delivered more quantitatively. Therefore, it has been used in studies to calculate coefficients of dye intercellular diffusion (Tucker et al., 1989).

When CF was microinjected (diffusion or iontophoresis) into the cytoplasm of Setcreasea staminal hairs, it permeated to the fifth cell along the chain of cells within 1 to 2 minutes. Transport appeared slower in older and larger cells and could be inhibited if a large amount of dye was delivered from the pipette tip into the cytoplasm. Transport was faster if permeation was allowed to continue in the dark; that is, without large amounts of UV excitation radiation focussed on the dye-containing cells.

Size limit for plasmodesmata permeation

To examine permeation selectivity of plasmodesmata, molecular probes primarily made up of fluorescein isothiocyanate (FITC) complexed with amino acids and peptides were microinjected into plant cells. The spread of these fluorescent molecules through the symplast was monitored. We found that molecules traveled rapidly when composed of FITC complexed to single amino acids with polar and aliphatic R groups, while those containing peptides traveled more slowly (Tucker, 1982). No dye molecules composed of amino acids with an aromatic side group passed from cell to cell. We concluded that the material occluding the cytoplasmic annulus constituted a diffusion barrier, by presenting a hydrophilic environment which allows passage of molecules with a maximum molecular weight of 700-800 daltons, but which retains those with aromatic side groups. Goodwin (1983) and Erwee and Goodwin (1984) reported similar results when FITC conjugates were microinjected into mesophyll cells of Elodea canadensis and Egeria densa. Goodwin (1983) reported that FITC leucyl-diglutamylleucine of molecular weight 874 appeared to be close to the limit for cell-to-cell passage. He estimated the pore diameter to be between 3.0 to 5.0 nm. Erwee and Goodwin (1984) reported that FITC conjugates with aromatic amino acids did not pass through the plasmodesmata. Terry and Robards (1987) studied the permeation of FITC conjugates through plasmodesmata of nectary trichome cells of Abutilon striatum and reported a permeation channel width of 3 nm. Some of the aromatic amino acid conjugates which did not permeate plasmodesmata in Setcreasea, Elodea, and Egeria did pass through Abutilon nectary trichome plasmodesmata. However, permeation of many of the aromatic amino acid-containing conjugates appeared to be impeded. The authors suggested that the effect of aromatic amino acids was to increase the size of the conjugate until the molecules become too large to pass through plasmodesmata.

Exclusion limit differences were reported for different tissues in Egeria densa by Erwee and Goodwin (1985). Probes of 665 MW could pass between cells of the shoot-apex and leaf epidermis, but not between cells of stem and root epidermis. Carboxyfluorescein of MW 376 would pass between cells of the stem and root epidermis. In addition, impermeable boundaries were noted between the epidermis and cortex, at nodes between expanding internodes, and between root cap cells and the rest of the root. In a separate study, Erwee et al. (1985) reported CF coupling between epidermal, spongy and palisade mesophyll and vascular cells, but not between epidermal and guard cells in leaves of Commelina cyanea. The lack of dye coupling between mature epidermal and guard cells has also been reported in Commelina communis leaves by Palevitz and Hepler (1985).

Kinetics of cell-to-cell communication

Tyree published in 1970 a paper, "A General Theory Of Symplastic Transport According To The Thermodynamics Of Irreversible Processes". Using his own experimental observations as well as information from the literature, he concluded that (1) the plasmodesmata constitute the pathway of least resistance for the diffusion of all small solutes; (2) diffusion will be the predominant transport mechanism across the pores for small solutes; and (3) solute distribution within the bulk cytoplasm of each cell ought to be a combination of diffusion and cyclosis.

Results from several quantitative analysis systems support this theory. Tyree and Tammes (1975) introduced fluorescein into the cytoplasm of a Tradescantia staminal hair cell through a wound and measured the movement of the dye out of that cell and along the respective chain of cells. Fluorescein passage, in terms of the cell number and the distance that these cells were located away from the wounded uptake cell, was observed to be proportional to the square root of time (consistent with Tyree's diffusion theory). These authors stated that $x^2 = 4k^2Dt$ where x is the distance in cm traversed in time t in s, and k is a factor determined by the ratio of the concentration of the detectable front, c_m, and the source concentration, c_o. Translocation proceeded in both the acropetal and basipetal directions at the same rate and the dye was not observed to pass into the vacuole. The calculated diffusion coefficient of fluorescein was calculated to be much smaller than the diffusion coefficient of fluorescein in water and the authors concluded that the plasmodesmata could be obstructed by as much as 99.3%. Barclay, Peterson, and Tyree (1982) performed similar experiments on trichomes of Lycopersicon esculentum and also found that translocation of fluorescein was proportional to the square root of time. Again, the diffusion coefficient of fluorescein through plasmodesmata was calculated to be smaller than through water.

The kinetics of cell-to-cell transport has been studied also in the staminal hairs of Setcreasea purpurea (Tucker and Spanswick, 1985; Tucker et al., 1989). In our first study, dye was allowed to continue to diffuse from the tip of the micropipette into the cells over the entire experiment. No effort was made to choose cells that were uniform in size and developmental stage along the chain of cells. In these experiments, coefficients could be calculated only when assuming that each cell pair had its own coupling coefficient. We found that transport along the chain of cells was much faster (through 5 cells in less that 2 min) than had been reported previously for Tradescantia (through 5 cells in 25 min). To extend these studies, a computer program was developed to generate concentration curves and compare them to data from microinjection experiments. In these experiments, which employed iontophoresis, CF was microinjected into the cytoplasm in a 5 to 10 s pulse, the pipette was removed and the transport along the chain of cells was videotaped. For these experiments, care was taken to select staminal hairs that were composed of cells uniform in size and developmental stage. For the computer program the equation

$$\Delta C_i = [K(C_{i-1} - C_i) - K(C_i - C_{i+1}) - L(C_i)](\Delta t)$$

was used. In this equation, ΔC_i = change of fluorescence intensity (V), including apparatus scaling factors, in cell i; C_i, C_{i-1}, C_{i+1} = fluorescence intensity (V) at time t for cell i, i-1, i+1, respectively; K = coefficient of intercellular junction diffusion in s^{-1}; L = coefficient of loss in s^{-1}; t = time step of calculation. The average diffusion coefficient of CF in the intercellular pores was calculated to be $5.34 \pm 1.52 \times 10^{-8}$ $cm^2 s^{-1}$, while the average loss across the vacuolar membrane was $9.44 \pm 1.69 \times 10^{-7}$ $um^{-2} s^{-1}$. The results of these experiments support Tyree's original symplastic diffusion theory and we have concluded that (1) symplastic transport of small hydrophilic molecules is governed by diffusion through intercellular pores (plasmodesmata) and intracellular loss (transport into the vacuole). Diffusion in the cell cytoplasm is never limiting; (2) each cell pair must be considered as its own diffusion system. Therefore, a diffusion coefficient cannot be calculated from an entire chain of cells; (3) the rate of passage through plasmodesmata in either direction is the same; and (4) diffusion through the intercellular pores is much slower than diffusion through similar pores filled with water.

Although the exact intraplasmodesmatal space through which molecules diffuse during cell-to-cell transport has not been identified in S. purpurea, it is assumed to be through the cytoplasmic annulus, as suggested in other plants (Lopez-Saez et al., 1966; Overall et al., 1982; Hepler, 1982). Pores (presumptive diffusive plasmodesmatal) have been observed in the cytoplasmic annulus of Azolla root plasmodesmata stained with tannic acid and ferric chloride (Overall et al., 1982). The chemical composition of the plasmodesmatal sieve material is unknown but it has been suggested to contain protein (Overall et al. 1982). It possibly contains gap junction-like proteins since antibodies to gap junction proteins have been reported to bind in the cell wall area of soybean cells (Meiners and Schindler, 1987).

Cytoplasmic streaming

Although the role that cytoplasmic streaming plays in the biology of a plant cell remains unknown, it is tempting to speculate that it assists cell-to-cell transport. Tyree (1970) stated, "solute distribution within the bulk cytoplasm of each cell ought to be by a combination of diffusion and cyclosis". Although it appears that cytoplasmic streaming is required to drive cell-to-cell transport in the large cells of Chara (Bostrom and Walker, 1976), no evidence exists for a similar symplastic transport requirement in smaller plant cells. Cande et al., (1973) reported that when cytoplasmic streaming in oat and maize coleoptile cells was completely inhibited with the microfilament-disrupting agent cytochalasin B, polar transport of auxin continued at a slightly reduced rate. In these experiments, the cytoplasm was drastically changed and could be seen as a piled up mass. Barclay et al. (1982) reported that when Lycopersicon esculentum trichomes were treated with cytochalasin B until the cytoplasmic structure collapsed, the rate of fluorescein transport was the same as that of untreated trichomes.

Using high-resolution, computer-enhanced video microscopy, cytoplasmic streaming in staminal hairs of Setcreasea purpurea was studied in our laboratory (Tucker and Allen, 1986). Cytoplasmic streaming was observed to be particles and organelles moving along well-defined pathways, in repeated and unequal saltatory steps, at different rates, and sometimes against the main direction of flow. Treatment of staminal hairs with azide or cyanide caused particle movement to stop, while low-temperature treatment caused changes in the cytoplasm and the particles to move in a disorganized fashion. In a separate study, it was noted that when streaming was stopped with either metabolic inhibitors or microfilament disrupting agents, passage of microinjected CF continued normally (Tucker, 1987). When these experiments were repeated using the diffusion equation computer analysis system, we found that when cytoplasmic streaming was completely inhibited, K values (coefficients of CF diffusion between cells) were similar to those of untreated staminal hairs. However, the L values (coefficients of intracellular loss) were lower, indicating that transport into the vacuole had decreased. It now was observed on the computer curves that the amount of CF remaining in the cytoplasm was higher than predicted from computer-generated kinetic curves, indicating that some of the CF had become bound to the cytoplasm.

Plasmolysis and dye coupling

With severe plasmolysis, the plasmodesmatal connections break and fluorescent dyes no longer pass from cell to cell. Barclay et al. (1982) found that when Lycopersicon esculentum trichomes cells were uncoupled by plasmolysis, coupling did not recur when cells were deplasmolysed for up to 4 hours. We found that when Onoclea sensibilis prothallia cells were plasmolysed, dye coupling was lost but would slightly recur 48 hours after deplasmolysis. Recovery might be due to the formation of new plasmodesmata since it appears that secondary formation of plasmodesmata in tissue culture takes about the same period of time (J. Monzer, personal communication). Erwee and Goodwin (1984) reported that plasmolysis of Egeria densa cells followed by deplasmolysis results in increased permeability that persists for at least 20 hours. Dyes of MW 1678 and dyes containing aromatic amino acids freely passed from cell to cell, and could not be blocked by microinjecting calcium.

Regulation of cell-to-cell communication

Plasmodesmata are presumed to be dynamic structures; that is, their diffusive pores open and close in response to stimuli, and thus regulate symplastic transport. One stimulus could be light and Racusen (1976) reported a 2-fold increase in cell coupling when oat coleoptilcs were irradiated with red light. Huisinga (1968) reported that when oat mesocotyls were exposed to red light, transport of fluorescein was inhibited, indicating that plasmodesmata were closed. These apparently conflicting results need to be clarified. Ishizawa and Esashi (1985) reported that ethylene enhanced the transport of fluorescein through the coleoptiles of rice seedlings after a 2 hour lag period. Strong fluorescence was seen in the phloem and the authors concluded that the ethylene-stimulated transport was symplastic. Data from my laboratory indicate that the inositol trisphosphate-diacylglycerol (IP_3-DAG) second messenger pathway may be involved in the closing of plasmodesmata. (For reviews of this pathway see Berridge, 1984; Berridge and Irvine, 1984.) In 90% of the experiments in which either IP_2 or IP_3 was microinjedted into the staminal hair cells, cell-to-cell diffusion of CF was inhibited (Tucker, 1988). Inhibition of symplastic transport was noted in 90% of the trials when cells were treated with either 1-oleoyl-2-acetyl-glycerol (OAG) or 1,3-dioleoyl glycerol, whereas only slight inhibition occurred when staminal hairs were treated with 1,2-dioctanoyl glycerol. This inhibition appeared to be transient since the effect diminished with time. For example, diffusion was blocked in 75% of trials treated with OAG for 0 to 30 min, 94% of trials treated for 30 to 60 min; 70% of trials treated for 60 to 90 min, and 10% of trials treated for 90 to 120 min. Calcium (may be considered as the third messenger in the IP_3-DAG pathway) has been reported to block cell-to-cell passage of fluorescein glutamylglutamic acid between Egeria leaf cells (Erwee and Goodwin, 1983). Treatment of soybean and Oedogonium cells with the calcium ionophore A23187 inhibited CF intercellular diffusion in the soybean (Baron-Epel et al., 1987) and induced a transient (1 min) electrical uncoupling in the Oedogonium. We found that microinjection of calcium-loaded BAPTA, [1,2-bis(2-aminophenoxy)ethane-N,N,N',N'-

tetraacetic acid], inhibited both cytoplasmic streaming and dye passage. In addition, when staminal hair cells were bathed in 50 ul of a high-calcium buffer (200 mM calcium - 1% triton X-100 - 10 mM KCl - 10 mM HEPES, pH 7) containing calcium ionophore A23187, cytoplasmic streaming stopped and microinjected CF did not diffuse into the neighboring cell. Thus, cell-to-cell diffusion of CF was blocked when staminal hairs of S. purpurea were either microinjected with IP_2, IP_3, or Ca^{2+} BAPTA buffers, or bathed with solutions containing diacylglycerols or the calcium ionophore A23187. Collectively, these results support an inositol trisphosphate - diacylglycerol second messenger pathway involvement in the regulation of cell-to-cell diffusion in plants.

Growth and Development

Evidence is circumstantial to support the widely held hypothesis that cell-to-cell communication via plasmodesmata is required for the proper growth and development of plants (Carr, 1976). For example, plasmodesmata are found between cells which make up a developing tissue or organ but are not found between cells that develop independently (e.g., microspores and sporophytes). We recently have found that if plasmodesmata in Onoclea sensibilis prothallia are broken by plasmolysis, the normal development into the heart shaped prothallia is altered. When cells are uncoupled, in terms of dye passage, each cell of the prothallia becomes totipotent and will develop into its own heart-shaped prothallia. The resultant mass of prothallia resembles a head of leaf lettuce with each leaf derived from a cell of the original prothallia. Similar results have been reported for Pteris vittata gametophytes (Nakazawa, 1963). These results clearly indicate that plasmodesmata are important in maintaining the proper growth and development of plants.

References

Barclay GF, Peterson CA, Tyree MT (1982) Transport of fluorescein in trichomes of Lycopersicon esculentum. Can J Bot 60: 397-402.
Baron-Epel O, Hernandez D, Jiang L-W, Meiners S, Schindler M (1987) Dynamic continuity of cytoplasmic and membrane compartments between plant cells. J Cell Biol 106: 715-721.
Berridge MJ (1984) Inositol trisphosphate and diacylglycerol as second messengers. Biochem J 220: 345-360.
Berridge MJ, Irvine RF (1984) Inositol trisphosphate, a novel second messenger in cellular signal transduction. Nature 312: 315-321.
Bostrom TE, Walker NA (1976) Intercellular transport in plants. Cyclosis and the rate of intercellular transport of chloride in Chara. J Exp Bot 27: 347-357.
Botha CEJ, Evert RF (1988) Plasmodesmatal distribution and frequency in vascular bundles and contiguous tissues of the leaf of Themeda triandra. Planta 173: 433-441.

Cande WZ, Goldsmith MHM, Ray PM (1973) Polar auxin transport and auxin-induced elongation in the absence of cytoplasmic streaming. Planta 111: 279-296.

Carr, DJ (1976) Plasmodesmata in growth and development. In Intercellular Communication in Plants: Studies on Plasmodesmata. Gunning BES, Robards AW, eds. Springer-Verlag, NY

Erwee MG, Goodwin PB (1983) Characterization of the Egeria densa Planch. leaf symplast. Inhibition of the intercellular movement of fluorescent probes by group II ions. Planta 158: 124-130.

Erwee MG, Goodwin PB (1984) Characterization of the Egeria densa leaf symplast: response to plasmolysis, deplasmolysis and to aromatic amino acids. Protoplasma 122: 162-168.

Erwee MG, Goodwin PB (1985) Symplast domains in extrastelar tissues of Egeria densa Planch. Planta 163: 9-19.

Erwee MG, Goodwin PB, van Bel AJE (1985) Cell-cell communication in the leaves of Commelina cyanea and other cells. Plant, Cell and Environment 8: 173-178.

Evert RF, Eschrich W, Heyser W. (1977) Distribution and structure of the plasmodesmata in mesophyll and bundle-sheath cells of Zea mays L. Planta 136: 77-89.

Goodwin PB (1983) Molecular size limit for movement in the symplast of the Elodea leaf. Planta 157: 124-130.

Goodwin PB, Erwee MG (1985) Intercellular transport studied by micro-injection methods. In: Botanical microscopy 1985. pp. 335-358, Robards AW, ed. Oxford University Press, Oxford.

Gunning BES, Overall RL (1983) Plasmodesmata and cell-to-cell transport in plants. BioSci 33: 260-265.

Gunning BES, Robards AW (eds) (1976) Intercellular communication in plants: studies of plasmodesmata. Springer-Verlag, New York

Hepler PK (1982) Endoplasmic reticulum in the formation of the cell plate and plasmodesmata. Protoplasma 111: 121-133.

Huisinga B (1968) Influence of red light on transport of fluorescein in the mesocotyls of dark-drown avena seedlings. Acta Bot Neerl 17: 390-392.

Ishizawa K, Esashi Y (1985) Ethylene-enhanced transport of uranine, a fluorescent dye, in rice seedling explants in relation to ethylene-stimulated coleoptile growth. Plant Cell Physiol 26: 237-244.

Lopez-Saez JF, Gimenez-Martin G, Risueno MC (1965) Fine structure of the plasmodesms Protoplasma 61: 81-84.

Meiners S, Schindler M (1987) The immunological evidence for gap junction polypeptides in plant cells. J Biol Chem 262: 951-953.

Nakazawa, S (1963) Role of the protoplasmic connections in the morphogenesis of fern gametophytes. Sci. Rep. Tokoku Univ. Ser. IV 25: 247-255.

Olesen P (1979) The neck constriction in plasmodesmata. Evidence for a peripheral sphincter-like structure revealed by fixation with tannic acid. Planta 144: 349-358.

Overall RL, Wolfe, JG, Gunning BES (1982) Intercellular communication in Azolla roots. I. Ultrastructure of plasmodesmata. Protoplasma 111: 134-150.

Palevitz BA, Hepler PK (1985) Changes in dye coupling of stomatal cells of Allium and Commelina demonstrated by microinjection of Lucifer yellow. Planta 164: 473-479.

Racusen RH (1976) Phytochrome control of electrical potentials and intercellular coupling in oat-coleoptile tissue. Planta 132: 25-29.

Robards AW (1968) Desmotubule - a plasmodesmatal substructure. Nature 218: 784.

Spanswick, RM (1975) Symplastic transport in tissues, in "Encyclopedia of Plant Physiology", New Series, Vol 2 Transport in Plants (Eds. U. Luttge and MG Pitman) Springer-Verlag, Berlin.

Tangl E (1879) Ueber offene communicationen zwischen den zellen des endosperms einiger Samen. Jb. wiss. Bot. 12: 170-190.

Terry BR, Robards AW (1987) Hydrodynamic radius alone governs the mobility of molecules through plasmodesmata. Planta 171: 145-157.

Thompson WW, Platt-Aloia K (1985) The ultrastructure of the plasmodesmata of the salt glands of Tamarix as revealed by transmission and freeze-fracture electron microscopy. Protoplasma 125: 13-23.

Tucker EB (1982) Translocation in the staminal hairs of Setcreasea purpurea. I. A study of cell ultrastructure and cell-to-cell passage of molecular probes. Protoplasma 113: 193-201.

Tucker EB (1987) Cytoplasmic streaming does not drive intercellular passage in staminal hairs of Setcreasea purpurea. Protoplasma 137: 140-144.

Tucker EB (1988) Inositol bisphosphate and inositol trisphosphate inhibit cell-to-cell passage of carboxyfluorescein in staminal hairs of Setcreasea purpurea. Planta 174: 358-363.

Tucker EB (1989) Inositol phosphates and diacylglycerols inhibit cell-to-cell transport. In: Inositol Metabolism in Plants, Morre DJ, Boss WF, Loewus FA ed. Alan Liss, Publishers

Tucker EB, Allen NS (1986) Intracellular particle motions (cytoplasmic streaming) in staminal hairs of Setcreasea purpurea. Effects of azide and low temperature. Cell Motility and Cytoskeleton 6: 305-313.

Tucker JE, Mauzerall D, Tucker EB (1989) Symplastic transport of carboxyfluorescein in staminal hairs of Setcreasea purpurea is diffusive and includes loss to the vacuole. Plant Physiol. 90: 1143-1147.

Tucker EB, Spanswick RM (1985) Translocation in the staminal hairs of Setcreasea purpurea. II Kinetics of intercellular transport. Protoplasma 128: 167-172.

Tyree MT (1970) The symplast concept. A general theory of symplast transport according to the thermodynamics of irreversible processes. J Theor Biol 26: 181-214.

Tyree MT, Tammes PML (1975) Translocation of uranin in the symplasm of staminal hairs of Tradescantia. Can J Bot 53: 2038-2046.

FUNCTION AND LOCALISATION OF MOVEMENT PROTEINS OF TOBACCO MOSAIC VIRUS AND RED CLOVER MOTTLE VIRUS

K Tomenius
Swedish University of Agricultural Sciences
Department of Plant and Forest Protection
P O Box 7044
750 07 Uppsala
Sweden

Introduction

Viral infection of a plant, in terms of cell to cell movement, is presently considered to be an active process, mediated by a virus-encoded function (Leonard and Zaitlin 1982; Taliansky *et al.* 1982a). Increasing knowledge about viral genomes and virus-encoded proteins, and of tobacco mosaic virus (TMV) in particular, has demonstrated that the 30 kD protein of TMV is involved in the local symplasmic spread of this virus (Meshi *et al.* 1987; Deom *et al.* 1987).

Plasmodesmata are intercellular channels between parenchyma cells through which the exchange of molecules and various metabolites takes place. Small pathogens, such as virus, also pass via plasmodesmata to adjacent healthy cells. The 30 kD protein of TMV has been localised to the plasmodesmata of infected tobacco cells by immunogold cytochemistry (Tomenius *et al.* 1987), suggesting that the protein operates there to facilitate the systemic spread of the virus. The molecular events involved in this transfer of TMV through plasmodesmata remain to be elucidated, but it may well be that there is an interaction between a host protein and the virus-encoded 30 kD protein.

Putative systemic movement proteins of other plant viruses, alfalfa mosaic virus, cauliflower mosaic virus and red clover mottle virus have also been identified and localised to the cell walls or plasmodesmata of infected tissues (Stussi-Garaud *et al.* 1987; Linstead *et al.* 1988; Shanks *et al.* 1989).

Systemic Movement Protein of TMV

The genome of TMV consists of a single-stranded RNA of about 6400 nucleotides, and the complete nucleotide sequences for the common strain vulgare and the tomato strain L have been analysed (Goelet *et al.* 1982; Takamatsu *et al.* 1983; Ohno *et al.* 1984). The TMV RNA encodes for at least four proteins, the 183 kD protein, a read-through product of the smaller 126 kD

NATO ASI Series, Vol. H 46
Parallels in Cell to Cell Junctions in Plants and Animals
Edited by A. W. Robards et al.

protein, both supposed to be active in the TMV RNA replication (Ishikawa *et al.* 1986) and for two other proteins, the 17.5 kD coat protein and the 30 kD putative movement protein (Beachy *et al.* 1976; Bruening *et al.* 1976; Beachy and Zaitlin 1977). The 126 and 183 kD proteins are translated directly from the genomic RNA and the smaller proteins, the coat protein and the 30 kD protein, are translated from two subgenomic mRNAs (Siegel *et al.* 1976; Hunter *et al.* 1976; Beachy and Zaitlin 1977).

Small amounts of the 30 kD protein were identified by its electrophoretic mobility in SDS-PAGE gels in extracts of TMV-infected protoplasts or tobacco leaves (Beier *et al.* 1980; Joshi *et al.* 1983). More definitive evidence for the identification of the 30 kD protein, *in vivo,* was obtained by immunoprecipitation of the protein from TMV-infected cell extracts using antibodies prepared against a synthetic polypeptide corresponding to the C-terminus of the 30 kD protein of the TMV OM (common) strain (Ooshika *et al.* 1984).

In protoplasts transcription and translation of the viral mRNA encoding for the 30 kD protein occur as transient events approximately 2 - 9 hours post infection. This is in contrast to other TMV-encoded proteins, which are produced continuously; thus the synthesis of the 30 kD protein is controlled at the level of mRNA synthesis (Watanabe *et al.* 1984). Joshi *et al.* (1983) detected a maximum of the 30 kD protein 21 hours after infection in TMV-infected leaves, and it was further localised to the plasmodesmata of tobacco mesophyll cells approximately 16 hours after infection. The amount of protein in the plasmodesmata appeared to reach a maximum after 24 hours and thereafter both the amount of label and the number of cross-reactive plasmodesmata seemed to decrease (Tomenius *et al.* 1987), suggesting that the 30 kD protein may play an early and transient role in establishing systemic infection before significant amounts of virus particles have accumulated.

Analysis of TMV Strains Defective in Cell to Cell Spread of Virus

A temperature-sensitive (ts) mutant, Ls-1, of the tomato (L) strain of TMV was isolated by Nishiguchi *et al.* (1978, 1980). Ls-1 replicates at the nonpermissive temperature, 32°C, but, in contrast to the L strain, cannot move from cell to cell at this temperature. Peptide mapping of the 30 kD proteins synthesised, *in vitro,* from TMV L and Ls-1, revealed only a minor difference in the 30 kD sequence (Leonard and Zaitlin 1982). Subsequent analysis of the protein sequence showed that an amino acid, proline in L was substituted with serine in the Ls-1 mutant (Ohno *et al.* 1983). The substituted 30 kD protein of the Ls-1 strain was supposed to be less stable, which may explain its defective function at the high, nonpermissive temperature. An amino acid substitution seems also to be responsible for the ts defect in cell to cell spread of the Ni 2519 mutant of TMV (Zimmern and Hunter 1983).

The function of the 30 kD protein was further analysed by Meshi *et al.* (1987) creating both point mutations and chimeric constructs of the TMV movement gene. When the Ls-1 gene was introduced into the parent strain L, the resulting mutant had the same defect in the function of cell to cell spread as was observed in Ls-1. Frame-shift mutants having different mutations in the gene encoding for the 30 kD movement protein were constructed and, although all mutants replicated in protoplasts, they were unable to infect tobacco plants. These results support the contention that the function of the 30 kD protein is to mediate in the cell to cell spread of the virus.

Inefficient cell to cell movement may also be correlated with a lower production than normal of the 30 kD protein. The low amount of the 30 kD protein produced by an attenuated strain of TMV, derived from TMV L, was related to three amino acid changes in the 126 kD protein, which is active in the process of virus replication. Alterations in the 126 kD protein were correlated with a decrease in the synthesis of the 30 kD mRNA, and the resultant lower concentrations of the 30 kD protein may have reduced the efficiency in cell to cell spread of virus in the infected plant (Watanabe *et al.* 1987).

The role of the TMV systemic movement proteins has also been investigated in functionally resistant plants in which viral replication occurs in initially infected cells, but the virus is unable to spread to adjacent healthy cells. This is the so called subliminal infection state (Sulzinski and Zaitlin 1982). An excellent example of this situation is found in tomato plants carrying the Tm-2 gene, where the TMV L strain is confined to initially infected cells. The resistance of Tm-2 tomatoes is expressed only in leaf tissue and not in protoplasts, and the Tm-2 gene is therefore supposed to interfere with the systemic movement of the virus (Motoyoshi and Oshima 1977). A spontaneous mutant of the L strain, Ltb1, systemically infects Tm-2 tomatoes (Motoyoshi 1984), and a nucleotide sequence analysis of the Ltb1 mutant revealed two base substitutions in the 30 kD movement gene, resulting in two amino acid changes in the movement protein, which were correlated with the ability of Ltb1 to overcome the Tm-2 resistance (Meshi *et al.* 1989).

Unfortunately the mode of interaction between the Tm-2 gene for resistance and the different movement proteins of TMV remains unresolved and information about the Tm-2 gene, or its gene products, are also unavailable.

The 30 kD TMV Protein and the Hypersensitivity Response

Hypersensitivity is generally characterised as a localised infection limited to a small spot surrounded by a necrotic area, although small amounts of virus may be present outside the necrotic area. The induction of the metabolic changes, which include production of mechanical barriers, defense proteins and enzymes, and which lead to necrosis is the result of host-virus interactions. A number of host-encoded proteins specific for the hypersensitivity reaction seem to be involved, but a possible gene product has not been isolated (Smart *et al.* 1987), and it is not known how the host-encoded proteins may interact with proteins active in promoting movement of virus from cell to cell.

The 30 kD protein of TMV was found to be reduced in the cell walls in plants of *Nicotiana tabacum*, cv Samsun NN, which reacts with localisation and necrosis to TMV infection, in contrast to a systemic host, cv Samsun. The protein also disappeared much faster in Samsun NN plants than in the systemic host, which was suggested to be caused by a range of changes induced by the necrogenesis (Moser *et al.* 1988).

Complementation of a Defective Virus Movement Function

Defective functions of the movement proteins in some hosts have been shown to be complemented by related virus strains, or by unrelated viruses belonging to different taxonomic groups (Atabekov *et al.* 1984). It was shown that the TMV L strain, which is unable to spread in Tm-2 tomatoes, could do so in the presence of potato virus X (PVX, potexvirus group), which systemically infects this host (Taliansky *et al.* 1982a).

Red clover mottle virus (RCMV, comovirus group) has a bipartite single stranded RNA, and the two RNA-components, designated B and M, are encapsidated separately in isometric particles (Oxelfelt and Abdelmoeti 1978). The type member of the comovirus group, cowpea mosaic virus, replicates and spreads only in the presence of both components. It was shown that the B-component replicates independently, but cannot move from cell to cell in the absence of the M-component, which codes for the coat proteins and a possible movement protein (Rezelman *et al.* 1982). The M-component of RCMV encodes for the same functions as the M-component of CPMV (Shanks *et al.* 1986). The B-component of RCMV was complemented in virus movement by sunhemp mosaic virus, a tobamovirus, and also by the TMV ts coat protein mutant, Ni 118 (Malyshenko *et al.* 1988). The TMV strain, Ls-1, which is ts in movement function, was complemented by PVX at the nonpermissive temperature. No comparison of the 30 kD protein of Ls-1 with a corresponding protein of PVX was made, but it was suggested that movement proteins of different viruses, for instance TMV and PVX, which systemically infect the same hosts, must have some important functions in common (Taliansky *et al.* 1982b).

Homologies throughout the entire 30 kD movement protein genes of different TMV strains have been found (Saito *et al.* 1988), and extensive similarities between a putative movement protein of tobacco rattle virus (tobravirus group) and TMV have also been found (Boccara *et al.* 1986). More limited similarities in the central region and at the C-terminus were found between the TMV 30 kD movement protein and the gene product 1 of cauliflower mosaic virus (Hull *et al.*

1986). Antibodies against the 30 kD protein of TMV L strain reacted as expected with the 30 kD protein in plasmodesmata infected with the TMV OM strain (Fig. 1), which confirms the amino acid sequence homology between their transport proteins (78%).

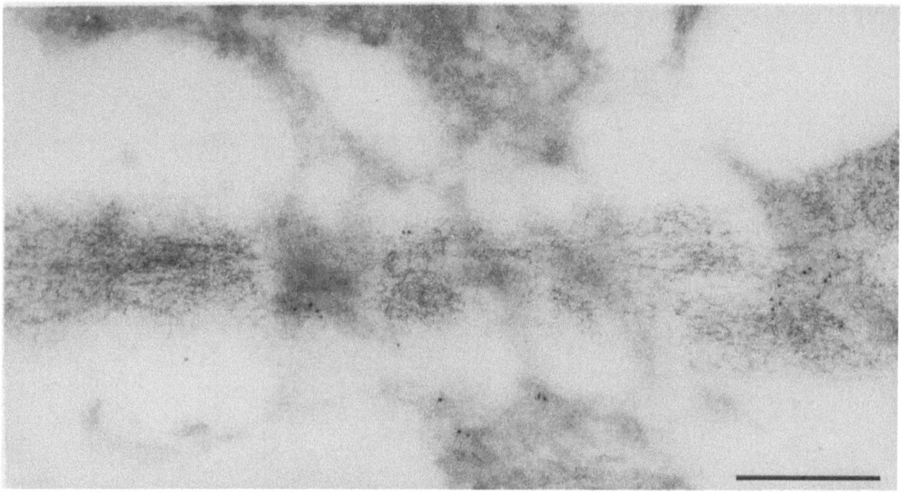

Fig. 1. Plasmodesma in a tobacco cell, infected with the Japanese TMV strain OM is labelled with colloidal gold coupled to antibodies raised against the 30 kD protein from a different TMV strain (L, tomato strain). Bar represents 200 nm.

The Short Distance Transport Form of Viral Infection

Virus-specific ribonucleoproteins (vRNPs) were proposed by Dorokhov *et al.* (1983) to be the transport form of TMV in short distance spread of infection. vRNPs contain genomic TMV RNA, subgenomic mRNAs and at least six different polypeptides including the coat protein subunits. vRNPs appear in the electron microscope as filamentous structures of different lengths and thickness. TMV may also be transported in the form of viral RNA, as mutants of TMV defective in the production of mature virions, or those that are unable to produce coat protein also move from cell to cell in parenchyma tissues (Siegel *et al.* 1962). It is also possible that infection may spread in the form of assembled particles, as virions have been observed by electron microscopy in the plasmodesmata of tissue infected with viruses from different virus groups including caulimovirus, comovirus, nepovirus, luteovirus, cucumovirus, tombusvirus and hordeivirus (Francki *et al.* 1985). Particles of red clover mottle virus were easily located inside plasmodesmata in infected pea plants (Tomenius *et al.* 1982, 1983) and sometimes the virus-like particles appeared in a row extending from a plasmodesma into the cytoplasm (Fig. 2a).

Localisation of Viral Movement Proteins

Virus-encoded proteins which promote virus spread from cell to cell, have been localised to the cell walls and plasmodesmata. In some cases the structure of the plasmodesmata is modified, possibly to promote virus movement, although this has not been proved.

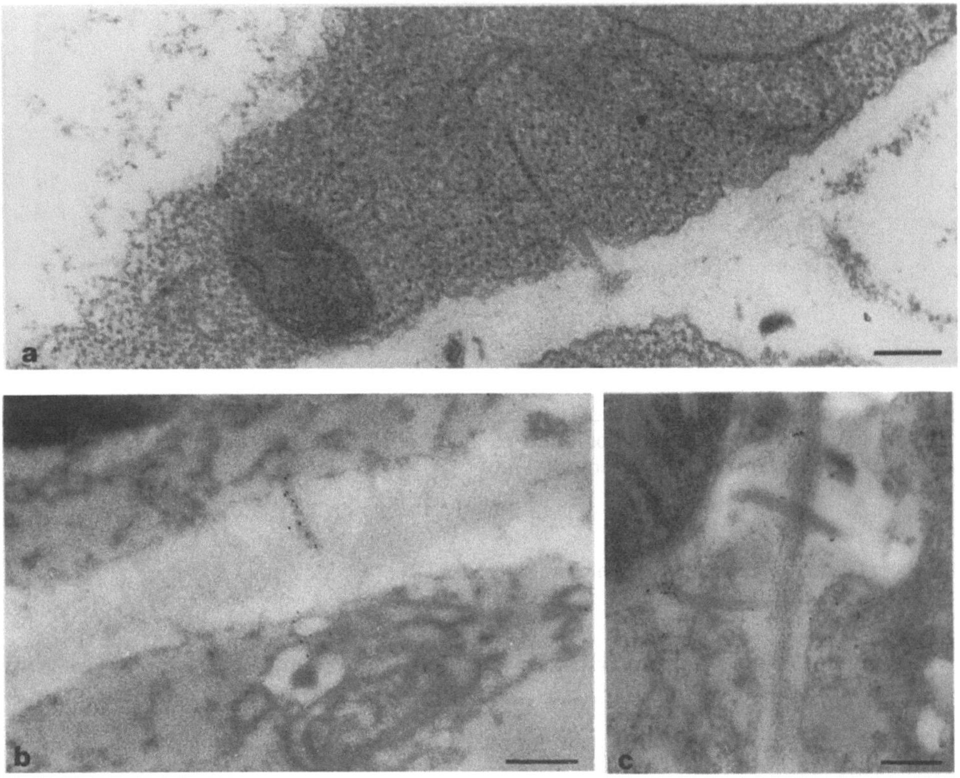

Fig. 2. (a-c). Virus-like particles close to a plasmodesma in RCMV-infected tissue (a). In (b) the label indicates the putative movement protein of RCMV, located inside a plasmodesma, and in (c) the movement protein label is associated with a modified plasmodesma. Bars represent 200 nm.

The inability of a virus to move from cell to cell in some hosts was proposed to be related to virus-induced changes in plasmodesmata frequency, size or structure (Sulzinski and Zaitlin 1982). An increased number of plasmodesmata as a result of incubating plants in the dark and faster movement of virus was found to be correlated (Wieringa-Brants 1981). On the other hand, a significantly lower number of plasmodesmata was found in mesophyll cells infected with the ts

TMV strain Ls-1 at the nonpermissive temperature in comparison with leaves held at 22°C compared to the parent strain, L, at the same temperatures. (Shalla *et al.* 1982).

The TMV OM strain 30 kD protein was first localised to the nuclei in infected protoplasts (Watanabe *et al.* 1986), and it was suggested that the virus-encoded movement function may overcome the host defense reactions by interfering with a possible *de novo* transcription involved in the defense mechanism, and that this should make adjacent cells susceptible to virus infection and allow virus to pass through plasmodesmata. The nuclei were separately investigated, in intact tissue, by immunogold cytochemistry with antibodies raised against a synthetic polypeptide corresponding to 16 amino acids of the C-terminus of the 30 kD protein. Only a very slight increase of the 30 kD protein label was present in nuclei in infected tissue compared to nuclei in healthy tissue (Tomenius *et al.* 1987). The 30 kD protein label was mainly located in the plasmodesmata and in the middle lamella of the cell wall close to plasmodesmata in infected tissue. Virtually no label was present in plasmodesmata in healthy tissue. The protein seemed to accumulate in the plasmodesmata before significant amounts of viral antigen began to pass through them. The 30 kD protein was often detected in the cytoplasm in the vicinity of a plasmodesma and, occasionally, in the intercellular space at the junction of two cells. No virus-induced modifications of the plasmodesmata in infected tissue were observed compared to plasmodesmata in healthy tissue, but the fine structure of plasmodesmata is not very well defined in tissue fixed and embedded only with very mild methods for subsequent treatment with antibodies.

The nonstructural protein P3 of alfalfa mosaic virus (AlMV), was present in the cell wall fraction of infected tobacco plants (Godefroy-Colburn *et al.* 1986), and it was suggested that the AlMV P3 protein was active in virus movement. The P3 protein was later localised to the cell walls, especially to the middle lamella, in infected cells by immunogold cytochemistry with antibodies raised against a synthetic peptide corresponding to the C-terminus of P3 (Stussi-Garaud *et al.* 1987). A few plasmodesmata were observed, but none of these contained label.

The gene 1 product (Pl) of cauliflower mosaic virus (CaMV), which has been proposed to function in cell to cell spread of virus, was detected with antibodies to a 15 amino acid long synthetic peptide corresponding to the C-terminus of the gene 1 product in the cell wall-enriched fraction of CaMV-infected turnips (Albrecht *et al.* 1988) The subcellular localisation of CaMV gene product 1 was made by immunogold cytochemistry with antibodies raised in rabbits to a lacZ-gene 1 fusion product (Harker *at al.* 1987). Label indicating the gene 1 protein was associated with the cell wall close to plasmodesmata in mesophyll cells and with the ends of phloem parenchyma cells in small vascular elements (Linstead *et al.* 1988). The antibody appeared to react with components of the extended cell wall surrounding the enlarged intersymplastic channel of the plasmodesmata. Virus particles were often visible in the modified plasmodesmata, and the localisation of the Pl protein to these plasmodesmata strongly indicates that the protein is involved in the systemic spread of CaMV. The mechanism promoting cell to cell movement is unknown. Furthermore it has yet to be established that the structural modifications of plasmodesmata are important for virus spread.

A possible virus movement protein of red clover mottle virus (RCMV, a comovirus), was recently investigated by Shanks *et al.* (1989). The complete nucleotide sequence of the putative movement RNA component of RCMV was determined earlier (Shanks *et al.* 1986) , and a comparison of the amino acid sequences of the translation products of this RCMV RNA, predicted from the nucleotide sequence with those of cowpea mosaic virus (CPMV, the type member of the comovirus group), showed that, in addition to two capsid proteins, RCMV encodes a nonstructural protein corresponding to the 48/58 kD proteins of CPMV. The 48/58 kD proteins are cleaved from a primary translation product by a protease to give a pair of proteins (48/58 kD) differing only at their N-terminus (Franssen *et al.* 1982). The function of the 48/58 kD proteins of CPMV is unknown, but either or both may be involved in the cell to cell spread of virus (Rezelman *et al.* 1982; Wellink *et al.* 1987).

Antibodies to the 48/58 kD proteins of RCMV were made with a 15 amino acid long oligopeptide corresponding to the C-terminus of the proteins. These antibodies were used to detect the 43 kD product in RCMV-infected protoplasts and for immunogold cytochemistry on sections of pea tissue infected with RCMV. Results from these immunocytochemical investigations demonstrated that the anti-48/58 kD antibodies were strongly bound to the plasmodesmata and located inside the channels (Fig.2b) or in tubular extensions of the channels into the cytoplasm. Cell wall protrusions were occasionally associated with the extended tubules. Many of the plasmodesmata in infected tissue were thus extensively changed (Fig. 2c) in comparison with those in healthy tissue, and the role of these modifications of the plasmodesmata in infected tissue, remains to be established.

Many other questions remain to be answered concerning the functions of the movement protein in plants; for instance the role of the movement protein in both the host range and in the resistance reactions of viruses in plants. The molecular events of plant virus interactions with the genes, or gene products, of host plants has now become the object of intensive research in this field. The outcome of these studies will be of importance both for the plant pathologists and cell biologists working on plasmodesmatal structure and function.

References

Albrecht H., Geldreich, A., Menissier de Murcia, J., Kirchherr, D., Mesnard, J., and Lebeurier, G. (1988). Cauliflower mosaic virus gene 1 product detected in a cell wall-enriched fraction. Virology 163, 503-508

Atabekov, J.G., and Dorokhov, Y.L. (1984). Plant virus-specific transport function and resistance of plants to viruses. Adv. Virus Res. 29, 313-364.

Beachy, R.N., and Zaitlin, M. (1977). Characterisation and *in vitro* translation of the RNAs from less-than-full-length, virus-related, nucleoprotein rods present in tobacco mosaic virus preparations. Virology 81, 160-169.

Beachy, R.N., Zaitlin, M., Bruening, G., and Israel. H.W. (1976). A genetic map for the cowpea strain of TMV. Virology 75, 498-507.

Beier, H., Mundry, K.W. and Issinger, O. (1980). *In vivo* and *in vitro* translation of the RNAs of four tobamoviruses. Intervirology 14, 292-299.

Boccara, M., Hamilton, W.D.O., and Baulcombe, D.C. (1986). The organisation and interviral homologies of genes at the 3' end of tobacco rattle virus RNA 1. EMBO J.5, 223-229.

Bruening, G., Beachy, R.N. Scalla, R., and Zaitlin, M. (1976). *In vitro* and *in vivo* translation of the ribonucleic acids of a cowpea strain of tobacco mosaic virus. Virology 71, 498-517.

Deom, C.M., Oliver, M.J., and Beachy, R.N. (1987). The 30-kilodalton gene product of tobacco mosaic virus potentiates virus movement. Science, Vol. 237, 389-393.

Dorokhov, Y.L., Alexandrova, N.M., Miroshnichenko, N.A., and Atabekov, J.G. (1983). Isolation and analysis of virus-specific ribonucleoprotein of tobacco mosaic virus-infected tobacco. Virology 127, 237-252.

Francki, R.I.B., Milne, R.G., and Hatta, T. (1985). Atlas of Plant Viruses. Vol. I-II. CRC Press. Boca Raton, Florida, USA.

Franssen,H., Goldbach, R., Broekhuijsen, M., Moerman, M., and van Kammen, A. (1982). Expression of middle component of cowpea mosaic virus: *in vitro* generation of a precursor to both capsid proteins by a bottom component RNA-encoded protease from infected cells. J. Virol. 41, 8-17.

Godefroy-Colburn, T., Gagey, M., Berna, A., and Stussi-Garaud, C. (1986). A non-structural protein of alfalfa mosaic virus in the walls of infected tobacco cells. J. Gen. Virol., 67, 2233-2239.

Goelet, P., Lomonossoff, G.P., Butler, P.J.G., Akam, M.E., Gait, M.J., and Karn, J. (1982). Nucleotide sequence of tobacco mosaic virus RNA. Proc. Natl. Acad. Sci. USA. Vol.79, 5818-5822.

Harker, C.L., Mullineaux, P.M., Bryant, J.A., and Maule, A.J. (1987). Detection of CaMV gene 1 and gene IV products *in vivo* using antisera raised to COOH-terminal beta-galaktosidase fusion proteins. Pl. Mol. Biol. 8, 275-287.

Hull. R., Sadler, J., and Longstaff, M. (1986) The sequence of carnation etched ring virus DNA:comparison with cauliflower mosaic virus and retroviruses. EMBO J. 5, 3083-3090.

Hunter, T.R., Hunt, T., Knowland, J., and Zimmern, D. (1976). Messenger RNA for the coat protein of tobacco mosaic virus. Nature 260, 759-764.

Ishikawa, M., Meshi, T., Motoyoshi, F., Takamatsu, N., and Okada, Y. (1986). *In vitro* mutagenesis of the putative replicase genes of tobacco mosaic virus. Nucleic Acids Res. 14:21, 8291-8305.

Joshi, S., Pleij, C.W.A., Haenni, A.L., Chapeville, F., and Bosch, L. (1983). Properties of the tobacco mosaic virus intermediate length RNA-2 and its translation. Virology 127, 100-111.

Leonard, D.A., and Zaitlin, M. (1982). A temperature-sensitive strain of tobacco mosaic virus defective in cell-to-cell movement generates an altered viral-coded protein. Virology 117, 416-424.

Linstead, P.J., Hills, G.J., Plaskitt, K.A., Wilson, I.G., Harker, C.L., and Maule, A.J. (1988). The subcellular location of the gene 1 product of cauliflower mosaic virus is consistent with a function associated with virus spread. J. gen. Virol. 69, 1809-1818.

Malyshenko, S.I., Lapchic, L.G., Kondakova, O.A., Kuznetzova, L.L., Taliansky, M.E., and Atabekov, J.G. (1988). Red clover mottle comovirus B-RNA spreads between cells in tobamovirus-infected tissues. J. gen. Virol. 69, 407-412.

Meshi, T., Motoyoshi, F., Maeda, T., Yoshiwoka, S., Watanabe H., and Okada, Y. (1989). Mutations in the tobacco mosaic virus 30 kD protein gene overcome Tm-2 resistance in tomato. The Plant Cell, vol. 1, 515-522.

Meshi, T., Watanabe, Y., Saito, T., Sugimoto, A., Maeda, T., and Okada, Y. (1987). Function of the 30 kD protein of tobacco mosaic virus: involvement in cell to cell movement and dispensability for replication. EMBO J. 6, 2557-2563.

Moser, O., Gagey, M., Godefroy-Colburn, T., Stussi-Garaud, C., Ellwart-Tschurtz, M., Nitschko, H., and Mundry, K.W. (1988). The fate of the transport protein of tobacco mosaic virus in systemic and hypersensitive tobacco hosts. J. gen. Virol. 69, 1367-1373.

Motoyoshi, F. (1984). A mutant of tobacco mosaic virus that kills tomato seedlings homozygous for genes Tm-2 and nv. Abstracts of the 6th International congress of virology, Sendai, Japan, p94.

Motoyoshi, F., and Oshima, N. (1977). Expression of genetically controlled resistance to tobacco mosaic virus infection in isolated tomato leaf mesophyll protoplasts. J. gen. Virol. 34, 499-506.

Nishiguchi, M., Motoyoshi, F., and Oshima, N. (1978). Behaviour of a temperature-sensitive strain of tobacco mosaic virus in tomato leaves and protoplasts. J. gen. Virol. 39, 53-61.

Nishiguchi, M., Motoyoshi, F., and Oshima, N. (1980). Further investigations of a temperature-sensitive strain of tobacco mosaic virus: Its behaviour in tomato leaf epidermis. J. gen. Virol. 46, 497-500.

Ohno, T., Aoyagi,M., Yamanashi, Y., Saito, H., Ikawa, S., Meshi, T., and Okada Y. (1984). Nucleotide sequence of the tobacco mosaic virus (tomato strain) genome and comparison with the common strain genome. J. Biochem. 96, 1915-1923.

Ohno, T., Takamatsu, N., Meshi, T., Okada, Y., Nishiguchi, M., and Kiho, Y. (1983). Single amino acid substitution in 30K protein of TMV defective in virus transport function. Virology 131, 255-258.

Ooshika, I., Watanabe, Y., Meshi, T., Okada, Y., Igano, K., Inoye, K., and Yoshida, N. (1984). Identification of the 30K protein of TMV by immunoprecipitation with antibodies directed against a synthetic peptide. Virology 132, 71-78.

Oxelfelt, P., and Abdelmoeti, M. (1978). Genetic complementation between natural strains of red clover mottle virus. Intervirology 10: 78-86.

Rezelman, G., Franssen, H.J., Goldbach, R.W., Ie, T.S., and van Kammen, A. (1982). Limits to the independence of bottom component RNA of cowpea mosaic virus. J. gen. Virol. 60, 335-342.

Saito, T., Imai, Y., Meshi, T., and Okada, Y. (1988). Interviral homologies of the 30K proteins of tobamoviruses. Virology 167, 653-656.

Shalla, T.A., Petersen, L.J., Zaitlin, M. (1982). Restricted movement of a temperature-sensitive virus in tobacco leaves is associated with a reduction in numbers of plasmodesmata. J. gen. Virol. 60, 355-358.

Shanks, M., Stanley, J., and Lomomossoff, G.P. (1986). The primary structure of red clover mottle virus middle component RNA. Virology 155, 697-706.

Shanks, M., Tomenius, K., Huskison, N., Barker, P., Wilson, I.G., Maule, A.J., and Lomonossoff, G.P. (1989). Identification and sub-cellular localisation of putative cell to cell transport protein from red clover mottle virus. Virology in press.

Siegel, A., Hari, V., Montgomery, I., Kolacz, K. (1976). A messenger RNA for capsid protein isolated from tobacco mosaic virus-infected tissue. Virology 73, 363-371.

Siegel, A., Zaitlin, M., and Sehgal. O.P. (1962). The isolation of defective tobacco mosaic virus strains. Proc. Natl Acad. Sci. USA. 48, 1845-1851.

Smart, T.E., Dunigan, D.D. and Zaitlin, M. (1987). *In vitro* translation of mRNAs derived from TMV-infected tobacco exhibiting a hypersensitive response. Virology 158, 461-464.

Stussi-Garaud, C., Garaud, J., Berna, A., and Godefroy-Colburn, Y. (1987). *In situ* localisation of an alfalfa mosaic virus non-structural protein in plant cell walls: correlation with virus transport. J. gen. Virol. 68, 1779-1784.

Sulzinski, M., and Zaitlin, M. (1982). Tobacco mosaic virus replication in resistant and susceptible plants: in some resistant species virus is confined to a small number of initially infected cells. Virology 121, 12-19.

Takamatsu, N., Ohno, T., Meshi, T., and Okada, Y. (1983). Molecular cloning and nucleotide sequence of the 30K and the coat protein cistron of TMV (tomato strain) genome. Nucleic Acids Res. 11:3767-3778.

Taliansky, M.E., Malyshenko, S.I., Pshennikova, E.S., and Atabekov, J.G. (1982a). Plant virus-specific transport function. II. A factor controlling virus host range. Virology 122, 327-331.

Taliansky, M.E., Malyshenko, S.I., Pshennikova, E.S., Kaplan, I.B., Ulanova, E.F., and Atabekov, J.G. (1982b). Plant virus-specific transport function. I. Virus genetic control required for systemic spread. Virology 122, 318-326.

Tomenius, K., Clapham, D., and Meshi, T. (1987). Localization by immunogold cytochemistry of the virus-coded 30K protein in plasmodesmata of leaves infected with tobacco mosaic virus. Virology 160, 363-371.

Tomenius, K., and Oxelfelt, P. (1982). Ultrastructure of pea leaf cells infected with three strains of red clover mottle virus. J. gen. Virol. 61, 143-147.

Tomenius, K., Clapham, D., and Oxelfelt, P. (1983). Localisation by immunogold cytochemistry of viral antigen in sections of plant cells infected with red clover mottle virus. J. gen. Virol. 64, 2669-2678.

Watanabe, Y., Emori, Y., Ooshika, I., Meshi, T., Ohno, T., and Okada, Y. (1984). Synthesis of TMV-specific RNAs and proteins at the early stage of infection in tobacco protoplasts: transient expression of the 30K protein and its mRNA. Virology 133, 18-24.

Watanabe, Y., Ooshika, I., Meshi, T., and Okada, Y. (1986). Subcellular localisation of the 30K protein in TMV-inoculated tobacco protoplasts. Virology 152, 414-420.

Watanabe, Y., Morita, N., Nishiguchi, M., and Okada, Y. (1987). Attenuated strains of tobacco mosaic virus. Reduced synthesis of a viral protein with cell to cell movement function. J. Mol. Biol. 194, 699-704.

Wellink, J., Jaegle, M., Prinz, H., van Kammen, A., and Goldbach, R. (1987). Expression of the middle component M RNA of cowpea mosaic virus *in vivo*. J. gen. Virol. 68, 2577-2585.

Wieringa-Brants, D.H. (1981). The role of the epidermis in virus-induced local lesions in cowpea and tobacco leaves. J. gen. Virol. 54, 209-212.

Zimmern, D., and Hunter, T. (1983). Point mutation in the 30-K open reading frame of TMV implicated in temperature-sensitive assembly and local lesion spreading of mutant Ni 2519. EMBO J. 2, 1893-1900.

PLASMODESMATA - VIRUS INTERACTION

W. J. Lucas*, S. Wolf*, C. M. Deom#, G.M. Kishore+ and R. N. Beachy#
*Department of Botany, University of California, Davis CA 95616
#Department of Biology, Washington University, St. Louis MO 63130
+Monsanto Research Center, Chesterfield, MO 63198

INTRODUCTION

The initial entry and replication of a virus in a host plant occurs in only a small number of cells. Expression of viral symptoms in the plant is a result of the spreading of competent viral genetic material (either as particles or as unencapsidated RNA/DNA) through the host plant. In nonhost plants, the virus either fails to replicate or replicates in initially infected cells but fails to move to neighbouring cells. Two general pathways have been identified by which viruses spread within plants. Rapid long-distance movement of viruses can occur either via the phloem or the xylem (Atabekov and Dorokhov, 1984). In such situations, viral entry is normally by means of insect vectors. Alternatively, the spread of viral infection can occur by means of cell-to-cell movement from the infected site to neighbouring cells and tissues. In this chapter we examine the various ways in which viruses may move between cells, given that viral particles are much larger than cellular metabolites which normally pass between physiologically competent cells.

PLASMODESMATA STRUCTURE IN RELATION TO VIRAL MOVEMENT

Plasmodesmata are narrow strands of cytoplasm penetrating through adjoining cell walls to interconnect plant cells. This system, which elevates a plant from a collection of individual cells to an interconnected community of living protoplasts, is widely known as the symplasm. Electron micrographs show that plasmodesmatal structure appears to be a membrane-lined pore of 20-80 nm in diameter (Fig. 1). The central region of the pore contains the axial component or desmotubule, which is associated with the endoplasmic reticulum. In some cases, the cytoplasmic pore may be constricted at the neck regions and appears to be partially occluded with closely-packed, globular subunits which are thought to be proteins (Fig. 1B). Using the technique of image reinforcement, maximum reinforcement was obtained with a presence of nine subunits in the neck constriction region (Olesen, 1979; see also Olesen, this volume). A

NATO ASI Series, Vol. H 46
Parallels in Cell to Cell Junctions in Plants and Animals
Edited by A. W. Robards et al.

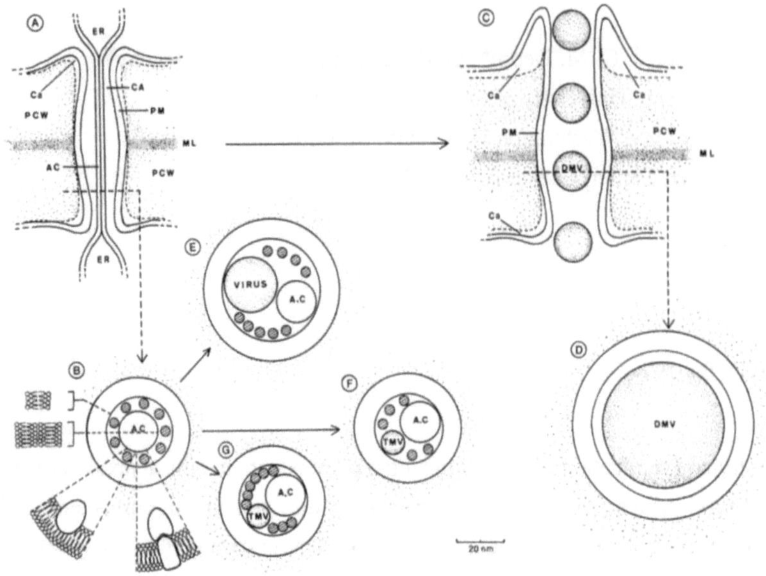

Fig. 1 Schematic representation of a normal plasmodesma (A and B), and a plasmodesma (C and D) in which the sub-structure has been modified by infection with dahlia mosaic virus (DMV). Virally-induced modifications of plasmodesmatal structure include, but may not be limited to removal of the axial component or desmotubule, controlled deposition of callose, and removal of cell wall material to increase the overall diameter (see C, D and E). Other viruses, like tobacco mosaic virus (TMV), may well interact with plasmodesmata to alter the arrangement of the putative proteins in the neck constriction region of the cytoplasmic annulus (cf B, G and F). Symbols are as follows: AC, axial component; CA, cytoplasmic annulus; Ca, callose deposits; ER, endoplasmic reticulum; ML, middle lamella; PCW, primary cell wall; PM, plamsa membrame. Note scale bar refers to B,D - G.

recent model of plasmodesmata, which was based on dye exclusion studies, suggests that the effective pathway for solute movement is through the spaces between the protein subunits, which partially occlude the cytoplasmic annulus (Terry and Robards, 1987). These spaces have an effective diameter of 1.5-2 nm and resemble the physical dimensions found in individual animal gap junctions (Lowenstein and Rose, 1978; see also Willecke et al., this volume).

Certain types of plant viruses are known to spread throughout the host by moving through plasmodesmata, and in some cases this movement is rather ubiquitous, as in the case of mosaic viruses (Esau, 1969; Gibbs, 1976). Kitajima and Lauritis (1969) presented electron microscopic evidence which indicated that dahlia mosaic virus (DMV) interacts with the plasmodesmata of an infected cell in such a way that the axial component is removed. In addition, the physical diameter of these plasmodesmata seems to be increased, and such modifications allow the DMV particles to pass through the plasmodesmata to neighbouring cells (Fig. 1C & D). Electron microscopic evidence of viral particles moving through plasmodesmata of other virally infected plant systems has also been reported (Allison and Shalla, 1974; Esau et al., 1967; Weintraub et al., 1976). Modifications of plasmodesmata were observed also by

Esau (1968), who showed that for plants infected by the beet yellows virus "both the longitudinal and the transectional views of plasmodesmata containing virus particles show no cores (desmotubules) and no endoplasmic reticulum associations".

The particles of most viruses are either globular (icosahedral with a diameter of 18-80 nm), helical, or filamentous rods (rigid or flexuous; diameters ranging from 10-25 nm and lengths of up to 2.5 μm). Very little is known about the size and the conformation of these viruses in the cytoplasm of the host plant, but it is clear that most viruses must act on the plasmodesmata so as to modify them to open the "gates" for the migration of viral genetic material into healthy cells. This viral infection process seems to offer an exciting way to modify the plasmodesmata, providing that one could achieve such a modification without the complications associated with viral infection *per se*.

MOLECULAR STUDIES OF VIRAL MOVEMENT

Molecular biology has aided considerably in both elucidating the viral mechanisms involved in conferring symplasmic mobility, as well as providing transgenic plants in which one of the putative viral movement proteins have been expressed (Deom et al., 1987). One important outcome of these studies has been the conceptual development that a special class of virally-encoded movement proteins (MP) interact with plasmodesmata, thereby causing sub-structural modifications which allow transmission between cells. We will illustrate this aspect of plasmodesmata-virus interaction by focusing on tobacco mosaic virus (TMV).

Nishiguchi et al. (1978, 1980) isolated a temperature-sensitive mutant of tobacco mosaic virus (TMV Ls1) which, under restrictive temperatures (32°C), replicated and assembled normally in both epidermal cells and protoplasts, but did not move from cell-to-cell in inoculated tobacco leaves. An analysis of two-dimensional tryptic peptide maps of proteins synthesized by *in vitro* translation of viral RNAs from a temperature-restrictive mutant and a closely related strain that could move from cell-to-cell, revealed a slight difference only in the 30 kD protein. One base substitution in the 30 kD protein cistron changed the amino acid <u>proline</u> in the temperature-resistant strain to <u>serine</u> in the Ls1 strain. Leonard and Zaitlin (1982) concluded that this 30 kD polypeptide must somehow be involved in mediating the cell-to-cell movement of TMV. A similar conclusion was recently drawn by Huisman et al. (1986) concerning the function of a closely homologous 30 kD protein of alfalfa mosaic virus.

Further support for the hypothesis that the 30 kD viral protein is involved in modifying the plasmodesmatal sub-structure comes from the observation that viruses defective in cell-to-cell movement, like TMV Ls1, can be complemented by a "helper" virus. When TMV Ls1 was coinoculated with a common strain, TMV Ls1 also spread through the symplasm (Meshi and Okada 1986). This observation that different viruses can function as helpers, or can complement a defect in a nonspreading virus in a non-host plant, suggests that the underlying mechanism of cell-to-cell movement may be common to many groups of plant viruses (Meshi and Okada, 1986).

A study of the intracellular localization of the 30 kD protein in TMV-inoculated tobacco protoplasts indicated that the 30 kD protein may associate with the cell nucleus, in order to function, and may not directly interact with the plasmodesmata (Watanabe et al., 1986). Tomenius et al. (1987) reported contradictory results for infected tobacco leaves, since they found, by using immunogold cytochemistry, that the 30 kD MP of TMV was localized to the plasmodesmata. However, these studies provided no direct evidence that the putative MP was changing the plasmodesmata. Actually, nothing was known about the mode by which this TMV 30 kD protein might potentiate cell-to-cell movement. Atabekov and Dorokhov (1984) proposed two very different mechanisms by which systemic spread might be achieved. Plasmodesmata are either modified by the virally encoded MP, or the viral "movement" protein functions by suppressing plant defense responses.

To explore further the mechanism whereby TMV gains access to the symplasm, a chimeric gene encoding the TMV 30 kD protein was introduced into tobacco plants via a modified Ti plasmid in *Agrobacterium tumefaciens* (Deom et al., 1987). By expressing the MP gene in transgenic plants, the function of the 30 kD MP could be studied in the absence of the expression of other viral genes. Deom et al. (1987) found that expression of the MP in these tobacco plants complemented the transport deficiency of the TMV Ls1 strain. Transgenic plants, infected with Ls1 and maintained at the nonpermissive temperature, potentiated cell-to-cell movement of the Ls1 virus in both inoculated and upper systemic leaves. This finding provided direct evidence that the 30 kD MP of TMV is necessary for virus movement. Further support for this concept of a MP was provided by Meshi et al. (1987) who constructed TMV mutants by using different types of frame-shifts at various positions on the MP region of the TMV genome. All such mutants were replication-competent but none showed systemic infectivity in tobacco plants. Furthermore, deletion from the TMV RNA of most of the region encoding for the MP did not prevent this mutant from replicating as well as producing the subgenomic mRNA for the coat protein.

EXPERIMENTAL INVESTIGATION OF THE SYMPLASMIC DOMAIN

A preliminary electron microscopy study on control and tobacco plants expressing the TMV MP showed no significant difference between their plasmodesmata. Axial components (desmotubules) and the cytoplasmic annulus were present in both plant types, indicating that a change in plasmodesmatal sub-structure, not easily detected with routine electron microscopic methods, may permit systemic spread of TMV (W.J. Lucas and T. Pesacreta unpublished results).

Techniques for microinjection of non-toxic membrane-impermeable fluorescent dyes (e.g., Lucifer Yellow CH), which were developed for tracing neurological interactions (Stewart, 1981) and studying gap junctions (Loewenstain and Rose, 1978; see also Pitts et al., this volume), have found, with only minor refinements, direct application in a large number of plant tissues (Fisher, 1988; Madore et al., 1986; Palevitz and Hepler, 1985; Tucker, 1982). The finding that fluorescein passes through cell junctions led to the synthesis of other fluorescent molecules for probing junctional permeability (Erwee and Goodwin, 1983; Terry and Robards, 1987). These new probes combined some of the desired features of fluorescein, such as high fluorescence, excitation by visible light, low toxicity toward cells, lack of nonjunctional membrane transport, with some non-fluorescent parameters, e.g., size and charge. Synthesis of fluorescent peptide probes of known molecular weight and radius, developed to probe the size exclusion limits of gap junctions (Flagg-Newton et al., 1979), has also been exploited by plant scientists to establish the extent of symplasmic permeability in plant tissues (Erwee and Goodwin, 1983; Terry and Robards, 1987; Tucker, 1982).

A limitation to the use of tracer dyes, and other chemical probes, is that their injection into the plant cell, by means of a micropipette, generally occurs into the large central vacuole. This problem was overcome by developing a technique in which the dye was pre-encapsulated in 0.2 μm liposomes. Pressure injection of these liposomes into the vacuole allowed them to contact

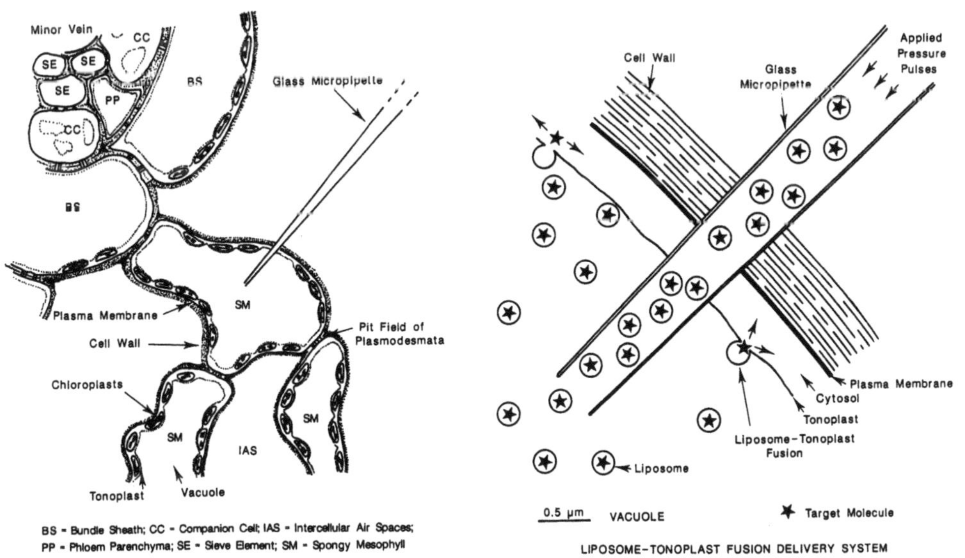

Fig. 2 Schematic representation of the technique employed to introduce fluorescently-labelled target molecules, *in vivo*, into the cytosol of tobacco mesophyll cells (see Wolf et al., 1989).

and subsequenty fuse with the tonoplast, thereby releasing the dye into the cytoplasm (Fig. 2; see also Madore et al., 1986). We recently employed this liposome delivery system to compare the plasmodesmatal size exclusion limits of control and transgenic tobacco plants that express the 30 kD MP gene (Wolf et al., 1989).

Lucifer yellow CH (LYCH), fluorescein isothiocyanate (FITC) labeled hexaglycine (F-gly$_6$) and FITC labeled dextrans (F-dextrans) with various molecular weights were used as fluorescent probes. Liposomes were loaded into the tips of glass micropipettes having a tip diameter of 0.5-1.0 µm. The capillary was sealed into a micropipette holder equipped with a luer port (WPI, New Haven CT, USA), and computer controlled pressure injection was achieved by means of a pneumatic PicoPump (WPI, model PV830). Pipette movement for cell impalement was controlled by an hydraulically driven micromanipulator (Narishige, Tokyo, Japan; model MO-102). Symplasmic movement of these dyes was monitored using epi-illumination (halogen lamp) on a Leitz Orthoplan Photomicroscope equipped with a blue (BP 390-490) excitation filter. Fluorescence was measured using an analytical photon detection system (Hamamatsu Photonics K.K., model C1966-20) mounted on the Leitz-photomicroscope. Images stored on a Sony video cassette recorder (model VO-5800) were processed through the Hamamatsu system and false color analysis was displayed on a Sony Triniton color video monitor (model PVM-1271Q) from which permanent photographic records could be obtained.

The molecular exclusion limits of plasmodesmata have been established to be 700 to 800 D in several plant systems (Erwee and Goodwin, 1983, 1984; Tucker, 1982; Tucker and Spanswick, 1985). As expected from experimental data available from other plant species, the size exclusion limits for cell-to-cell mobility in control tobacco plants was in the 750 D range (Table 1). However, tobacco plants in which the 30 kD MP was expressed allowed large F-dextran molecules to move from cell to cell. Movement of 9,400 D F-dextran was observed in 95% of the injection experiments, but F-dextran of 17,200 D did not move even 20 minutes after injection (Table 1).

To ensure that fluorescent movement was not a consequence of aberrant metabolism, in the transgenic tobacco plants, which resulted in cleavage of small fluorescein-dextran fragments, leaf extracts from F-dextran injected tissues were chromatographed. By using the high sensitivity of the photon analytical mode of the Hamamatsu system, the small quantities of 9,400 D F-dextran were readily analyzed by fluorescence detection. No FITC, or smaller molecular weight F-dextran was found in these experiments. These results, along with the fact that F-dextran (17,200 D) did not move even 20 minutes after injection, indicated that the F-dextran probes were not metabolized by the mesophyll cells of the transgenic tobacco plants.

Collectively, the microinjection studies and fluorescence-chromatography data support the hypothesis that the MP is responsible for the detected change in size exclusion limit of the

Table 1. Mobility of fluorescent probes through plasmodesmata connecting mesophyll cells of transformed tobacco plants. Data are presented as the percent of injections which showed movement of the specific probe, as judged 2 min after injection. (Values in parenthesis represent number of injections.) Data from Wolf et al. (1989).

Probe	MW (D)	Transformant	Genotype MP	% Injections indicating dye movement
LYCH	457	277	+	100 (5)
		306	-	100 (5)
F-Gly$_6$	749	277	+	100 (8)
		306	-	50 (10)
F-Dextran	3,900	277	+	100 (6)
		306	-	14 (7)
F-Dextran	9,400	277	+	95 (20)
		306	-	0 (12)
F-Dextran	17,200	277	+	0 (6)
		306	-	0 (6)

tobacco mesophyll plasmodesmata. The modification in size exclusion limit is significant, representing more than an order of magnitude increase. Furthermore, these results represent the first experimental verification that a putative viral MP actually interacts with plasmodesmata in a physiological sense (see Wolf et al., 1989).

SEQUENCE AND STRUCTURAL DATA ON VIRAL MOVEMENT PROTEINS

To date the MP genes from four tobamoviruses have been cloned and sequenced. Table 2 shows the characteristics of the genes and the predicted proteins which range in size from 264-283 amino acids. The predicted amino acid sequences indicate that the proteins have a high content of charged amino acids, over half of which are located in the C-terminal third of the protein.This C-terminal third of the protein can be further divided into regions (Saito et al., 1988); a large region rich in basic amino acids is flanked by two smaller regions rich in acidic amino acids. As seen in Fig. 3, the N-terminal two-thirds of the TMV MP is predominantly hydrophobic, while the C-terminal third of the protein is hydrophilic. The hydropathic profiles of the ToMV, CGMMV, and SHMV movement proteins are very similar to that of TMV (data not shown). These movement proteins have an overall positive net charge (basic) with the C-terminal third of each protein being more basic than the remainder of the protein. The similarities between the charge distribution and hydrophobic profiles of the tobamovirus MPs

Table 2. Characteristics of the MP genes and proteins of four tobamoviruses. (Abbreviations: TMV, tobacco mosaic virus; ToMV, tomato mosaic virus; CGMMV, cucumber green mosaic virus; SHMV, sunn-hemp mosaic virus.)

Virus	Gene size (nt)	Protein size (aa)	Calculated MWt	References
TMV	807	268	30.0 kD	Goelet et al., 1982
ToMV	795	264	29.5 kD	Takamatsu et al., 1983
CGMMV	795	264	29.0 kD	Saito et al., 1988
SHMV	852	283	30.0 kD	Meshi et al., 1982

suggest that the hydropathic properties of these proteins might be important for structure and/or function.

The percentage of overall amino acid identity between the four MPs is illustrated by the data presented in Table 3. Here we should stress that higher degrees of identity can be found in shorter regions of the sequences. The most significant overall homology occurs between the TMV and ToMV movement proteins (79%). The homology is greater than 90% in the N-terminal four fifths, but decreases to approximately 38% in the C-terminal portion of these proteins. Interestingly, the divergence in the C-termini of the TMV and ToMV MPs does not interfere with the ability of the TMV MP, present in transgenic plants, to complement the temperature sensitive defect in the MP of ToMV Ls1 strain (Deom et al., 1987).

A comparison of the secondary structural predictions (Chou and Fasman, 1974) for the tobamovirus MPs indicates tentative structural similarities between these proteins (Fig. 4). There is the greatest degree of structural similarity between TMV and ToMV, which is

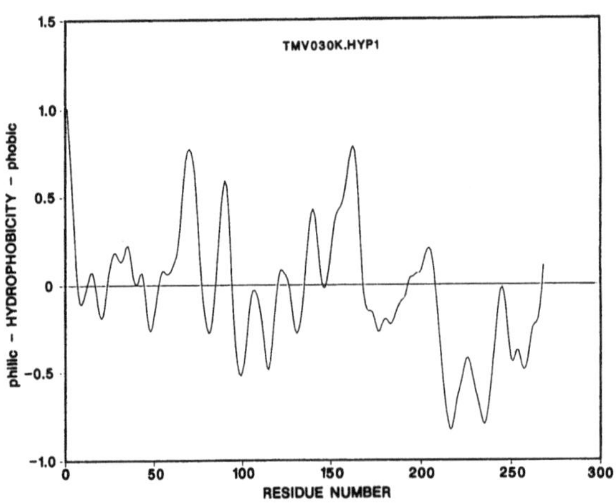

Fig. 3. Hydropathy plot for the 30 kD movement protein of TMV.

Table 3. Amino acid identity (%) between four tobamovirus movement proteins. Identities were determined by the University of Wisconsin Genetics Computer Group sequences analysis program, Gap (gap weight=1, gap length=0).

Virus	TMV	ToMV	CGMMV	SHMV
TMV	100	79	41	34
ToMV	-	100	41	37
CGMMV	-	-	100	41
SHMV	-	-	-	100

indicative of the above described high degree of amino acid homology. Considerable similarities also are predicted between TMV, ToMV and SHMV. Structural similarities between the CGMMV MP and the other tobamoviruses appear to be more prevalent in the C-terminal two-thirds of the respective proteins. Obvious similarities that are predicted in the four movement proteins include the turn structures between amino acids 70-90 and in the C-terminal one-third of the proteins. Interestingly, the turns predicted between amino acids 70-90 are flanked by hydrophobic regions.

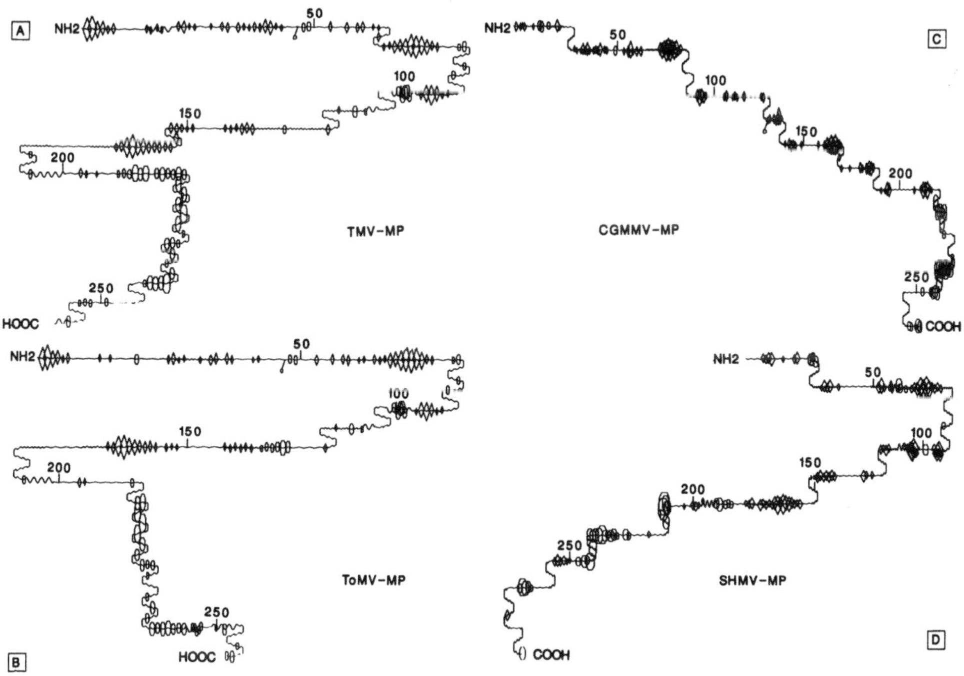

Fig. 4. Secondary structural predictions according to Chou and Fasman (1974) for the tobamovirus movement proteins. The University of Wisconsin Genetics Computer Group sequence analysis programs for the VAX were used to obtain these predictions. Symbols represent: diamond, hydrophobicity; hexagon, hydrophilicity; sine wave, alpha-helix; sharp saw-tooth wave, beta-sheet; dull saw-tooth wave, no prediction.

Many plant viruses encode proteins of Mr in the range of 30 kD, some of which have been proposed to be equivalent in function to the TMV MP. Experimental evidence localizing three of these putative MPs in virus infected tissue has now been obtained. The 38 kD protein encoded by cauliflower mosaic virus (CaMV), a DNA virus, has been found associated with plasmodesmata by immunogold-labeling of thin sections of infected leaves (Linstead et al., 1988). This is interesting in light of the observations that caulimovirus infections are characterized by alterations of the cell walls (Conti et al., 1972). The 32 kD protein of alfalfa mosaic virus (AlMV) has been localized to the middle lamella of the cell wall (Stussi-Garaud et al., 1987), yet the protein was apparently not associated with any specific cell wall structure. In contrast, the P3 protein (Mr 35 kD) of cucumber mosaic virus (CMV) has been localized to the nucleoli of infected tissue using immunogold probes (Mackenzie and Tremaine, 1988). Interestingly, AlMV and CMV have tripartite RNA genomes and the 32 kD and P3 proteins are considered to be analogues. The apparent paradox between the cellular location of these proteins and their putative function has yet to be resolved.

A number of plant viruses may not encode movement proteins. For example, the luteoviruses and the monopartite geminiviruses are limited to phloem parenchyma cells. These viruses may lack the required MP to enter into nonvascular tissue. It is interesting in this respect that movement of potato leafroll virus (PLRV), a luteovirus, into leaf mesophyll cells of *Nicotiana clevelandii* is accentuated in the presence of a helper virus like the potato Y potyvirus (Barker, 1987). Equally interesting is the finding that transgenic tobacco plants expressing the TMV MP and inoculated with PLRV allow the virus to spread into leaf mesophyll cells where the virus replicates. In control plants PLRV spreads much more slowly than in transgenic plants, and is phloem limited (C. Holt, M. Kaniewska and R. Beachy, unpublished data).

DYNAMICS OF PLASMODESMATA

The mechanism(s) by which virally-encoded movement proteins might function to modify plasmodesmata has yet to be established. It is generally accepted that viruses and/or their nucleic acids can move through plasmodesmata, but their possible effect on these structures remains conjectural. However, since we now realize that the plasma membrane is a dynamic structure that is constantly being turned over (Steer, 1988), we can begin to better appreciate the ways by which the plant (and viruses) may modify plasmodesmata. In an earlier section (see Fig. 1), we presented possible ways in which virally-induced sub-structural changes may occur to allow for the passage of viruses. At present, almost nothing is known about the nature of the putative proteins that may act to regulate the size exclusion limits of the cytoplasmic annulus. It would appear that the number of said protein inclusions is not constant between species, but may range from as low as 6 to around 20, with the "norm" being close to 9 (Olesen, 1979; Overall et al., 1982; Terry and Robards, 1987; Thomson and Platt-Aloia, 1985). Furthermore, except for the particle distribution data obtained from freeze-fracture studies by Thomson and Platt-Aloia (1985), there is no information in the published literature pertaining to the cellular mechanism(s) responsible for targeting these proteins to the neck constriction region. In this regard, the differences in the structure of plasmodesmata and gap junctions indicate a divergence in development which may mean that the two systems have no residual homologous constituents (but see Meiners and Schindler, 1987; Schinder at al., this volume). Clearly, it is going to be a real challenge to develop a protocol to isolate and characterize these vital plasmodematal proteins.

Studies on the tissue-specific constancy, or otherwise, of the arrangement of these structures (proteins) needs to be carefully documented. Of equal importance, there is a real need for information on the dynamics of the plasma membrane and axial component within the plasmodesma. In this latter case it may well be possible to use fluorescence photobleaching techniques (see Wade et al., 1986) to obtain information on the dynamics of membrane flow between cells (Baron-Epel et al., 1988). Data of this nature, along with similar studies conducted on viral-infected, as well as transgenic species which express the various virus-specific putative movement proteins will establish whether these proteins, and their interaction with the membranous constituents of the plasmodesma (see Fig. 1B), are dynamic in nature. It will also be of interest to determine whether these plasmodesmatal proteins are capable of being modulated by environmental stresses such as heat and cold, light intensity, nutrient deficiencies, etc., as well as by bacterial and fungal infections.

Cytological and physiological studies have established that plants have the genetic information to enable them to truncate plasmodesmata (Palevitz and Hepler, 1985; Wille and Lucas, 1984). The expression of this information is under the influence of a tissue specific promoter, in that it only occurs in developing guard cells, where plasmodesmatal function results in symplasmic isolation of the guard cells from the surrounding epidermal tissue (Palevitz and Hepler, 1985; Wille and Lucas, 1984). Limited evidence is also available which suggests that plants may be able to form plasmodesmata, de novo, between existing cell walls (Burgess, 1972; Jones, 1976; Binding et al., 1987; see also Monzer, this volume). If new plasmodesmata can be formed, viruses may have discovered ways to manipulate the molecular events associated with secondary plasmodesmatal formation. Control over gene expression, to produce secondary plasmodesmata deficient in the axial component, would potentiate systemic movement of the virus.

A final area worthy of study is the role played by callose, i.e., the (1->3)-ß-D-glucan which has long been known to be involved in the defence mechanism of the plant. A recent immunocytological study, conducted by Northcote et al. (1989), revealed that the deposition of callose is restriced generally to the immediate wall area surrounding the plasmodesmata. Olesen (this volume) considers that the synthesis and degradation of callose may form the main control component that restricts the neck region of the plasmodesma. Thus, it would appear that symplasmically mobile viruses may have developed ways in which to inactivate the synthesis of callose, or to increase the relative activity of the (1->3)-ß-D-glucanase over the synthase.

Development of an understanding of the molecular and biochemical regulation of the synthesis and action of the TMV MP will provide an excellent means to further investigate plasmodesmatal function. Of equal importance, understanding the role of these putative movement proteins in cell-to-cell movement of viruses may aid in the selection of traits conferring resistance to viral infection.

ACKNOWLEDGEMENTS

This work was supported, in part, by grants from the National Science Foundation to R.N.B. (DMB 87017012) and W.J.L. (DMB 8703624).

REFERENCES

Allison AV, Shalla TA (1974) The ultrastructure of local lesions induces by potato virus x: A sequence of cytological events in the course of infection. Phytopathology 64:784-793

Atabekov JG, Dorokhov YL (1984) Plant virus-specific transport function and resistance of plants to viruses. Adv Virus Res 29: 313-364

Barker H (1987) Invasion of non-phloem tissue in *Nicotiana clevelandii* by potato leafroll lutevirus is enhanced in plants also infected with potato Y potyvirus. J Gen Virol 68:1223-1227

Baron-Epel O, Hernandez D, Jiang L-W, Meiners S, Schindler M (1988) Dynamic continuity of cytoplasmic and membrane compartments between plant cells. J Cell Biol 106:715-721

Binding H, Witt D, Monzer J, Mordhorst G, Kollmann R (1987) Plant cell graft chimeras obtained by co-culture of isolated protoplasts. Protoplasma 141:64-73

Burgess J (1976) The occurrence of plasmodesmata-like structure in a non-division wall. Protoplasma 74:449-458

Chou PY, Fasman GD (1974) Conformational parameters for amino acids in helical, ß-sheet, and random coil regions calculated from proteins. Biochemistry 13:211-222

Conti GG, Vegetti G, Bassi M, Favali MA (1972) Some ultrastructure and cytochemical observations on chinese cabbage leaves infected with cauliflower mosaic virus. Virology 47:694-700

Deom CM, Oliver MJ, Beachy RN (1987) The 30-kilodalton gene product of tobacco mosaic virus potentiates virus movement. Science 237:389-394

Erwee MG, Goodwin PB (1983) Characterization of the *Egeria densa* Planch leaf symplast: Inhibition of the intercellular movement of fluorescent probes by group II ions. Planta 158:320-328

Erwee MG, Goodwin PB (1984) Characterization of the *Egeria densa* leaf symplast: Response to plasmolysis, deplasmolysis and to aromatic amino acids. Protoplasma 122: 162-168

Esau K (1968) Viruses in Plant Hosts. University of Wisconsin Press, Madison 225pp

Esau K (1969) The Phloem. Handbuch der Pflanzenanatomie. Gebrueder Borntraeger Berlin pp505

Esau K, Cronshaw J, Hoefert LL (1967) Relation of beet yellows virus to the phloem and to movement in the sieve tube. J Cell Biol 32:71-87

Fisher DG (1988) Movement of lucifer yellow in leaves of *Coleus blumei* Benth. Plant Cell Env 11:639-644

Flagg-Newton J, Simpson I, Loewenstein WR (1979) Permeability of the cell-to-cell membrane channels in mammalian cell junction. Science 205:404-407

Gibbs A (1976) Viruses and Plasmodesmata. In BES Gunning and AW Robards (eds): Intercellular Communication in Plants: Studies on Plasmodesmata. pp 149-164. Springer-Verlag Berlin Heidelberg New York

Goelet P, Lomonossoff GP, Butler PJG, Akam ME, Gait MJ, Karn J (1982) Nucleotide sequence of tobacco mosaic virus RNA. Proc Natl Acad Sci USA 79:5818-5822

Huisman MJ, Sarachu AN, Ables F, Broxterman HJG, Van Voltendoting L, Bol JF (1986) Alfalfa mosaic virus temperature-sensitive mutants. III. Mutants with a putative defect in cell-to-cell transport. Virology 154:401-404

Jones MGK (1976) The origin and development of plasmodesmata. In BES Gunning, Robards AW (eds): Intercellular Communication in Plants: Studies on Plasmodesmata. pp 81-105. Springer-Verlag Berlin Heidelberg New York

Kitajima EW, Lauritis JA (1969) Plant virons in plasmodesmata. Virology 37:681-685

Leonard DA, Zaitlin M (1982) A temperature-sensitive strain of tobacco mosaic virus defective in cell-to-cell movement generates an altered viral-coded protein. Virology 117:416-424

Linstead PJ, Hills GJ, Plaskitt KA, Wilson IG, Harker CL, Maule AJ (1988) The subcellular location of the gene I product of cauliflowe mosaic virus is consistent with a function associated with virus spread. J Gen Virology 69:1809-1818

Loewenstein WR, Rose B (1978) Calcium in (junctional) intercellular communication and a thought on its behavior in intracellular communication. Ann New York Acad Sci 307: 285-307

MacKenzie DJ, H. Tremaine JH (1988). Ultrastructure location of non-structural protein 3A of cucumber mosaic virus in infected tissue using monoclonal antibodies to a cloned chimeric fusion protein. J Gen Virol 69:2387-2395

Madore MA, Oross JW, Lucas WJ (1986) Symplastic transport in *Ipomoea tricolor* source leaves: demonstration of functional symplastic connections from mesophyll to minor veins by a noval dye-tracer method. Plant Physiol 82:432-442

Meiners S, Schindler M (1987) Immunological evidence for gap junction polypeptide in plant cells. J Biol Chem 262:951-953

Meshi T, Ohno T, Okada Y (1982) Nucleotide sequence of the 30 K protein cistron of cowpea strain of tobacco mosaic virus. Nucleic Acid Res 10: 6111-6117

Meshi T, Okada Y (1986) Systemic movement of viruses. In T Kosuge, Nester EW (eds): Plant-Microbe Interaction: Molecular and Genetic Perspectives. pp 285-304. Macmillan New York

Nishiguchi M, Motoyoshi F, Oshima N (1978) Behavior of a temperature-sensitive strain of tobacco mosaic virus in tomato leaves and protoplasts. J Gen Virol 39:53-61

Nishiguchi M, Motoyoshi F, Oshima N (1980) Further investigation of a temperature-sensitive strain of tobacco mosaic virus: Its behavior in tomato leaf epidermis. J Gen Virol 46:497-500

Northcote DH, Davey R, Lay J (1989) Use of antisera to localize callose, xylan and arabinogalactan in the cell-plate, primary and secondary walls of plant cells. Planta 178:353-366

Olesen P (1979) The neck constriction in plasmodesmata: Evidence for a peripheral sphincter-like structure revealed by fixation with tannic acid. Planta 144:349-358

Overall RL, Wolfe J, Gunning BES (1982) Intercellular communication in *Azola* roots. I. Ultrastructure of plasmodesmata. Protoplasma 111:134-150

Palevitz BA, Helper PK (1985) Changes in dye coupling of stomatal cells of *Allium* and *Commelina* demonstrated by microinjection of Lucifer yellow. Planta 164:473-479

Saito T, Imai Y, Meshi T, Okada Y (1988) Interviral homologies of the 30K proteins of tobamoviruses. Virology 167:653-656

Steer MW (1988) Plasma membrane turnover in plant cells. J Exp Bot 39:987-996

Stewart WW (1981) Lucifer dyes-highly fluorescent dyes for biological tracing. Nature 292:17-21

Stussi-Garaud C, Garaud J, Berna A, Godefroy-Colburn T (1987). In situ location of alfalfa mosaic virus non-structural protein in plant cell walls: correlation with virus transport. J Gen Virol 68:1779-1784

Takamatsu N, Ohno T, Meshi T, Okada Y (1983) Molecular cloning and nucleotide sequence of the 30 K and the coat protein cistron of TMV (tomato strain) genome. Nucleic Acid Res 11: 3767-3778

Terry BR, Robards AW (1987) Hydrodynamic radius alone governs the mobility of molecules through plasmodesmata. Planta 171:145-157

Thomson WW, Platt-Aloia K (1985) The ultrastructure of the plasmodesmata of the salt glands of *Tamarix* as revealed by transmission and freeze-fructure electron microscopy. Protoplasma 125:13-23

Tomenius K, Claphan D, Meshi T (1987) Localization by immunogold cytochemistry of the virus-coded 30K protein in plasmodesmata of leaves infected with tobacco mosaic virus. Virology 160:363-371

Tucker EB (1982) Translocation in the staminal hairs of *Setcreasea purpurea*. I. A study of cell ultrastructure and cell-to-cell passage of molecular probes. Protoplasma 113:193-201

Tucker EB, Spanswick RM (1985) Translocation in the staminal hairs of *Setcreasea purpurea*. II. Kinetics of intercellular transport. Protoplasma 128:167-172

Wade HW, Trosko JE, Schindler M (1986) A fluorescence photobleaching assay of gap junction-mediated communication between human cells. Science 232:525-528

Watanabe T, Ooshika I, Meshi T, Okada Y (1986) Subcellular localization of the 30K protein in TMV-inoculated tobacco protoplasts. Virology 152:414-420

Weintraub M, Ragetli HWJ, Leung E (1976). Elongated virus particles in plasmodesmata. J Ultrastruct Res 56:351-364

Wille AC, Lucas WJ (1984) Ultrastructure and histochemical studies on guard cells. Planta 160:129-142

Wolf S, Deom CM, Beachy RN, Lucas WJ (1989) The movement protein of tobacco mosaic virus modifies the size exclusion limit of plasmodesmata. (Submitted to Science)

Subject Index

3-methylcholanthrene 122
12-O-tetradecanoylphorbol 13-acetate (TPA) 116
16k channel protein 56
16k protein 13ff
26 kDa protein
- of liver gap junctions 4
30Kd protein
- of TMV 247
70Kd protein 48
abutilon 221ff
access resistances 89
acetyl choline receptor 13
actin 45
action potential 90
african violet extract 175
agrobacterium tumefaciens 262
alfalfa mosaic virus 253, 261
alpha-actinin 45
amaranthus 198ff
anesthetics 69
ankyrin 45
aniline blue
- dynamics of160
- induced fluorescence159
antibody
- to liver gap junction proteins 7
antigenic cross-reactivity 2
aplysia 66
apoplast 53,198,221
arabadopsis thaliana 173
arachidonic acid 74
artichoke leaf extract 176
ascidians 129
ATPases 17,55
avocado extract 176
Azolla 228
bacteriorhodopsin 44
BALB/c 3T3 116ff
BAPTA buffers 244
beet yellows virus 261
beta 210,213,214
blastocoel expansion
- cell-to-cell coupling during 106
border cells104
brain tissue - maturation of 9
C3H10T1/2 cells 118ff
C4 photosynthesis 210
C_4 plants 198ff
calcium 71,125,243
calcium ionophore A23187 244
callose 162ff,269
calmodulin 70ff
CAMP 125
CAMP kinase 125
CAMP-dependent protein kinase 75
cananga 198ff
capparis 226
carboxyfluorescein 240
carcinogenesis 115,120
cardiac gap junction 67
cauliflower mosaic virus 250
caulimovirus 268
CDNA
- for connexin 6
cell adhesion molecules 56
cell cycle 140
cell line 121
cell plate 193
cell proliferation 1
cell transformation 118
cell-to-cell coupling
- during blastocoel expansion 106

cell-to-cell propagation of electrical
impulses 1
cellular differentiation 23ff
cellular oncogenes 120
cellular periphery 179
CGMP 75
channel gating 2
channel proteins 14
channels 54,63
chemical synapse 18
chimeras 185
chironomus 70
chlordane 117
ciliated cells 135
citrus 225ff
clover mottle virus 254
*colchicum*226
coleus 198ff
commelina 212ff
communication compartment 57,103,132
companion cells 205
compartments
- communicating 103
- developmental 103
conductances
- of gap junctions 87
- single channel 87
connexins
3,13ff,26,32,38,43,56,65ff,124,171ff
- expression of 7
- model of 23
- structural diversity of 10
connexin gene family 23ff
connexin transcripts 23ff
- expression of 28
connexin-related molecules 14
connexons 1,171
control
- of junctional communication 60
coupling coefficient 71
coupling strength 71
cowpea mosaic virus 250,254
crayfish 69
cross-reactivity 2
croton oil 123
crystallin proteins 35
cucumber exctract 176
cucumber mosaic virus 268
cyclic AMP 116,122
cyclic nucleotides 75
cytochalasin B 242
cytoplasmic annulus 227
cytoplasmic bridges 68
cytoplasmic streaming 242
dahli mosaic virus 260
daisy extract 175
DDT 117
dermal-epidermal boundary 57
dermis 57
desmotubule 148,226
development 129
developmental compartments 103
dieldrin 117
differentiation
- regulation of 1
differentiation 129
diffusible morphogens104
diffusion 214
- of cytoplasmic regulatory signals 107
diffusion barrier
- to tracer movement104
direct intercellular signal exchange 1
divergence 269
drosophila 68,70,78

dye
- lucifer yellow 190
dye-coupling 129
dye-uncoupling 133
electrical impulses
- cell-to-cell propagation of 1
electrical-coupling 129
electronic interaction 89
embryos 129
embryogenesis - of mouse 60
embryonic cells 69ff
endoplasmic reticulum 222
endothelial-stromal cell coupling 57
epidermal cells 57
epidermal communication compartments 58ff
epidermal core 108
epidermal growth control 58
epidermal proliferative control 58
epidermal proliferative unit 58
epidermis 58,109
epitopes 174
expression 68
- of connexin transcripts 28
- of connexins 7
expression of difference 60
extracellular space 230
extrafloral nectaries 221
F-type H^+-ATPases 17
F_1F_0 ATPases 56
F9 mouse embryo carcinoma cells 29
fatty acids 69
fibers 205
fibrocytes 88
floral nectaries 221
fluorescein isothiocyanate 239
fluorescence photobleaching 269
fluorescent tracer dyes 238
fluorescent dyes 209
fluorochrome microinjection 186
free cytoplasmic Ca^{++} 71
frequency histograms 94
freund's adjuvant 176
fruit, leaf and vegetative samples 176
fundulus 71
fusion pore 18
gap junction channels 14ff,54,67ff
- model for 16
gap junction formation
- specificity of 55
gap junction proteins 23ff
gap junctional conductances 87
gap junctional coupling and metabolic stress 106
gap junctional intercellular communication 115
gap junctions
- and tissue growth 107
gap junctions1,13ff,54ff,87,115,171,269
gating 68,73
geminiviruses 268
germ layers 137
glial fibrillary acid protein 9
glucan synthase 163
glucocorticoids 116.122
glucocorticoids 122
gossypium hirsutum 224ff
green bell pepper extract 176
growth factor 59
half plasmodesmata 185
halomethanes 74
halothane 74
healing over 70
heart connexin 178

hepatocytes 77
heptanol 74
heterologous communication 120ff
hibiscus 221ff
hill coefficient 71
homeostasis 115
homologous communication 120ff
human fibroblasts 121
human hepatocytes 121
human keratinocytes 121
hydropathic profiles
hydrophobic regions 267
hypersensitivity response 249
immunoelectron microscopy 4
immunogold cytochemistry 262
immunological relatedness 172
immunoreactivity
- of gap junctions 23ff
inositol triphosphate 125
inositol trisphosphate-diacylglycerol 243
inside-outside voltage dependence70
intercalated disk 63
intercellular bridges 53ff
intercellular communication 23ff
intercellular signal exchange 1
intracellular acidification 73
intracellular free Ca^{++} 70
intracellular pH 69,71
iontophoresis dyes 238
iris extract 175
junction formation 56
junctional channels
- opening and closing of 90
junctional communication 53
- control of 60
junctional conductance 70ff
junctional coupling
- in differentiation105
- in insect epidermis102
junctional permeability 66
junctional resistance and permeability
- correlation between 108
kinetics
- of cell-to-cell communication 240
l-glutamate106
lactuca sativa 233
lens fibres 35ff
lettuce leaf extract 176
lilly extract 175
lipid bilayers 70ff
lipophilic molecules 74
liposomes 263
- delivery system 204
liver tissue 4
lonicera japonica 224
low angle X-ray diffraction 3
lucifer yellow 57,66,130
luteoviruses 268
lycopersicon 209,242
MAC 207
- monoclonal antibody 155
maturation
- of brain tissues 9
mechanism
- of specificity 55
median cavities 228
membrane-impermeable fluorescent dyes 263
mesophyll and bundle sheath cells 201
metabolic activities 1
metabolic stress 106
mezerein 117
microbody-like organelles 203
microbody-like organelles 207

microinjection studies 264
milkweed bug - *Oncopeltus* 103
minor veins 199
model
- of gap junctional channel 16
- of connexins 23
molecular exclusion limits of plasmodesmata 264
molecular modelling 15
molecular probes 233
molluscs 129
morphogenesis 61,101
morphogens 101ff
mouse embryogenesis 60
mP26 35
myelin basic protein 9
myocytes 88
Na^{+} dependent uptake 106
nacreous walls 201
nacreous-walled sieve elements 208
neck regions 227
nectar 221
nectaries 221
neonatal rat heart cells 88
neuron specific enolase 9
neurotransmitter release 18
nexuses 1
NIH 3T3 cells 120ff
n-propyl gallate 180
NRK 121
octanol 74
octyl glucoside 41
oncogene 120
oncogene expression 120
openings and closing
- of junctional channels 90
P-protein 201
passive electrotonic coupling 93
patch clamp 10
patterns
- of junctional communication 56
pectinase 179
pectinase permeabilized 181
per 78
petal extracts 175
petunia extract 175
PH dependence 73
PH sensitivity 72
phenobarbital 116ff
phloem limited 268
phorbol esters 76,116ff
phosphoinositide second messenger 163
phosphorylation 75
photoreaction centre 13
pit-fields 181
plakoglobin 45
plant viruses 260
plasma membrane
- dynamic structure of 268
plasmodesmata 54,163,172,221,261,269
"basic type" 146
- C_4 grasses 152
- connection with ER 148
- cytoplasmic annulus 148
- cytoplasmic sleeve 148,152
- dimensions of 150
- freeze-fracture 152
- neck constriction 152
- plasma membrane tube 154
- plasmodesmatal canal 226
- plasmodesmatal frequencies 208ff
- plasmodesmatal proteins 269
- sphincter concept 156,162ff
- sphincter dynamics 159

- sphincter fluorescence 159
- sphincters in grasses 156
- structural analogies with animal junctions 166
plasmodesmograms 198ff
plasmolysis and dye coupling 242
polycyclic hydrocarbons 117
polyoma virus-BHK21 123
polyoma cells 120
populus 210ff
pore conductance 68
pore-plasmodesmata connections 201ff
potassium ferrocyanide 148
potyvirus 268
precursor half channels 54
prenectar 224
primary lens fibers 35
protein kinase C 75ff,125
proteoglycan 56,78
proteoliposomes
- reconstituted 43
proton translocation 18
protophloem 202
protoplast cultures 186
quail and chick embryo fibroblasts 121
rat F cells 121
rat liver connexin 172ff
reconstituted proteoliposomes 43
red clover mottle virus 250
red bell pepper extract 176
regulation
- of cell-to-cell communication 243
- of differentiation 1
regulatory mechanisms 79
repeated epilation 59
resistance
- during segment growth 108
retinoic acid 116,122
rhodopsin 13
rose extract 175
salamander 70
salsola kali 146ff
second messenger 58,69
secondary plasmodesmata 185
- formation of 193
secondary structural predictions
- for tobamovirus MPs 266
secretory reticulum 230
selectivity 67
setcreasea purpurea 237
sieve elements 201ff
signal exchange
- direct intercellular 1
signal recognition particles (SRP) 36
signal transduction 77
single channel conductance 87
single channel events 93
single channel recording 67
soybean connexin 175
soybean root cells 172
spartina townsendii 154
specific resistance
- of gap junctions 87
specificity
- mechanism of 55
- of gap junction formation 55ff
spectrin 45
sphinctered junctions 230
sporobolus rigens 154
squid 70
stalk cell 224
staminal hairs 237
structural diversity
- of connexins 10

subconductance states 87
suberin lamella 152
substomatal chambers 226
sunhemp mosaic virus 250
SV40 T antigen
symplasmic mobility 261
symplast 53,198,221
synapse 18
syncytium 87
systemic infectivity 262
tamarix 233
tannic acid contrast enhancement 152
temporal pattern
- of junctional permselectivity 109
tenebrio 66,109
TMV movement gene 249
TMV systemic movement proteins 249
tobacco mosaic virus 247
tobacco rattle virus 250
tobamoviruses 265
tomato extract 176
topological model
- of connexins 23ff
TPA 59
tracheary elements 207
tradescantia 209
transformed cells 119ff
transformed phenotypes 115
transgenic plants 261ff
transjunctional voltages 70ff
triads - of skeletal muscle cells 167
trochoblasts 133
truncate plasmodesmata 269
tumorigenic rat tracheal epithelial cells 123
tumour cells 115
tumour promoter 116
tumour promoter cells phenobarbital 116
tumour promotion 116
tumour suppression 116,122
tnitary conductance 67
unsaturated lipids 74
v-fos 120
v-myc120
v-ras 120
v-src 120
vacuolar H^+-ATPases 17
vascular parenchyma cells 203
vein endings 207
veins 198
vicia faba 222ff
vigna unguiculata 224
vinca 226
vinculin 45
viral movement proteins 254,261
viral particles moving 260
virus-specific ribonucleoproteins 251
voltage dependence 70,88
x-ray diffraction 3
zea mays 158
zinc iodide 148
zostera nana 208

NATO ASI Series H

Vol. 1: **Biology and Molecular Biology of Plant-Pathogen Interactions.**
Edited by J. A. Bailey. 415 pages. 1986.

Vol. 2: **Glial-Neuronal Communication in Development and Regeneration.**
Edited by H. H. Althaus and W. Seifert. 865 pages. 1987.

Vol. 3: **Nicotinic Acetylcholine Receptor: Structure and Function.**
Edited by A. Maelicke. 489 pages. 1986.

Vol. 4: **Recognition in Microbe-Plant Symbiotic and Pathogenic Interactions.**
Edited by B. Lugtenberg. 449 pages. 1986.

Vol. 5: **Mesenchymal-Epithelial Interactions in Neural Development.**
Edited by J. R. Wolff, J. Sievers, and M. Berry. 428 pages. 1987.

Vol. 6: **Molecular Mechanisms of Desensitization to Signal Molecules.**
Edited by T. M. Konijn, P. J. M. Van Haastert, H. Van der Starre, H. Van der Wel, and M. D. Houslay. 336 pages. 1987.

Vol. 7: **Gangliosides and Modulation of Neuronal Functions.**
Edited by H. Rahmann. 647 pages. 1987.

Vol. 8: **Molecular and Cellular Aspects of Erythropoietin and Erythropoiesis.**
Edited by I. N. Rich. 460 pages. 1987.

Vol. 9: **Modification of Cell to Cell Signals During Normal and Pathological Aging.**
Edited by S. Govoni and F. Battaini. 297 pages. 1987.

Vol. 10: **Plant Hormone Receptors.** Edited by D. Klämbt. 319 pages. 1987.

Vol. 11: **Host-Parasite Cellular and Molecular Interactions in Protozoal Infections.**
Edited by K.-P. Chang and D. Snary. 425 pages. 1987.

Vol. 12: **The Cell Surface in Signal Transduction.**
Edited by E. Wagner, H. Greppin, and B. Millet. 243 pages. 1987.

Vol. 13: **Toxicology of Pesticides: Experimental, Clinical and Regulatory Perspectives.**
Edited by L. G. Costa, C. L. Galli, and S. D. Murphy. 320 pages. 1987.

Vol. 14: **Genetics of Translation. New Approaches.**
Edited by M. F. Tuite, M. Picard, and M. Bolotin-Fukuhara. 524 pages. 1988.

Vol. 15: **Photosensitisation. Molecular, Cellular and Medical Aspects.**
Edited by G. Moreno, R. H. Pottier, and T. G. Truscott. 521 pages. 1988.

Vol. 16: **Membrane Biogenesis.** Edited by J. A. F. Op den Kamp. 477 pages. 1988.

Vol. 17: **Cell to Cell Signals in Plant, Animal and Microbial Symbiosis.**
Edited by S. Scannerini, D. Smith, P. Bonfante-Fasolo, and V. Gianinazzi-Pearson. 414 pages. 1988.

Vol. 18: **Plant Cell Biotechnology.**
Edited by M. S. S. Pais, F. Mavituna, and J. M. Novais. 500 pages. 1988.

Vol. 19: **Modulation of Synaptic Transmission and Plasticity in Nervous Systems.**
Edited by G. Hertting and H.-C. Spatz. 457 pages. 1988.

Vol. 20: **Amino Acid Availability and Brain Function in Health and Disease.**
Edited by G. Huether. 487 pages. 1988.

NATO ASI Series H

Vol. 21: **Cellular and Molecular Basis of Synaptic Transmission.** Edited by H. Zimmermann. 547 pages. 1988.

Vol. 22: **Neural Development and Regeneration. Cellular and Molecular Aspects.** Edited by A. Gorio, J.R. Perez-Polo, J. de Vellis, and B. Haber. 711 pages. 1988.

Vol. 23: **The Semiotics of Cellular Communication in the Immune System.** Edited by E.E. Sercarz, F. Celada, N.A. Mitchison, and T. Tada. 326 pages. 1988.

Vol. 24: **Bacteria, Complement and the Phagocytic Cell.** Edited by F.C. Cabello und C. Pruzzo. 372 pages. 1988.

Vol. 25: **Nicotinic Acetylcholine Receptors in the Nervous System.** Edited by F. Clementi, C. Gotti, and E. Sher. 424 pages. 1988.

Vol. 26: **Cell to Cell Signals in Mammalian Development.** Edited by S.W. de Laat, J.G. Bluemink, and C.L. Mummery. 322 pages. 1989.

Vol. 27: **Phytotoxins and Plant Pathogenesis.** Edited by A. Graniti, R.D. Durbin, and A. Ballio. 508 pages. 1989.

Vol. 28: **Vascular Wilt Diseases of Plants. Basic Studies and Control.** Edited by E.C. Tjamos and C.H. Beckman. 590 pages. 1989.

Vol. 29: **Receptors, Membrane Transport and Signal Transduction.** Edited by A.E. Evangelopoulos, J.P. Changeux, L. Packer, T.G. Sotiroudis, and K.W.A. Wirtz. 387 pages. 1989.

Vol. 30: **Effects of Mineral Dusts on Cells.** Edited by B.T. Mossman and R.O. Bégin. 470 pages. 1989.

Vol. 31: **Neurobiology of the Inner Retina.** Edited by R. Weiler and N.N. Osborne. 529 pages. 1989.

Vol. 32: **Molecular Biology of Neuroreceptors and Ion Channels.** Edited by A. Maelicke. 675 pages. 1989.

Vol. 33: **Regulatory Mechanisms of Neuron to Vessel Communication in Brain.** Edited by F. Battaini, S. Govoni, M.S. Magnoni, and M. Trabucchi. 416 pages. 1989.

Vol. 34: **Vectors as Tools for the Study of Normal and Abnormal Growth and Differentiation.** Edited by H. Lother, R. Dernick, and W. Ostertag. 477 pages. 1989.

Vol. 35: **Cell Separation in Plants: Physiology, Biochemistry and Molecular Biology.** Edited by D.J. Osborne and M.B. Jackson. 449 pages. 1989.

Vol. 36: **Signal Molecules in Plants and Plant-Microbe Interactions.** Edited by B.J.J. Lugtenberg. 425 pages. 1989.

Vol. 37: **Tin-Based Antitumour Drugs.** Edited by M. Gielen. 226 pages. 1990.

Vol. 38: **The Molecular Biology of Autoimmune Disease.** Edited by A.G. Demaine, J-P. Banga, and A.M. McGregor. 404 pages. 1990.

Vol. 39: **Chemosensory Information Processing.** Edited by D. Schild. 403 pages. 1990.

Vol. 40: **Dynamics and Biogenesis of Membranes.** Edited by J.A.F. Op den Kamp. 367 pages. 1990.

Vol. 41: **Recognition and Response in Plant-Virus Interactions.** Edited by R.S.S. Fraser. 467 pages. 1990.

NATO ASI Series H

Vol. 42: **Biomechanics of Active Movement and Deformation of Cells.**
Edited by N. Akkaş. 524 pages. 1990.

Vol. 43: **Cellular and Molecular Biology of Myelination.**
Edited by G. Jeserich, H. H. Althaus, and T. V. Waehneldt. 565 pages. 1990.

Vol. 44: **Activation and Desensitization of Transducing Pathways.**
Edited by T. M. Konijn, M. D. Houslay, and P. J. M. Van Haastert. 336 pages. 1990.

Vol. 45: **Mechanism of Fertilization: Plants to Humans.**
Edited by B. Dale. 710 pages. 1990.

Vol. 46: **Parallels in Cell to Cell Junctions in Plants and Animals.**
Edited by A. W. Robards, W. J. Lucas, J. D. Pitts, H. J. Jongsma, and D. C. Spray. 296 pages. 1990.

MIX
Papier aus verantwortungsvollen Quellen
Paper from responsible sources
FSC® C105338

If you have any concerns about our products,
you can contact us on
ProductSafety@springernature.com

In case Publisher is established outside the EU,
the EU authorized representative is:
Springer Nature Customer Service Center GmbH
Europaplatz 3, 69115 Heidelberg, Germany

Printed by Libri Plureos GmbH
in Hamburg, Germany